BEAUTY IN THE BEAST

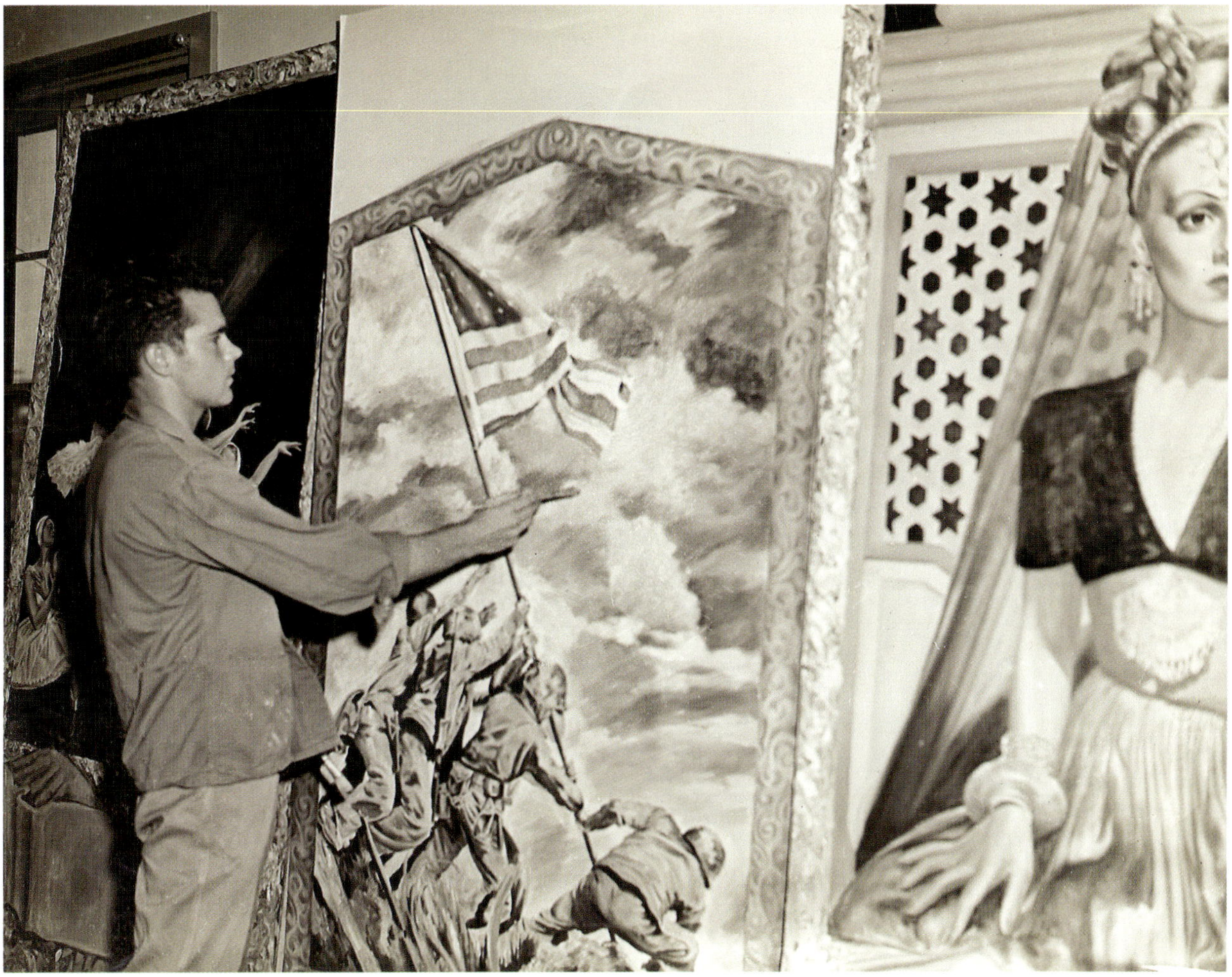

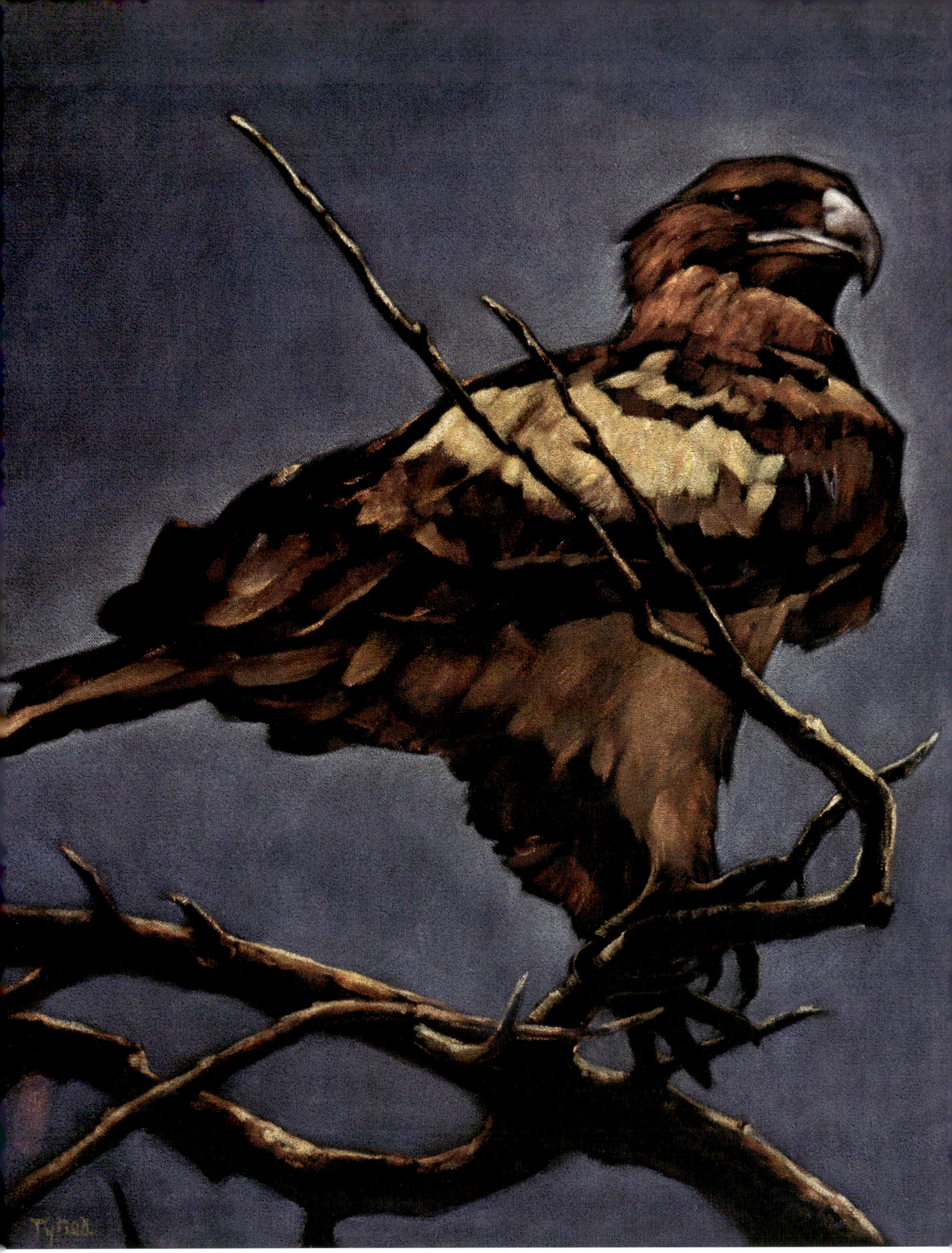

BEAUTY *in the* BEAST

Flora, Fauna, and Endangered Species of Artist Ralph Burke Tyree

CJ Cook

with Paige Herbert

SOUTH PACIFIC DREAMS PUBLISHING
San Diego
2023

Published by
South Pacific Dreams Publishing
www.SouthPacificDreams.com

Library of Congress Control Number: 2023908878
ISBN: 978-0-9984224-6-6

Interior book design by Claudine Mansour Design

Printed in South Korea by Think, Inc., El Dorado Hills, CA

(Previous photograph) Private Tyree, U.S. Marine Corps, paints a large mural in the officers' mess hall in Samoa. Tyree Family.

(Previous painting) *Juvenile Bald Eagle*, Ralph Burke Tyree. 1974, oil on velvet, 18" x 14", author's collection.

(Following painting) *Toco Toucan*, Ralph Burke Tyree. 1976, oil on velvet, 18" x 24", Tyree family.

(Painting opposite Contents) *Yellow Hibiscus*, Ralph Burke Tyree. 1968, oil on velvet, 16" x 20", author's collection.

To the lovely animals and plants of our blue planet.

"Our task must be to free ourselves by widening our circle of compassion to embrace all living creatures and the whole of nature and its beauty."

—ALBERT EINSTEIN

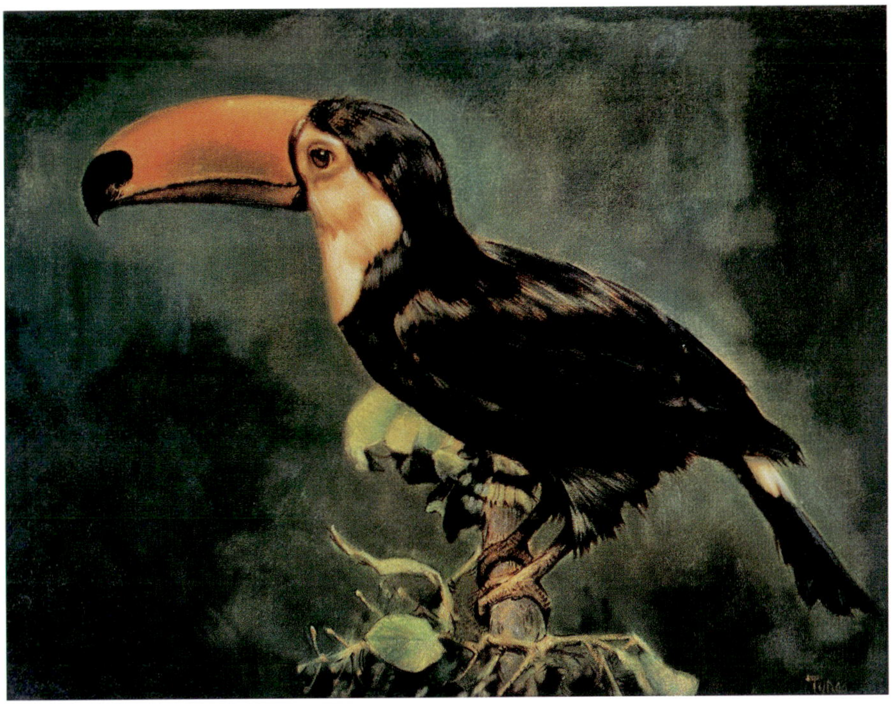

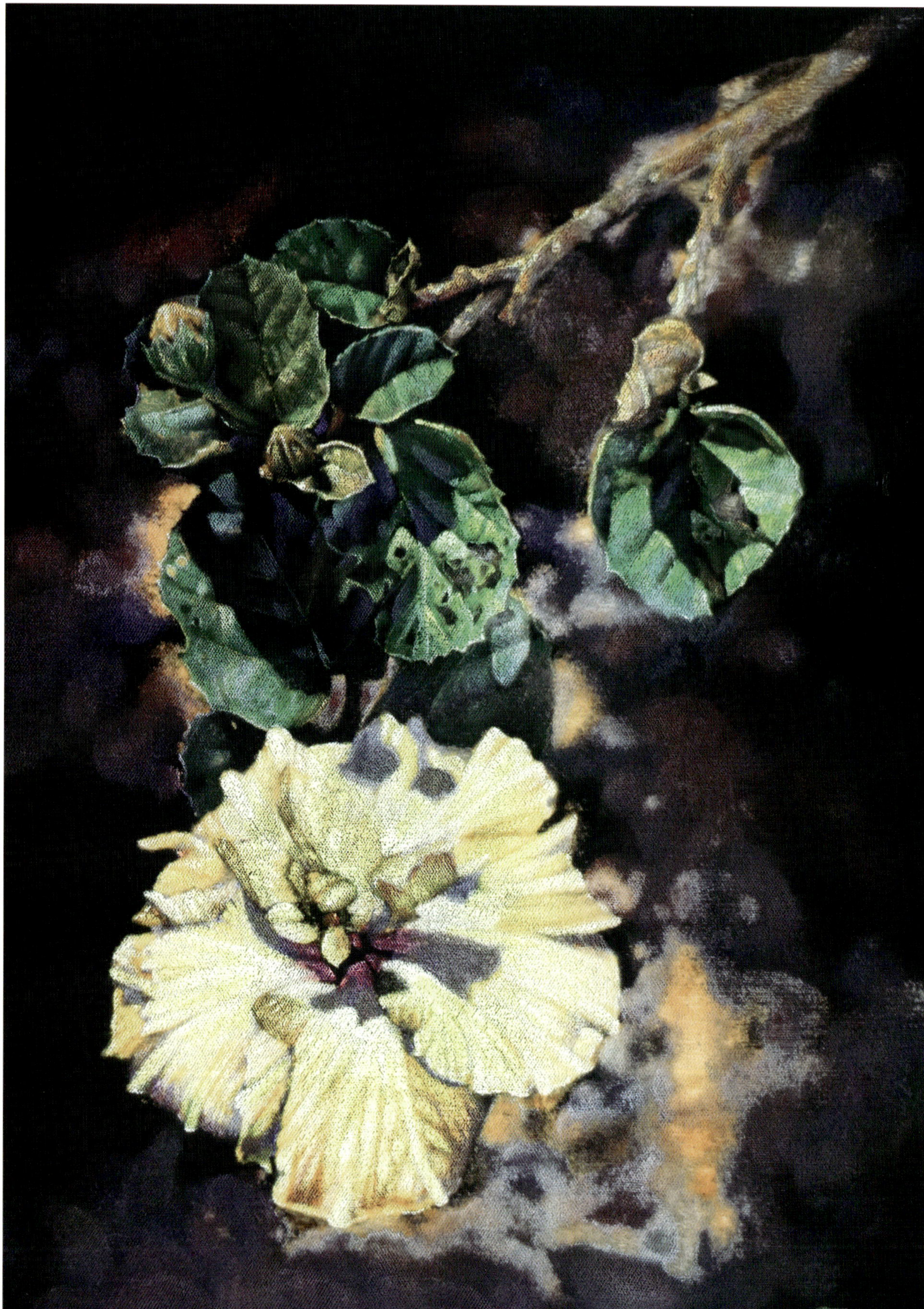
Tyree

Contents

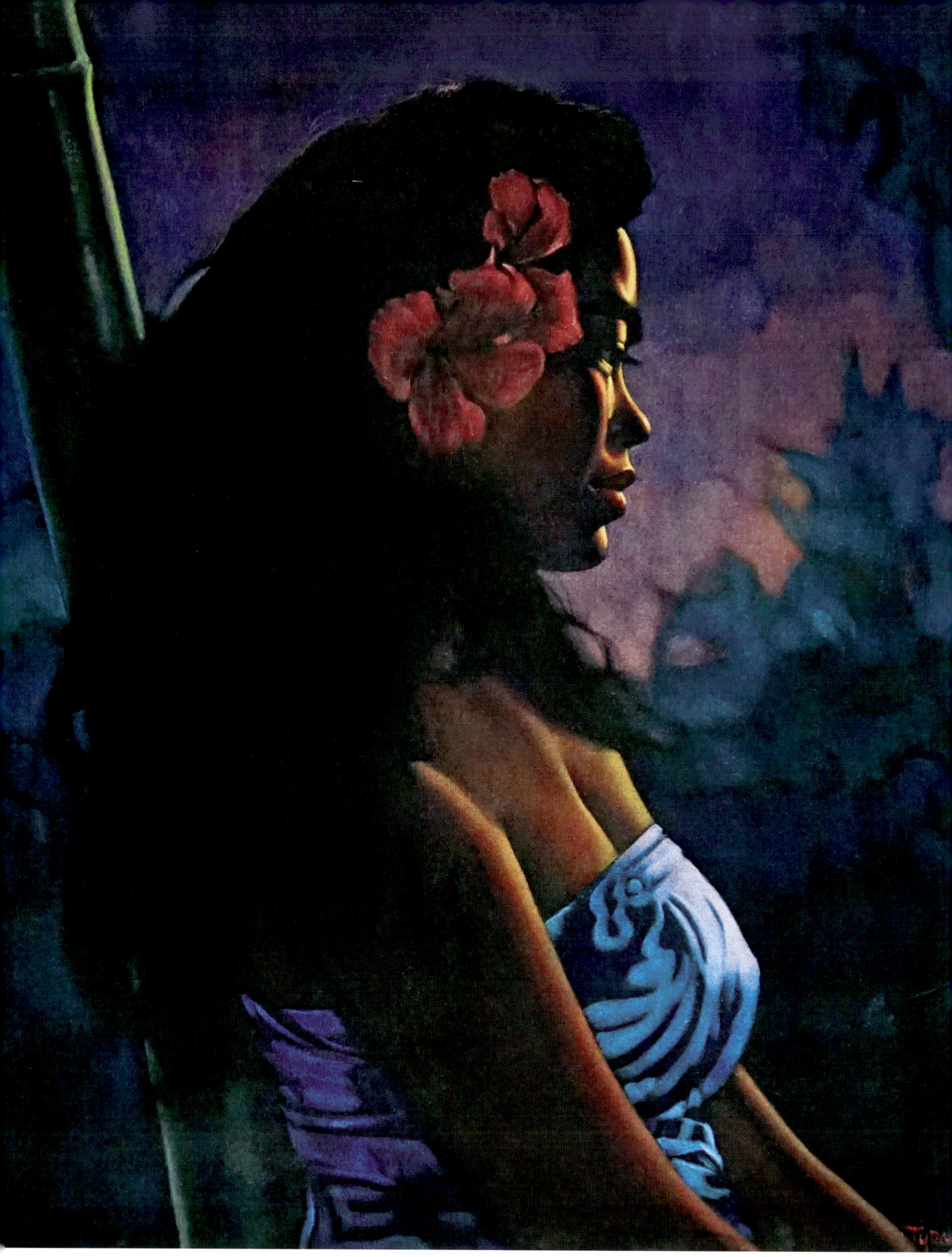

Preface

In 2008, I purchased my first black velvet painting by Ralph Burke Tyree, a prolific artist of the twentieth century. He was a quiet, gentle pied piper with a serious case of wanderlust and a family willing to follow. He summarized his fanaticism for painting, saying,

> The way I look at it . . . everybody needs a stabilizer . . . Painting is creative. It doesn't matter if it is music or art—it gets you away from the mad rush. Man must get away, be alone, and find himself in this type of society today. He's crowded in—there are too many people. He gets in a corner—he needs solitude.

A dreamer, he painted idealized people in picturesque South Pacific landscapes, the faces of wizened island men, and both familiar and exotic animals. His portraits, whether of humans or animals, captured their peaceful and sympathetic spirits. *Beauty in the Beast* is the story of Tyree's creative genius in painting flora, fauna, and endangered species.

After growing up in California's Central Valley, Tyree moved to the San Francisco Bay Area for his art education, attending the California College of Arts and Crafts and the California School of Fine Arts. Soon after his graduation, the attack on Pearl Harbor occurred, and seven weeks later, he joined the Marines. After completing basic training in San Diego, he deployed to Samoa. On this island in the South Pacific, Private Tyree befriended his commanding general, who subsequently designated him the resident artist of the Marine base. With that, he began his art career painting portraits of the officers' loved ones and murals in the mess halls.

Tyree returned to California after the war, married his high school sweetheart, started a family with her, and began to work as a professional artist. During this time, he depicted the idyllic beauty of the South Pacific's landscapes and people, often painting sensual island *wāhine* (Hawaiian for women) in beach or jungle settings (figure 0.1). While he occasionally worked with pastels and canvas, he primarily used oil paints and board. His love for the South Pacific even led him to move his family to the region where they lived for years in Guam, Oahu, Maui, and the Big Island of Hawaii. Over his 30-year career, he often traveled to other island paradises like Palau, Fiji, Tahiti, Samoa, and the Solomon Islands.

Mid-career, Tyree switched medium and began painting with oils on fine French silk black velvet to give his paintings added depth and texture. Thanks to Edgar Leeteg, this technique became

Figure 0.1 *Red Hibiscus, Hawaiian Wahine*, Ralph Burke Tyree. 1976, Oil on velvet, 20" x 16", author's collection.

popular, especially in South Pacific-themed restaurants. Leeteg was an American painter who had relocated to Tahiti and was the first modern artist to use oil paints on black velvet. With the rise of Tiki culture in the 1960s, Tyree became the most prolific South Pacific and Tiki artist of the century. He developed a highly efficient, technically complex method to paint on the velvet. One of his favorite sayings was, "Leonardo da Vinci was a great craftsman. There was a time when artists had to be good craftsmen. No one cared if he had a soul or studied in Paris; the question was, 'Can he paint well?'." With this mindset, Tyree worked hard at his profession and his craft.

Tyree had long been captivated by the plant world, depicting his wāhine surrounded by tropical foliage with their hair accented by a flower or three. Throughout the late 1960s, tropical flora, notably passion flowers, magnolia, and hibiscus, became a central element of his paintings, and in 1970, endangered animals entered his repertoire. For the next decade, his primary interest was painting the majestic creatures of the world with oils on board and velvet, bringing public awareness to their declining numbers. Sadly, he suffered a heart attack and died at 57 in 1979. This book explores the works of these last ten years of his life when animals became his focus.

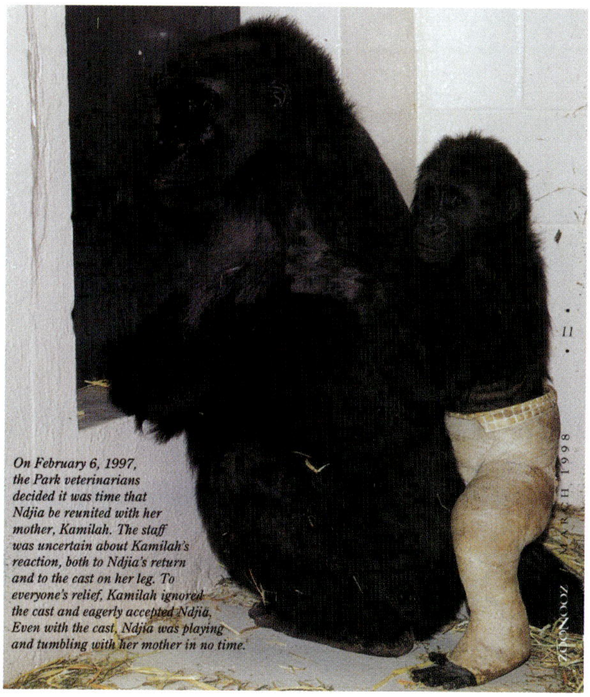

Figure 0.2 Four weeks after surgery Ndjia and her mom reunite, March 1998, San Diego Zoo, *ZooNooz*.

Like Tyree, I am a plant and animal lover. Many photographs that supplement his paintings in this book are from my collection, taken by friends and myself. I care deeply about our animal brethren, especially the vulnerable and endangered. Fortunately, I can support my local San Diego Zoo, which has played and continues to play a pivotal role in efforts to save species such as the California condor, featured in chapter 6. In 1998, I even had the chance to assist the efforts of pediatric orthopedic surgeons and zoo veterinarians to correct a severe knee deformity in a two-year-old lowland gorilla (*Gorilla gorilla*) named Ndjia. In Africa, this gorilla is protected but still faces dangers posed by poachers, Ebola, and habitat destruction, causing its numbers to decline so much that the International Union for Conservation of Nature (IUCN) Red List categorizes it as critically endangered. With Ndjia, her knee deformity gradually straightened after careful femoral growth plate surgery, a month in a body cast (a first for any gorilla), and a slow recovery (figure. 0.2). As the years passed, she became a viable breeding mate and a proud mama herself (figure 0.3). This successful gorilla growth plate surgery brought me great delight and pride.

After purchasing one of Tyree's paintings, I became fascinated by this man, his art, and the process of oil painting on velvet. My interest led me to write *Tyree: Artist of the South Pacific* (2017), which won two Gold Awards from the Independent Book Publishers Association in 2018 for Cover Design and

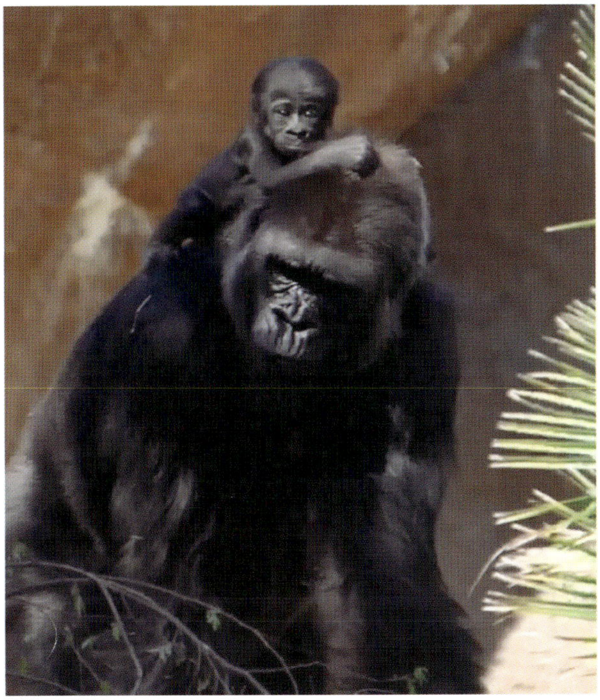

Figure 0.3 Ndjia and her baby, 20 years later, 2018, DVM San Diego Zoo and Los Angeles Zoo, courtesy of Jeff Zuba.

Biography. After writing the Tyree biography, I began investigating the man who started it all and soon released *Leeteg: Babes, Bars, Beaches, and Black Velvet Art* (2021).

Where to next, though? I had more to say about Tyree, especially his animal art, which only had one dedicated chapter in my previous book. There was more to be told and more paintings to present, thus, this book's genesis. With this book, as with my first, I had the support of the Tyree family in further exploring this fantastic man and artist, as well as finding more of his animal paintings. Tyree's granddaughter, Paige Herbert (daughter of Marda), is both my editor and coauthor. She brings her art history background to the book, conducting visual analyses and adding stories about the animals.

I am sure there are more Tyree animal paintings in the public sector—perhaps a gorilla, cheetah, or rhino, but without a master list of the animals he painted, it remains a mystery. I hope that this book will encourage folks to share their Tyree paintings, which I will post on my website (www.SouthPacificDreams.com), Facebook (cjcook), and Instagram (@cjcook1767).

With that, please enjoy the art of Ralph Burke Tyree, featuring the flora and fauna he loved so much.

Mahalo,
CJ Cook

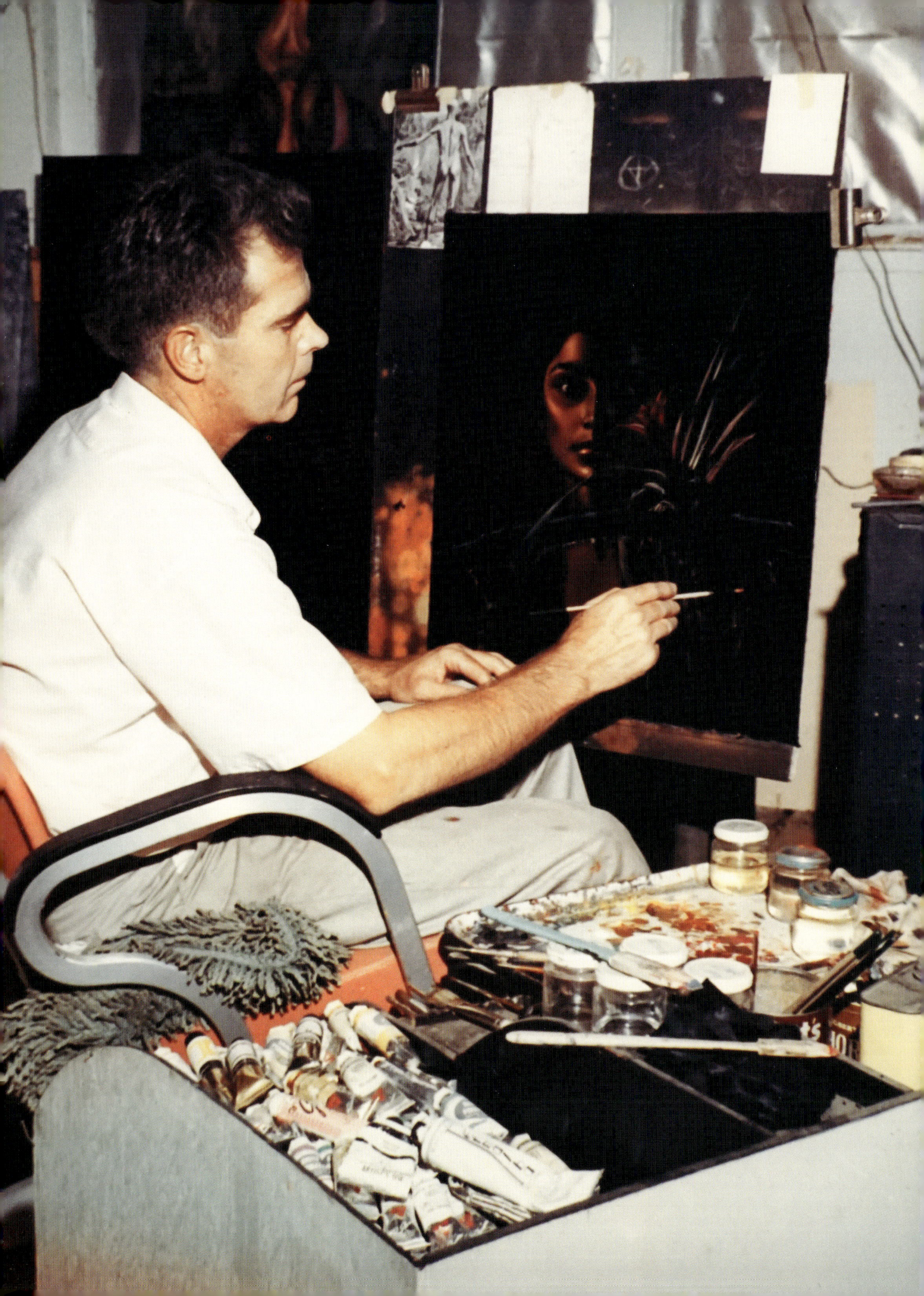

Foreword

Ralph Burke Tyree, My Father

My dad was a very private person. To say he preferred being at home painting, preparing panels, or finishing frames to going out and socializing is a huge understatement. While I was growing up, Dad's studios were always at home, either in a separate building or a room in our house, and that is usually where he could be found, painting in solitude. I recall a few close friends and patrons who would visit him in his studio over the years. Some would watch as he painted and converse for hours on end. I do not doubt that if they'd had the chance to meet, CJ Cook and my father would have become friends, sharing a love of art, nature, and history. CJ would have been one of those rare studio visitors in addition to being an avid collector and admirer of Dad's work.

I met CJ when he was starting his first book, *Tyree, Artist of the South Pacific*. His dedication to tell Dad's story and uphold his legacy moved me. That book focused on my father's early life and career and his vast catalog of Polynesian subjects, most prominently his nudes and portraits. My father was a prolific artist, and this book, Beauty in the Beast, pays tribute to his later works, which were almost exclusively animals with an emphasis on endangered species. It also features much of the flora and fauna he painted earlier in his career.

Beauty in the Beast is also a synchronistic effort that came together when I discovered CJ was seeking assistance with this book and my daughter Paige, an art historian, was looking for her next writing or editing project. Although her grandfather passed many years before she was born, working on the book allowed Paige to get to know him in an incredibly unique way by looking at his life and body of creative work from the perspective of an art historian and of a grandchild. CJ and Paige have created an impressive and timely book that accurately represents my father's work and his desire to raise awareness and preserve the animals we share with this planet. His paintings continue to serve that noble purpose with their depth and exquisite beauty. They come to life in this comprehensive book, which ties his artwork from the 1940s through the 1970s to current animal and habitat preservation themes.

My father valued living in harmony with nature, leaving broad open spaces untouched and respecting the earth's Indigenous peoples, plants, and animals. He would often go on walks through the vineyards and fields surrounding the places we lived in California, and sometimes I would tag along. As we walked, he knew which sounds to call the jackrabbits out of hiding so we could watch them bounding through the rows of crops, marveling at their speed and the strength of their spectacular

Figure 0.4 Ralph Burke Tyree painting a bromeliad as seen in figure 3.41. Circa 1965, Tyree family.

long-distance jumps. From our years in Hawaii, I have enduring memories of Dad showing us how to feed our pet goat branches of ti leaves without getting bitten ourselves and of him walking alongside my mom and brothers, balancing a gigantic bunch of bananas over one shoulder with a machete in his other hand, heading up the path to our home on a hill overlooking the ocean surrounded by sugar cane fields far from the nearest little village.

He instilled a deep love and respect for nature in all seven of us kids. He encouraged our curiosity on his own and made learning about animals fun. When my youngest brother Marc was born, Mom stayed in the hospital with him for a few days, and back at home, Dad read us our bedtime stories. As my brother Jeff and I drifted off to sleep, Dad filled our heads with his fantastical interpretations of Rudyard Kipling's *Just So Stories*, imagining how the rhinoceros got its wrinkly skin, why the camel grew a hump, or how armadillos learned to roll themselves up. He conjured up wildly inventive stories about spiders so enormous they could throw dinner plates across the room, instilling his fascination with animals in us through his imaginative tales and the books he chose to read to us.

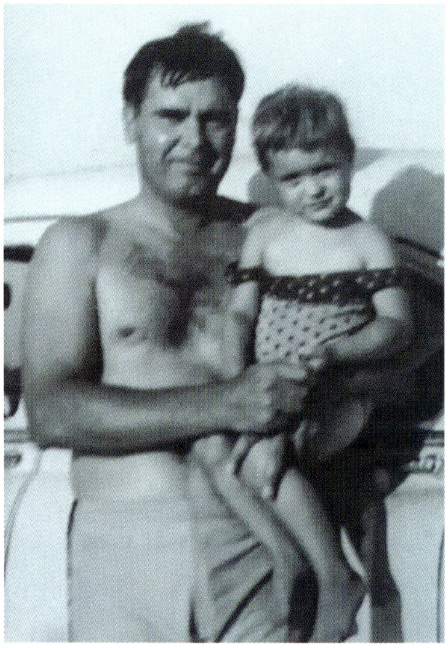

Figure 0.5 Ralph Tyree with daughter, Marda, age one. Tyree Family.

As a child in the 1960s, we did not go on vacations (we moved), and family outings were uncommon, but one place Dad took us several times was the San Francisco Zoo. I was proud to have my elephant-shaped key to unlock the exhibits. Once in a while, he would take us to a tiny local zoo with animals behind bars in concrete-lined cages and talk about how they should be housed in natural enclosures similar to their native environments, free to live closer to how they would exist in the wild. Dad dreamed of designing and constructing a vast wildlife refuge in Africa one day and discussed his vision around the family dinner table many times, speaking passionately about the need for conservation and sharing his knowledge and excitement about the extraordinary animals he was painting.

Although he made it apparent in many ways, my father expressed his deep reverence for the earth and its creatures most profoundly through his paintings, which capture their essence, visually preserving what he loved. I trust this book will inspire you with a similar sense of awe for the natural world that connects us all. Enjoy.

—Marda Tyree Herbert

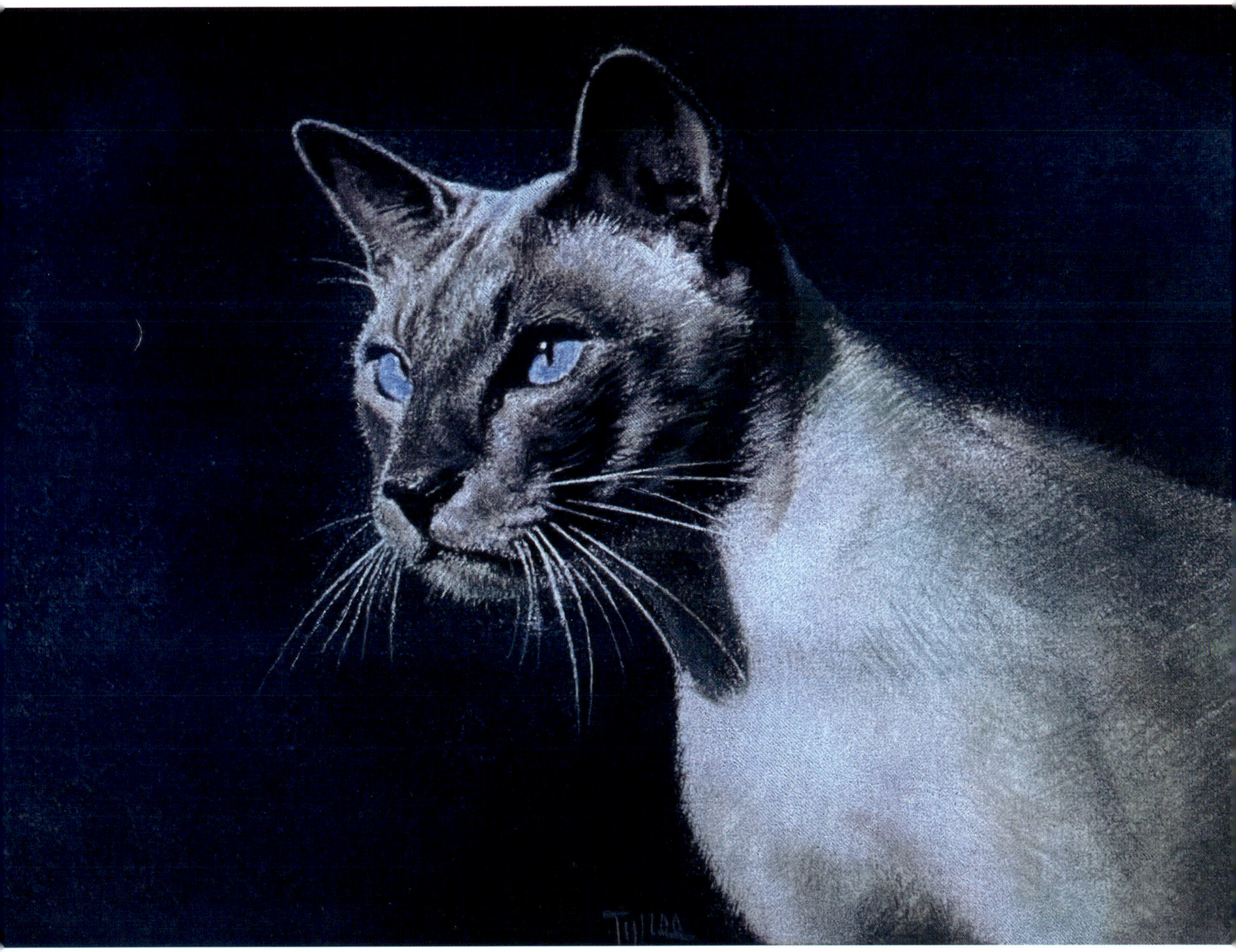

Figure 0.6 *Siamese Cat*, Ralph Burke Tyree. 1978, oil on board, 12" x 16", author's collection.

BEAUTY IN THE BEAST

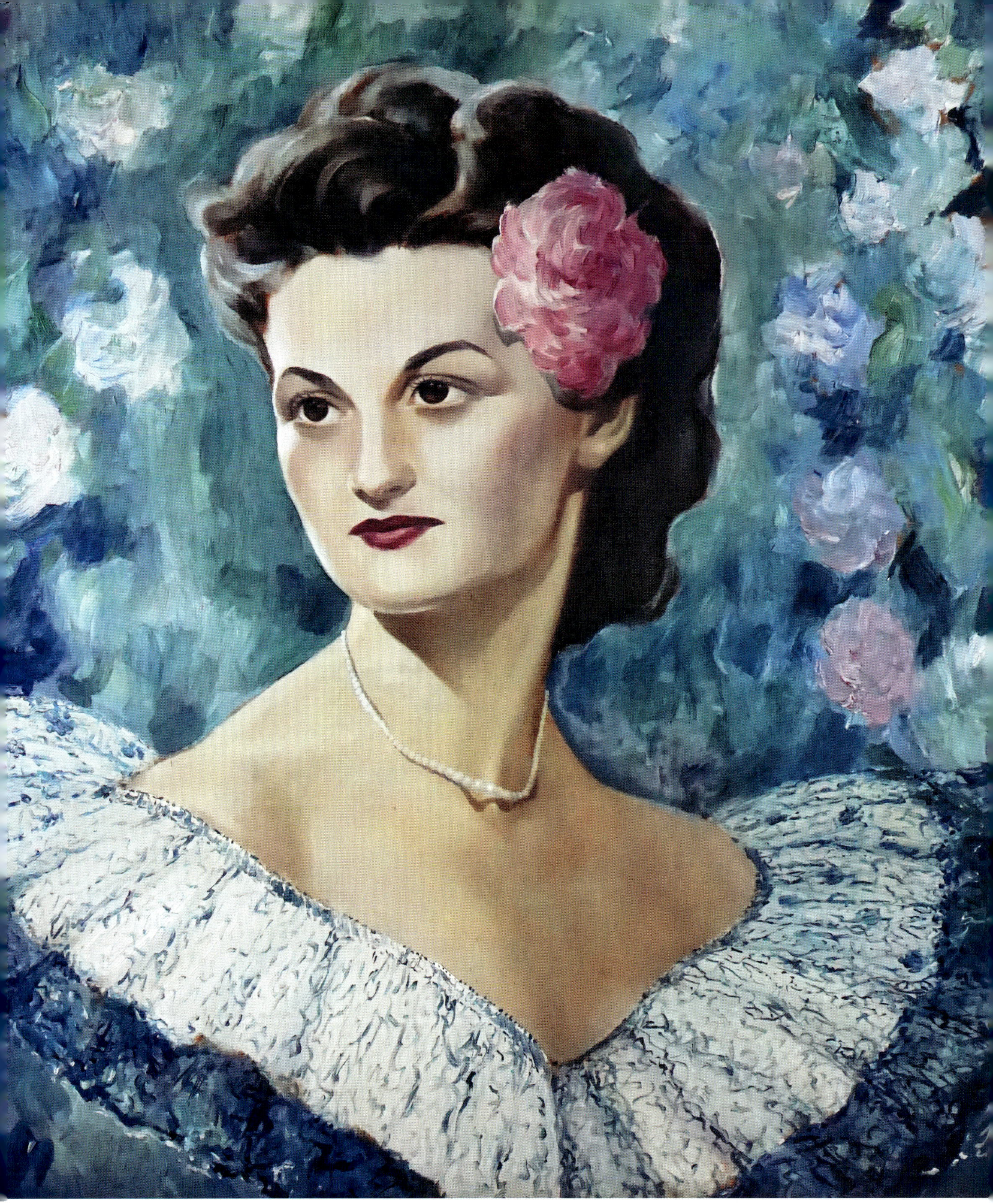

Figure 1.1 *Officer's Wife*, Ralph Burke Tyree. Samoa, 1944, oil on board, 24″ x 21″, author's collection.

CHAPTER 1

Marines and Margo

"My favorite thing to do is go where I've never been."

—ANONYMOUS

Ralph Burke Tyree was born in Irvine, Kentucky, on June 30, 1921. He was the seventh of eight children born to Sally Turpin and Charles Green Tyree. When Tyree was one, the family moved to the town of Delhi, about 20 miles south of Modesto in California's Central Valley, where his father opened a small general store. Tyree and his siblings attended grade school in Delhi and nearby Livingston (figure 1.2).

As a young lad, Tyree was creative and bold. He loved to paint and draw and was enchanted by the adventure stories of authors like Jack London and Zane Grey, dreaming of touring the South Pacific as they had. After Tyree entered high school in 1934, athletics and art dominated his spare time (figure 1.3). While he was good at both, it was art that most intrigued him, and after high school, Tyree attended junior college in Modesto, California, majoring in art.

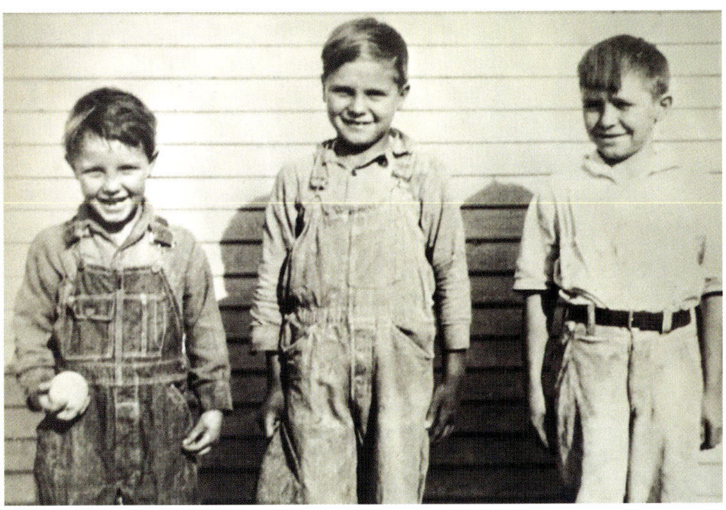

Figure 1.2 Tyree, center, with brothers Jean and Charles. Circa 1927, Tyree family.

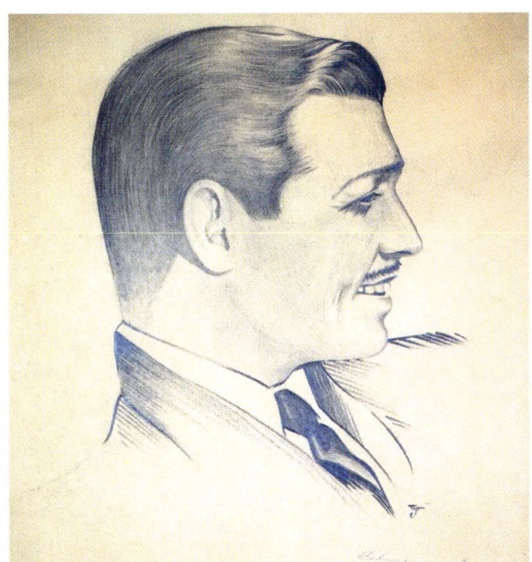

Figure 1.3 *Clark Gable*, Ralph Burke Tyree. 1937, pencil, 10" x 8", Tyree family.

Before long, Tyree won a scholarship to attend the California College of Arts and Crafts in Oakland (now the California College of Arts) for a portrait he painted of his high school sweetheart and future wife, Margo Almeida. He studied in Oakland for a year before enrolling in the San Francisco School of Fine Arts for more training. In 1939, after completing his formal art education, Tyree began to show his work, first in the Philadelphia Scholars' exhibit and the following year in San Francisco's well-known de Young Museum. He also joined Walt Disney Studios in Los Angeles to gain hands-on experience as a graphic advertising artist. Little did he know, barely a year later, world events would suddenly alter his career path.

Private Tyree

On December 7, 1941, the Japanese surprised the United States with an all-out aerial attack on the naval base at Pearl Harbor in Honolulu, Hawaii. Tyree was 20 years old and, like many Americans, passionate about his country. By late January 1942, he joined the Marines and reported for basic training at the Marine Corps Recruit Depot in San Diego, California (figure 1.4).

Private Tyree (figure 1.5) and his fellow Marines of the 22nd Regiment set sail on the *SS Lurline* for American Samoa on July 19, 1942, arriving 13 days later. Initially, he was a rifleman in the Corps but

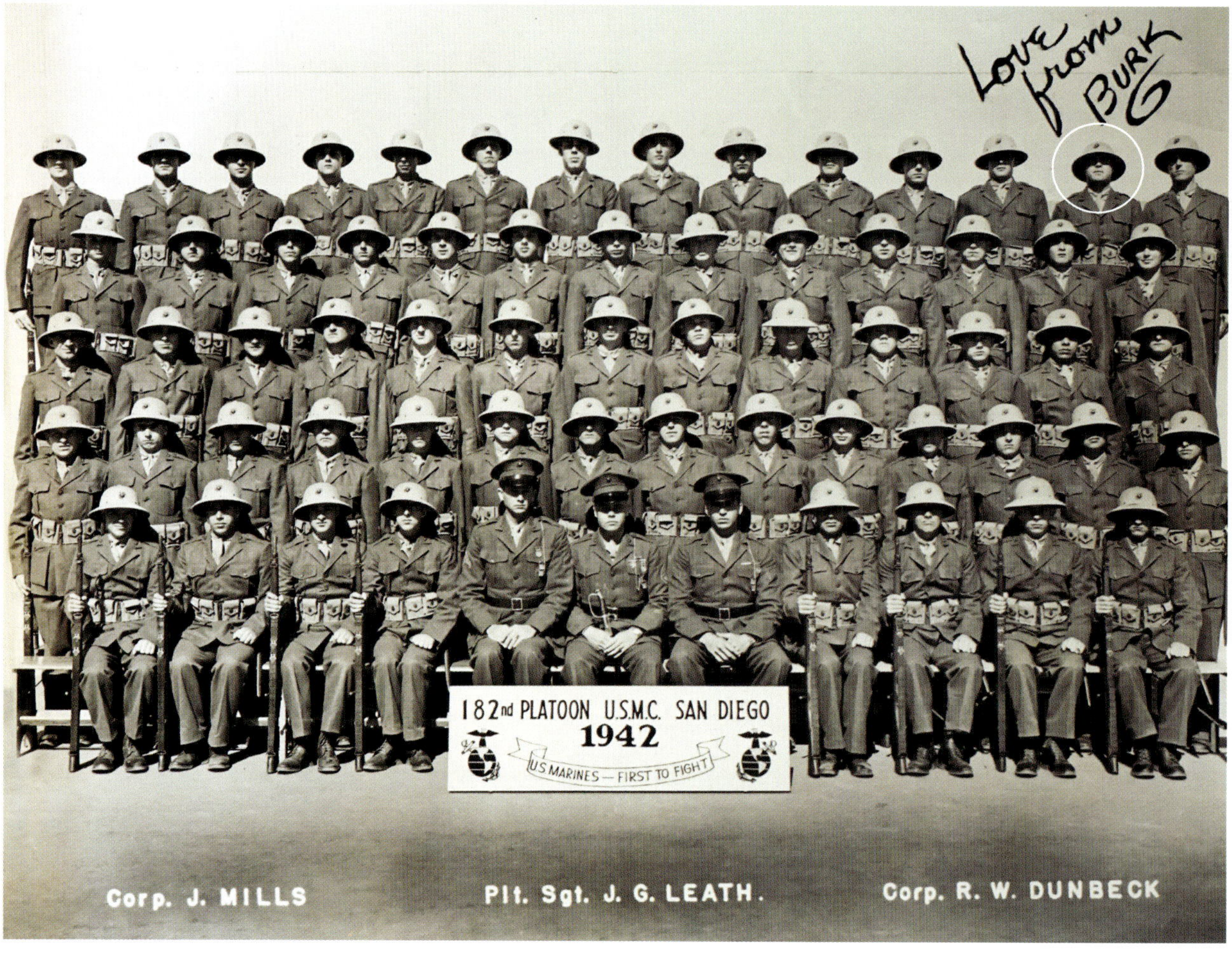

Figure 1.4 Tyree (top row circled, second from the right) with his USMC 182nd Platoon, March 1942, San Diego, Tyree family.

fortunately for him, the Marine officers soon recognized his artistic talents. Three months later, he was reassigned to make maps as a Draftsman. Before long, his designation again changed, this time to Sign Painter.

While working as a Sign Painter, Tyree's artistic skills caught the eye of his commander, General Charles Price. General Price took a liking to the young private and made him his Portrait Painter at the headquarters of the Samoan Area Defense Group (figure 1.6).

Tyree's new responsibilities included creating morale-boosting murals in the officer's mess hall, illustrating menus, painting portraits of the officers and their wives, and other artistic endeavors, in addition to his regular duties.

Many of the officers asked him to paint portraits of the officers' loved ones left back home (figure 1.1). By adorning the ladies' hair with

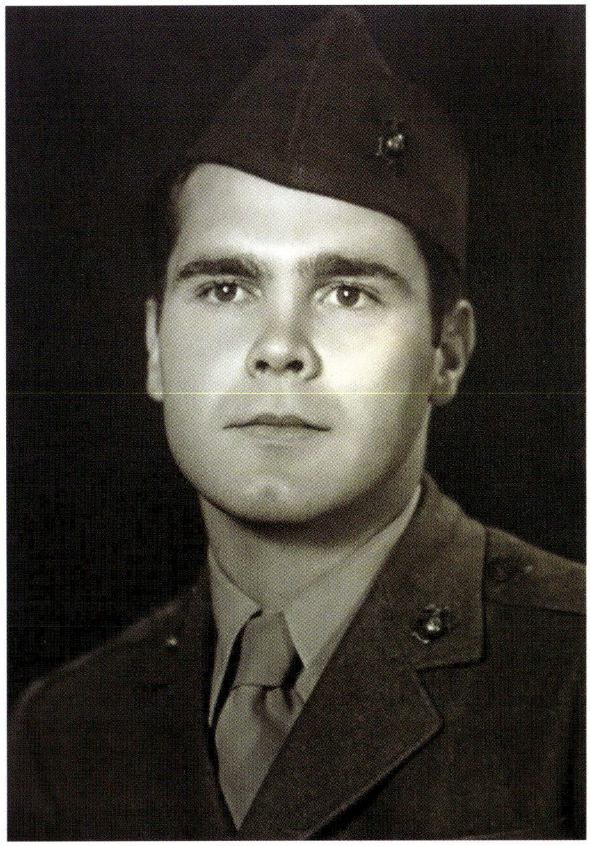

Figure 1.5 Private Ralph Burke Tyree, US Marine Corps. 1942, Tyree family.

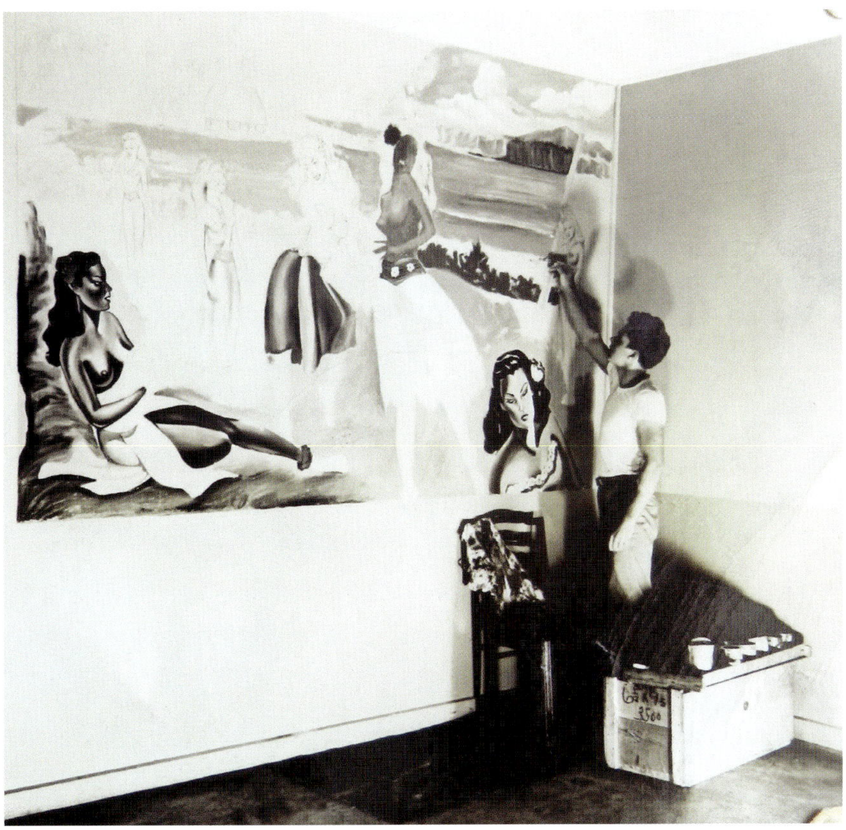

Figure 1.6 Private Tyree paints a mural in the officer's mess hall in Samoa, US Marine Corps, 1943, Tyree family.

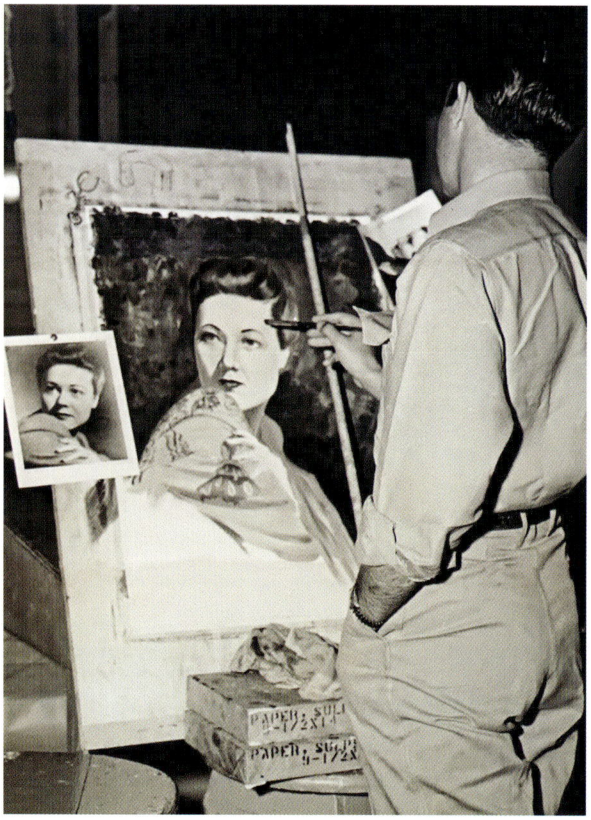

Figure 1.7 Tyree painting officer's wife in his Samoan studio. 1943, Tyree family.

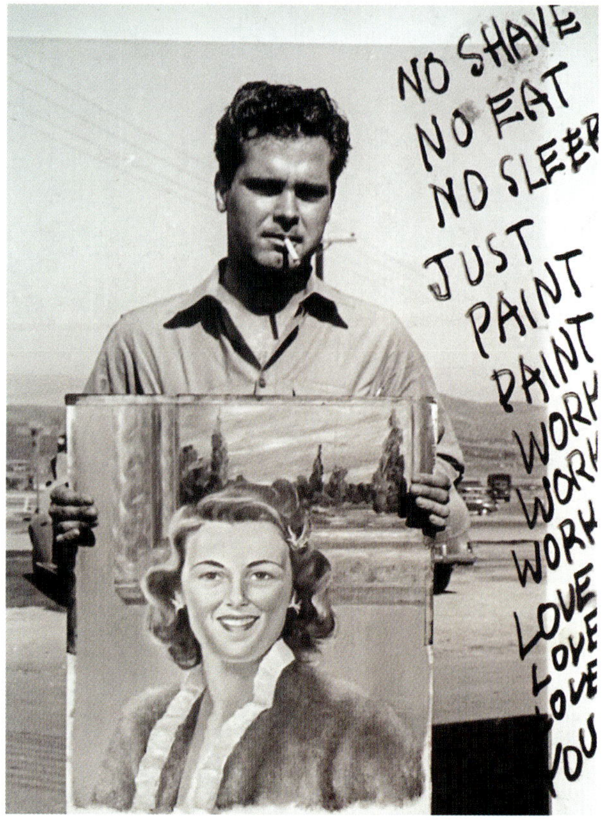

Figure 1.8 Photo sent to Margo from Samoa expressing his busy work schedule. 1942, Tyree family.

flowers, he connected these stateside lovelies to the tropical environment of the officers. He set up a makeshift studio near Commander Price's office where he could work on them (figure 1.7). All this painting quickly improved his artistic skills even further. Tyree would jump to his general's demands and work hard day and night, completing the regular duties of an enlisted Marine, creating maps and signs, and painting, writing on a photograph he sent to Margo (figure 1.8),

"No shave, no eat, no sleep, just paint, paint, work, work, work, love, love, love, you."

During his two-year stint in Samoa, Tyree fell in love with the people and islands of the South Pacific. Amidst so much destruction and death, he saw beauty. It was also at this time that he began to dabble with animal sketches, as can be seen in figure 1.9, a pencil drawing of water buffalo pulling his fellow Marines in a Samoan cart. In 1943, this piece was featured in the book *Marines at War*, edited by Aimee Crane.

Margo, His Sweetheart

While away in the South Pacific, Tyree's girlfriend back home was Marguerite Almeida, better known as Margo (figure 1.10). Born in Livermore, California, on October 1, 1922, she first met a young Tyree, who was one year her senior, during their time as students at Livingston High School. When Margo

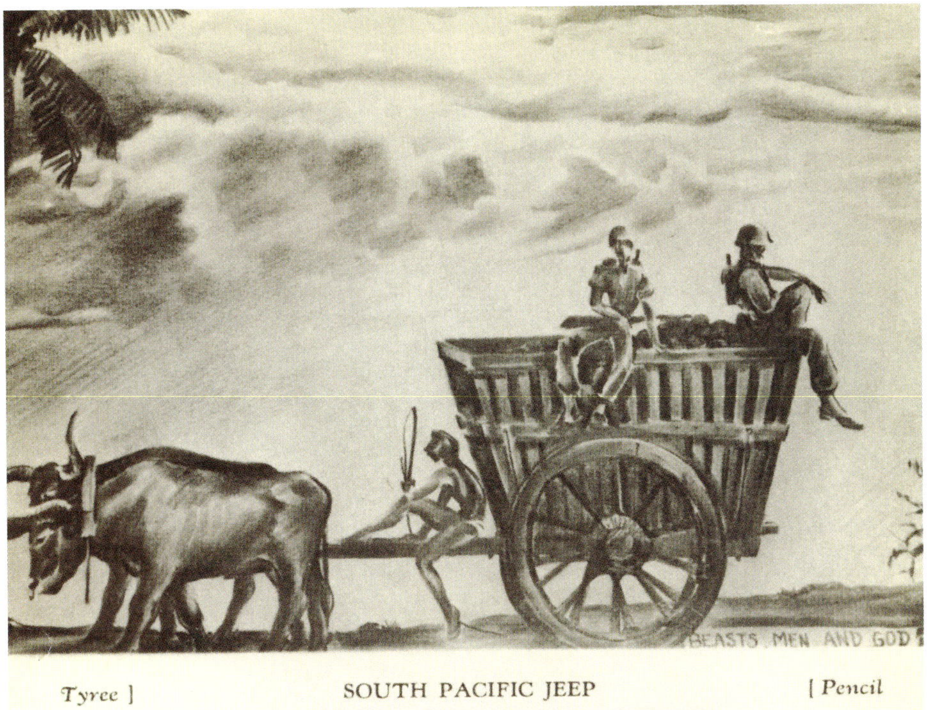

Figure 1.9 *Beasts, Men, and God*, Ralph Burke Tyree. 1943, pencil, from *Marines at War*, edited by Aimee Crane.

transferred to Turlock High, he coincidently made the same transfer the following year for better art classes, or at least that was his excuse. Tyree and Margo remained close friends after they graduated, and as previously mentioned, she posed as his model for the charcoal portrait that secured his scholarship to a prestigious art school. Their relationship continued after the war broke out and Tyree deployed to Samoa. He kept their romance alive by sending a series of bimonthly colored pencil drawings and love letters (figures 1.11–1.13).

Tyree was in Samoa for two years, returning to the US on April 8, 1944, when his boss, General Price, was reassigned to San Diego. Once again, he was put to work using his artistic talents, this time at the Camp Pendleton Marine base near Oceanside, California. The following year, on June 27, 1945, shortly before the war ended, Corporal Ralph Burke Tyree married Margo, his sweetheart of ten years (figure 1.14).

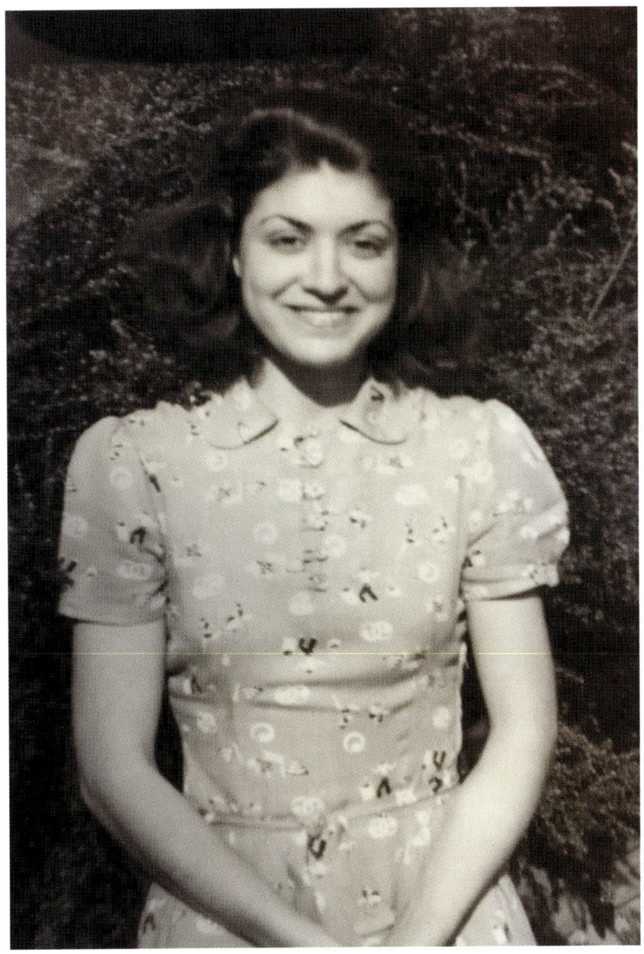

Figure 1.10 Margo Almeida as a high school senior. 1939, Tyree family.

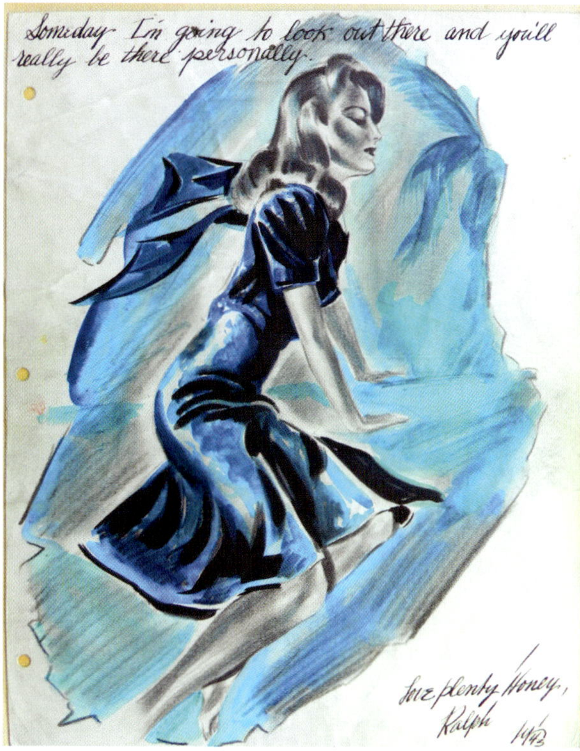

Figure 1.11 Love note featuring a woman in blue, "Someday I'm going to look out there and you'll really be there personally," Ralph Burke Tyree. Samoa, 1943, watercolor, 11" x 9", Tyree family.

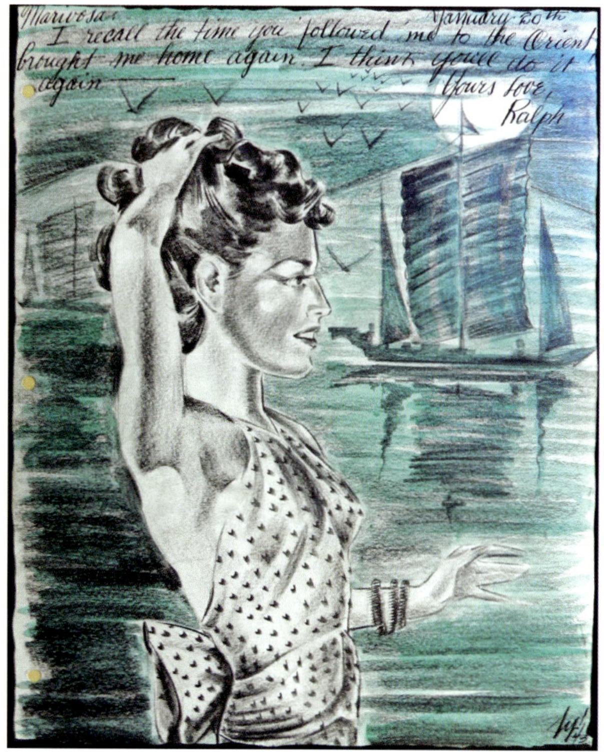

Figure 1.12 Love note, "Marivosa, you followed me to the Orient, brought me home again," Ralph Burke Tyree. Samoa, 1943, colored pencil, 11" x 9", Tyree family.

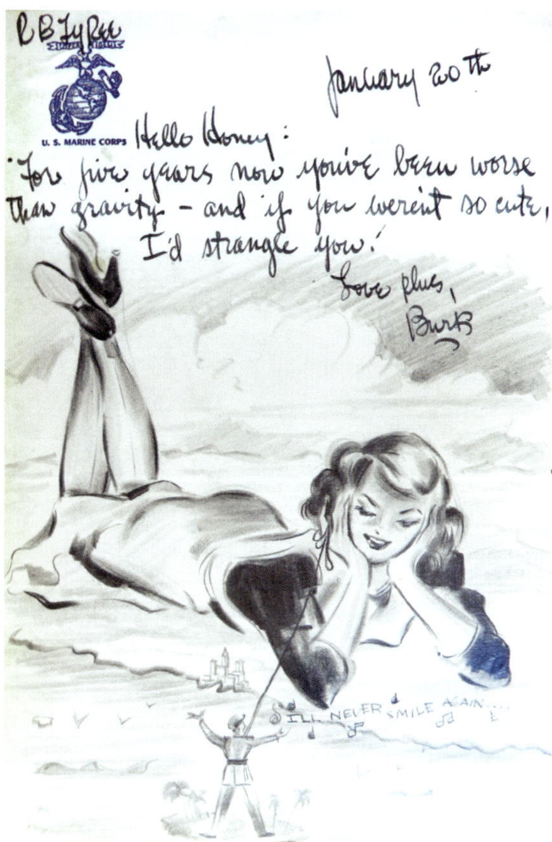

Figure 1.13 Love note sent to Margo, "Hello Honey: for five years now you've been worse than gravity—and if you weren't so cute, I'd strangle you! Love you, Burk," Ralph Burke Tyree. Samoa, 1944, pencil, 10" x 8", Tyree family.

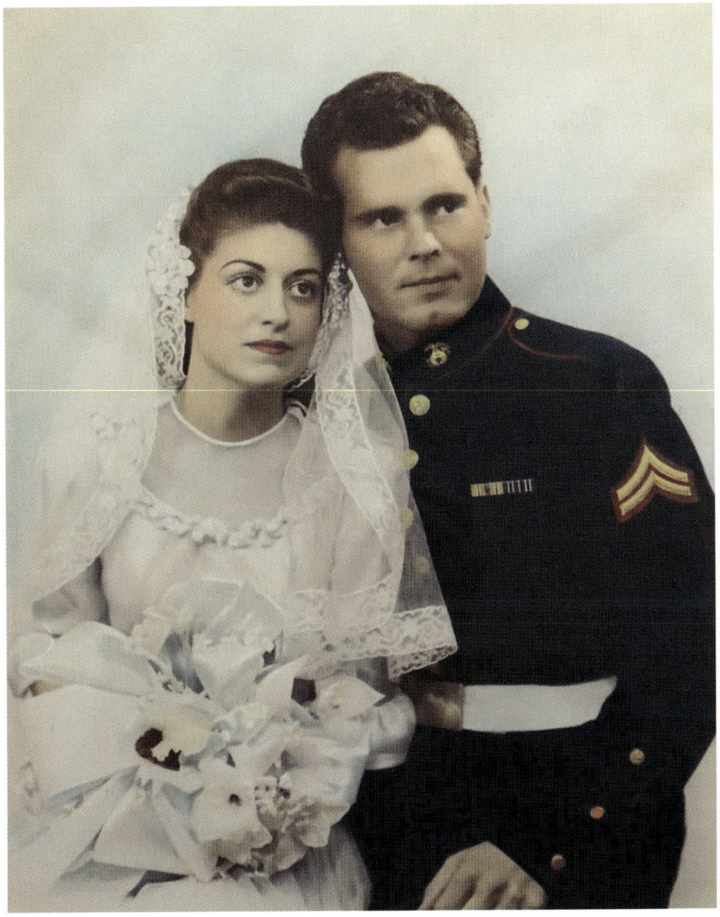

After being honorably discharged from the Marine Corps in January 1946, the 24-year-old corporal moved from Oceanside to Oakland with his new wife. Over the next six years, the family grew with baby-boomer efficiency as Margo gave birth to four sons, Greg, Michael, Steve, and Danny, one about every 15 months. During these years, the family moved back and forth between the San Francisco Bay Area and the Central Valley. This was a busy time for Tyree, who painted sporadically while also managing odd jobs to support his rapidly expanding family (figure 1.15).

Figure 1.14 Corporal Ralph Burke Tyree and his sweetheart, Margo's wedding portrait, June 27, 1945. Tyree family.

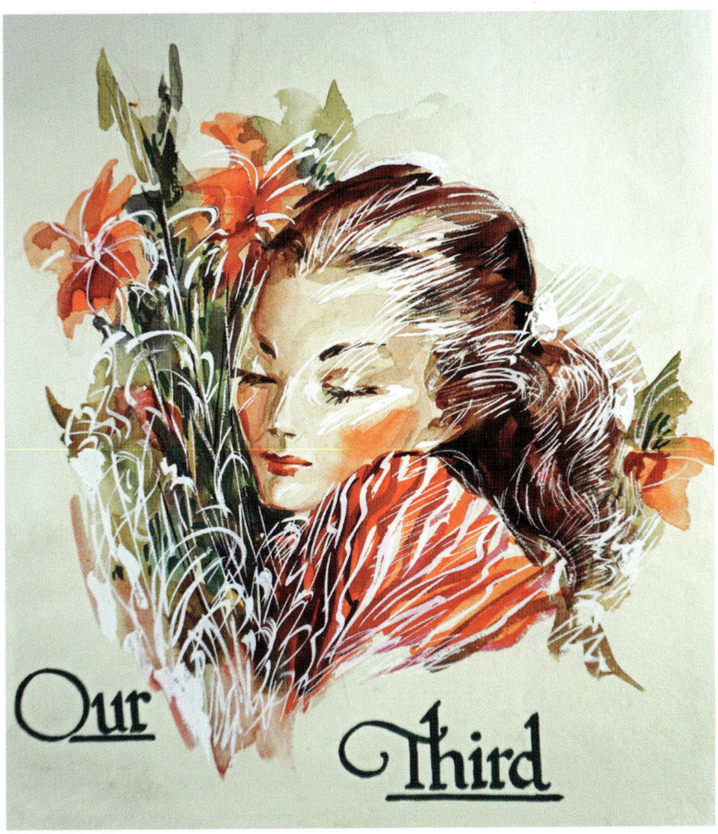

Figure 1.15 *Our Third*, painting for Margo on their third wedding anniversary, Ralph Burke Tyree. 1948, oil on board, 16" x 12", Tyree family.

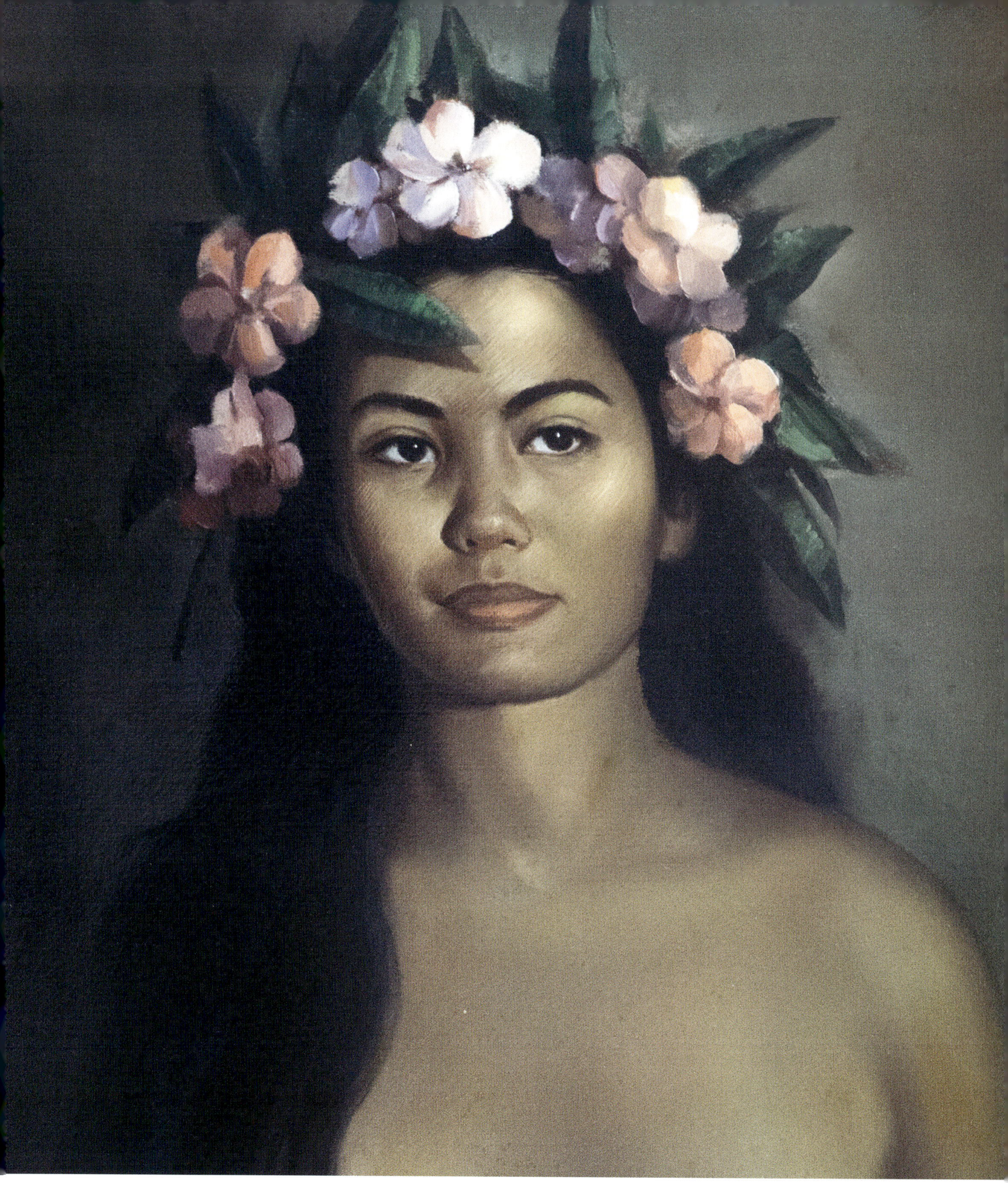

Figure 2.1 *Wahine with Plumeria Lei Po'o*, Ralph Burke Tyree. 1960, oil on board, private collection.

CHAPTER 2

Some Flowers in Her Hair

"Be Sure to Wear Some Flowers in Your Hair"

—WRITTEN BY JOHN PHILLIPS AND SUNG BY SCOTT MCKENZIE, 1967

A meaningful change came early in 1952 when Tyree, yearning to return to the South Pacific, accepted a civilian position drafting maps for the Navy on the island of Guam. Six months after he arrived on the island, Margo and their four boys joined him (figure 2.2).

This job brought more spare time and better pay, officially allowing Tyree to launch his professional art career. He set up a home studio and opened his first art business, South Sea Oils (figures 2.3 and 2.4). Primarily using oil paints on board, he preferred painting portraits of island wāhine (figure 2.1). Inspired by his Samoan experience, he always included flowers in the subject's hair or lush, tropical backgrounds (figure 2.5), sometimes painting the exotic flora on its own en plein air (figure 2.6). This inclusion of flora, at times impressionistic, at others realistic, accentuated his subjects and established the location as the tropics. Margo frequently acted as his model, often posing in a sarong, showing off her figure on a beach, near a stream, or over the ocean cliffs of Guam.

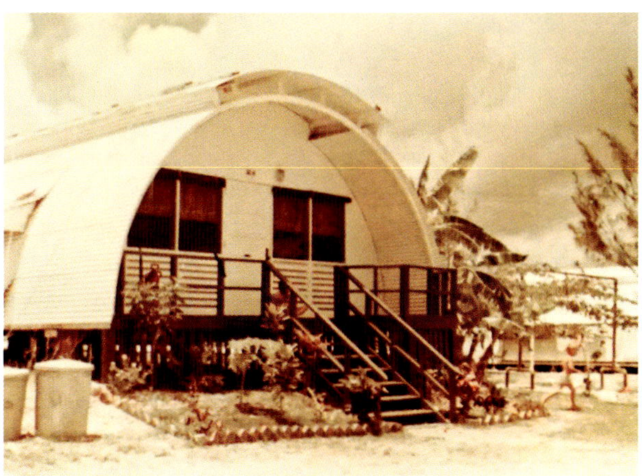

Figure 2.2 Tyree family Quonset hut, their home in Guam from 1952–55. Tyree family.

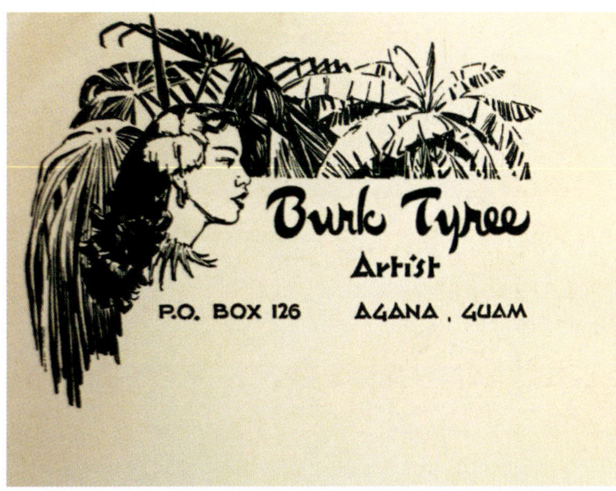

Figure 2.3 Tyree's stationery featuring palms and flowers, Ralph Burke Tyree. 1953, pen and ink, private collection.

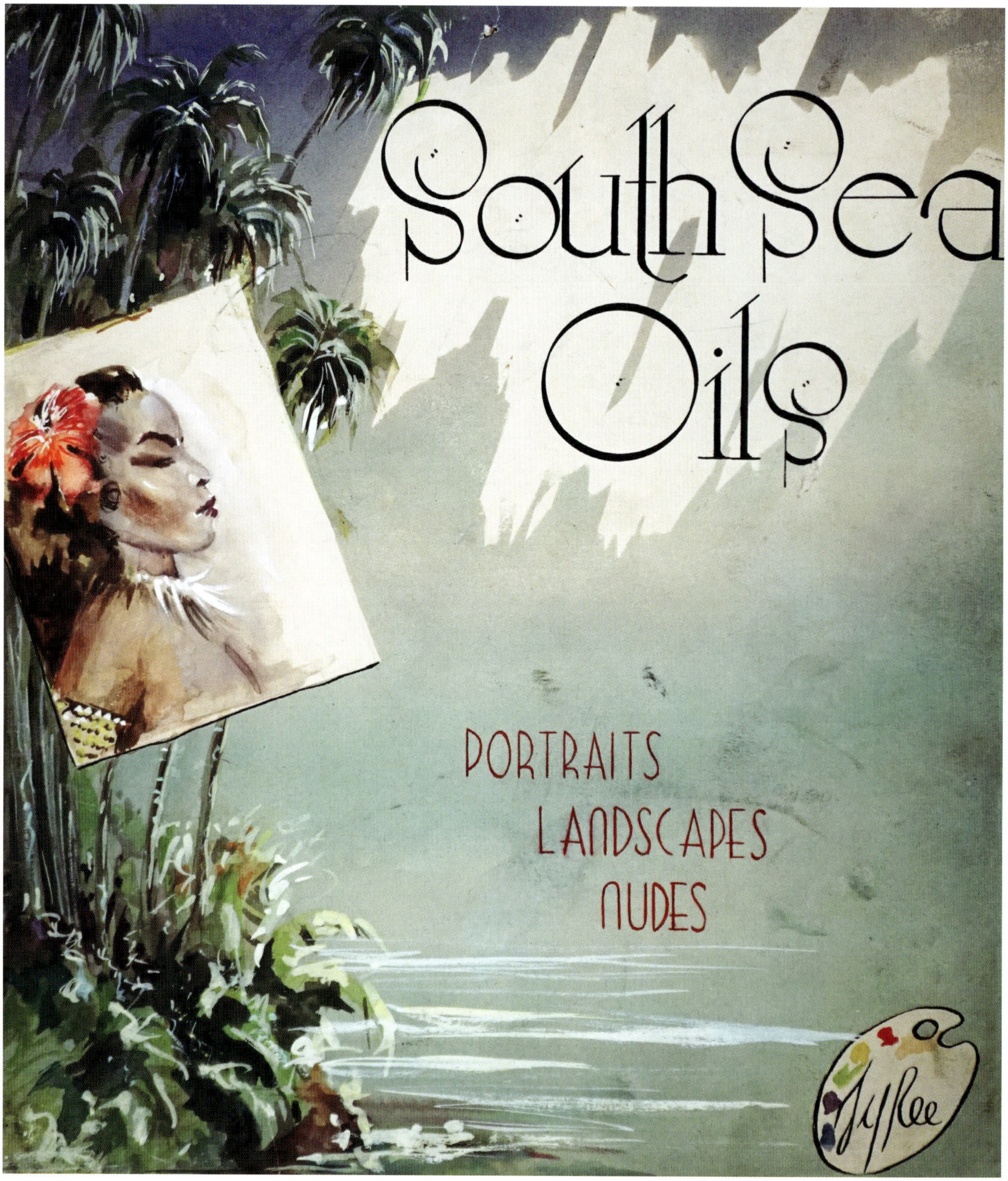

Figure 2.4 Advertisement for his art business, South Sea Oils: Portraits, Landscapes, and Nudes, Ralph Burke Tyree. Guam, 1952, oil on board, 16" x 12", Tyree family.

After three years in Guam working for the Navy and establishing his painting business, the Tyree family moved to Honolulu (figure 2.7). Margo raised the children, now five in total, with the arrival of their first and only daughter, Marda. Meanwhile, Tyree worked as an artist full-time, his career blossoming with the tourist trade in Hawaii. During this time, Tyree continued to focus on his favorite subject matter, Hawaiian beauties adorned with flowers.

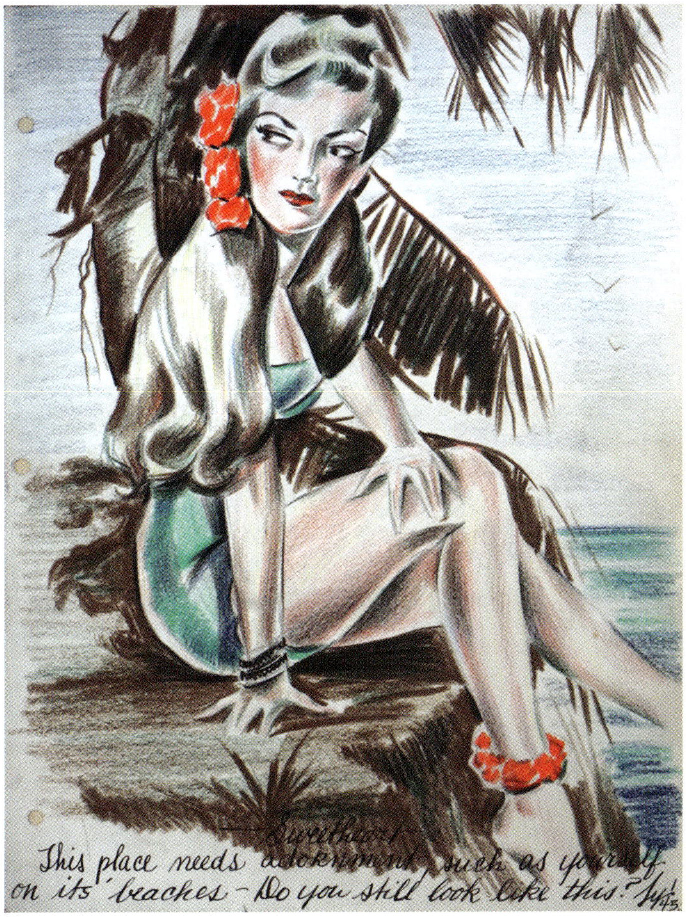

Figure 2.5 Tyree love letter featuring Margo with flowers in her hair, "Sweetheart: This place needs adornment such as yourself on its beaches- Do you still look like this?" Ralph Burke Tyree. 1943, colored pencil and watercolor, 11" x 9", Tyree family.

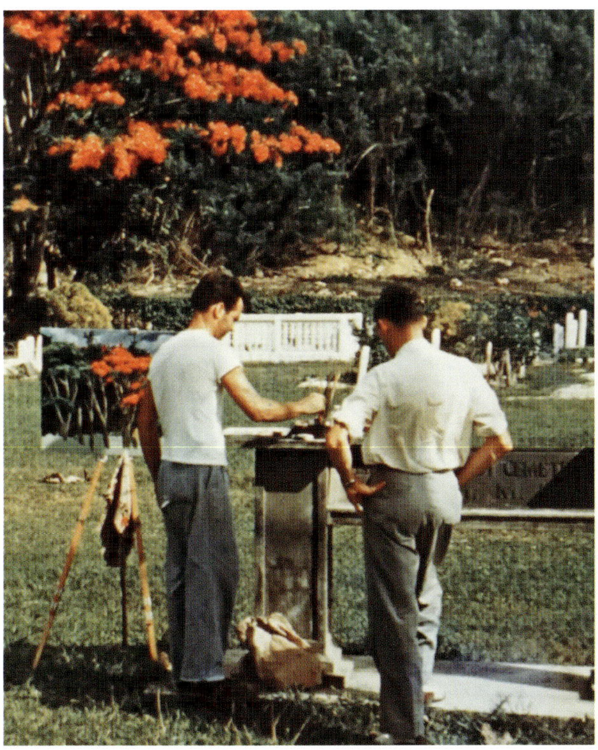

Figure 2.6 Tyree painting poinciana tree en plein air. 1954, Guam, Tyree family.

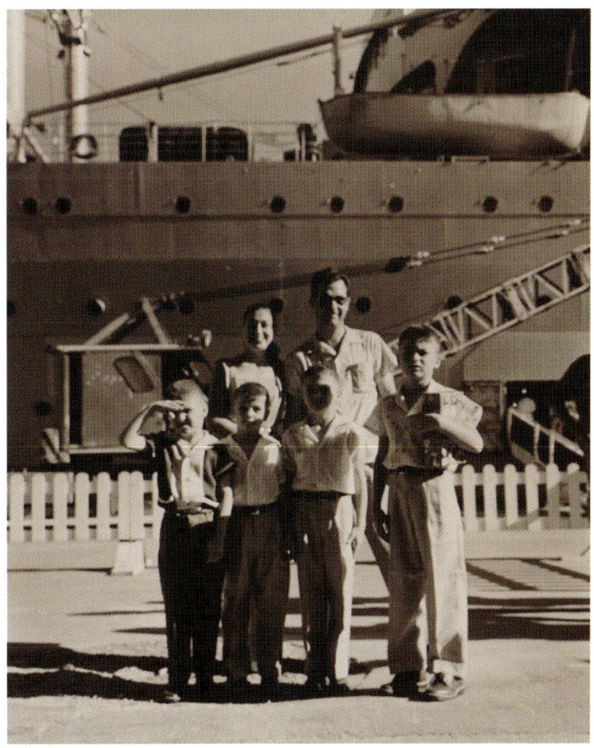

Figure 2.7 Tyree family heading from Agana, Guam to Honolulu, Hawaii, December 1955. Tyree family.

Leis and Lei Po'o

A **lei** is a wreath of objects like kukui nuts, shells, or flowers strung together as a necklace, common throughout Polynesia (figure 2.8). Usually, the lei is given to tourists and dignitaries upon arrival and sometimes departure as a greeting and a symbol of aloha. Floral leis are most often made with plumeria, orchid, tuberose, and pikake flowers strung with *maile* and *ti* leaves. In the tropics, a suitor will adorn the object of his affection with a lei instead of giving them a bouquet. The prettiest of all floral accouterments is the lei *po'o,* or head lei, made with flowers, feathers, leaves, or raffia braided together (figure 2.9). Sometimes called a *haku* lei, which means braided, the proper name is lei po'o (figures 2.10–2.13). While a lei of flowers laced on a line is simple to make, it takes a lot of time and patience to create a lei po'o. Hawaiians primarily use these beautiful time-consuming arrangements at weddings and formal parties (figure 2.14).

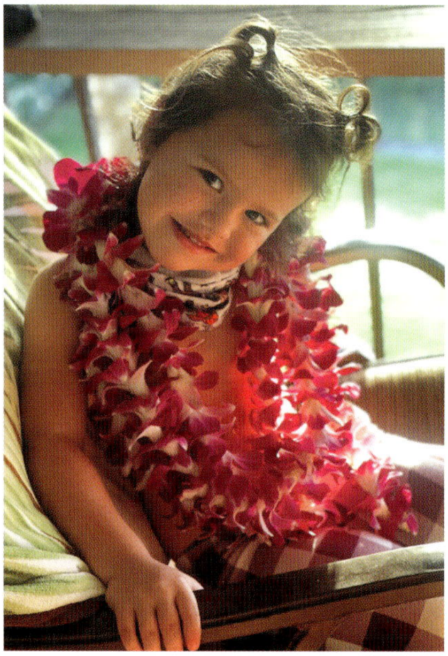

Figure 2.8 Scotlyn, my granddaughter, wearing leis, 2022. Author's photo.

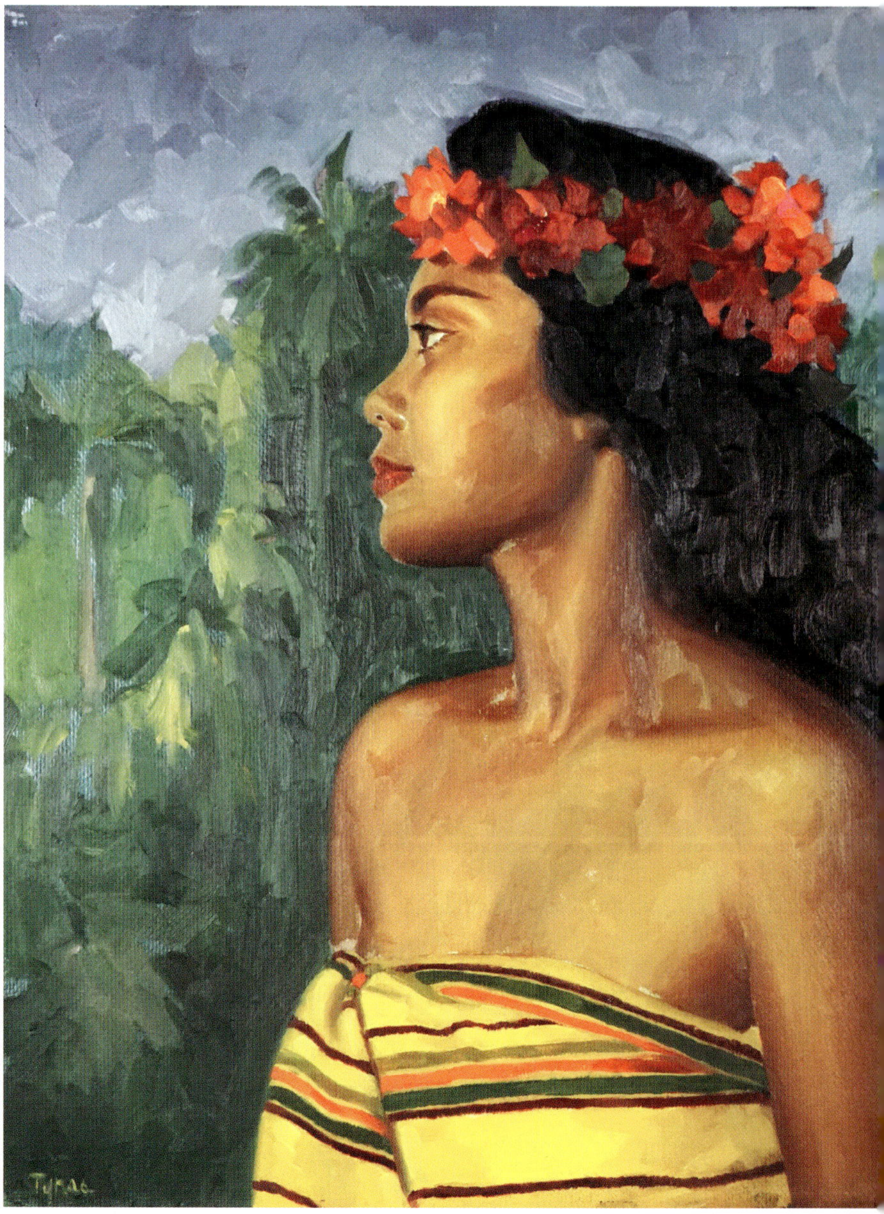

Figure 2.9 *Hawaiian Maiden with Lei Po'o,* Ralph Burke Tyree. 1965, oil on board, 16" x 13", author's collection.

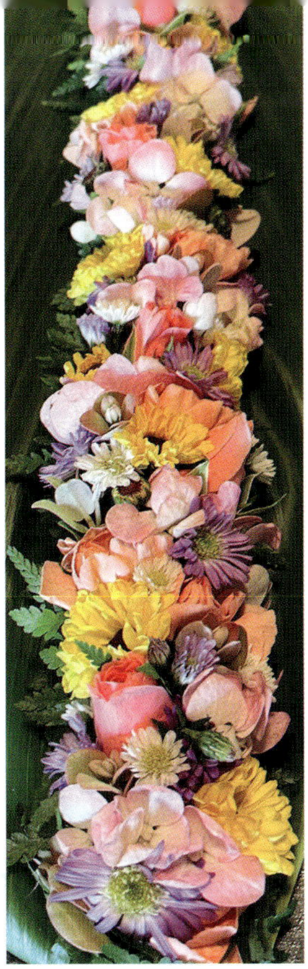

Figure 2.10 Bridal lei po'o created by Mia Thompson. Oahu, 2021, author's photo, courtesy of the floral artist.

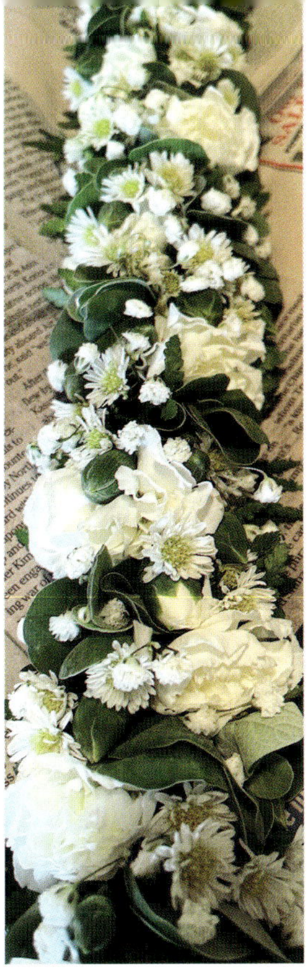

Figure 2.11 Bridal lei po'o created by Mia Thompson. Oahu, 2021, author's photo, courtesy of the floral artist.

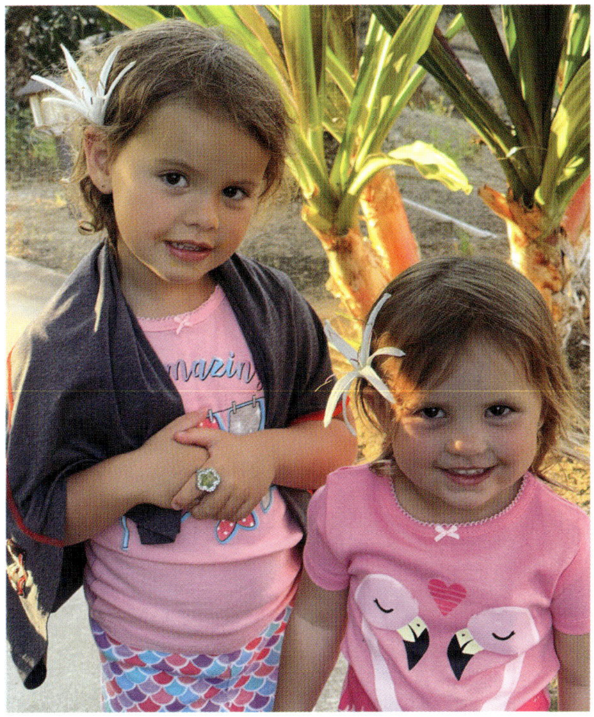

Figure 2.15 Scotlyn and my other granddaughter, Sadie, sporting spider lilies behind their right ears. 2022, author's photo.

Flowers are also often worn individually, over the ear (figure 2.15). Polynesian women use these flowers in their hair to symbolize availability for dating. A flower behind the right ear says, "I'm available," but behind the left ear means, "Stay away. I'm taken."

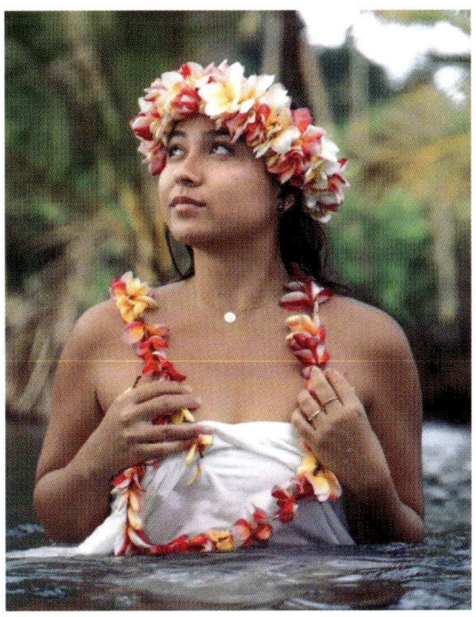

Figure 2.12 Lei and lei po'o worn by artist and model Tinaia Lucas (@Leipoo_Tahiti). Tahiti, 2021, Instragram, photo by Manu Tefan, courtesy of the floral artist.

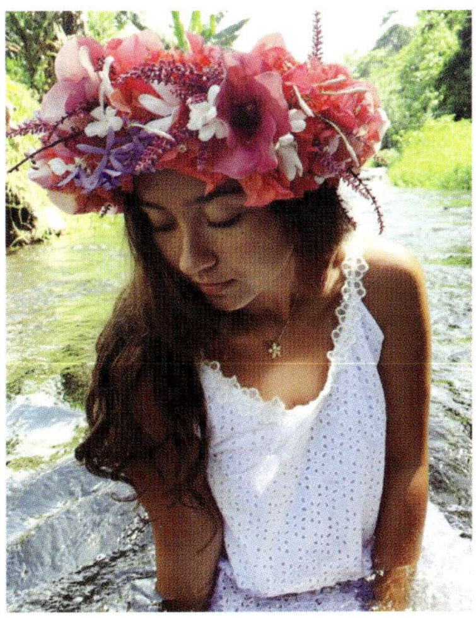

Figure 2.13 Lei po'o worn by artist and model Tinaia Lucas (@Leipoo_Tahiti). Tahiti, 2021, Instagram, photo by Manu Tefan, courtesy of the floral artist.

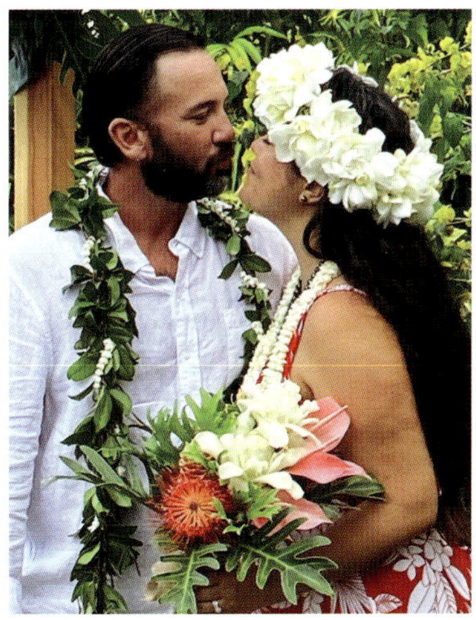

Figure 2.14 Wedding ceremony with the bride wearing a po'o and the groom sporting a maile lei. Floral art by Mia Thompson. Oahu, 2018, courtesy of the floral artist.

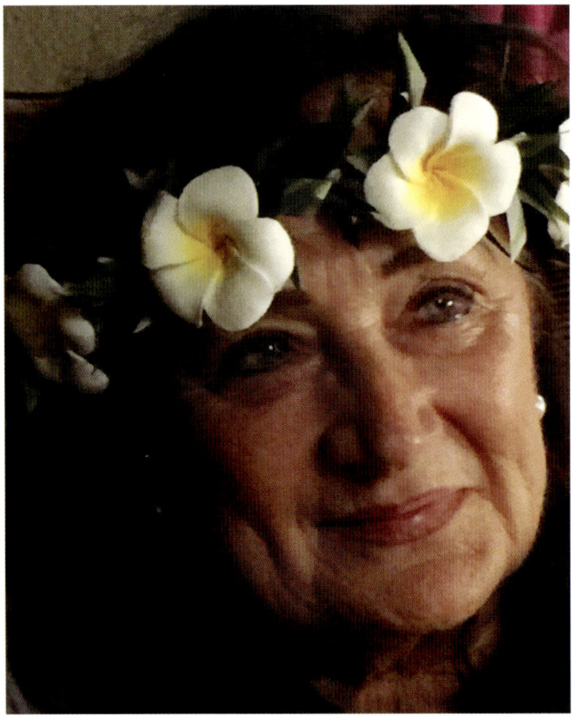

Figure 2.16 Wahine with plumeria lei po'o. Maui, 2019, author's photo.

Plumeria

First of Tyree's favorite flowers was the **plumeria** (*Plumeria*), also known as frangipani (figures 2.16–2.22). Most species of this genus of fragrant flowering trees in the *Apocynaceae* family are deciduous trees that shed their leaves annually. These small trees only reach about 20 to 30 feet in height and are widespread in subtropical and tropical regions worldwide. The flowers become extremely fragrant at night, luring pollinating insects despite their lack of nectar. It is the traditional flower of the lei and lei po'o, but the flowers are short-lived—the fragrance is gone within a couple of days, and the flowers turn brown. More than four hundred cultivars of this colorful and sweet-scented plant have been commercially developed through selective breeding.

Most plumeria can be propagated easily by drying stem cuttings and planting them in well-drained soil. However, they are easily susceptible to rot, as I have learned with many attempts in my California backyard. Even once a young plant has rooted, overwatering, especially if the soil is not draining well, frequently causes the plant's demise. With its permeable volcanic soil and frequent rain showers, Hawaii is the perfect environment for this fragrant flora to thrive.

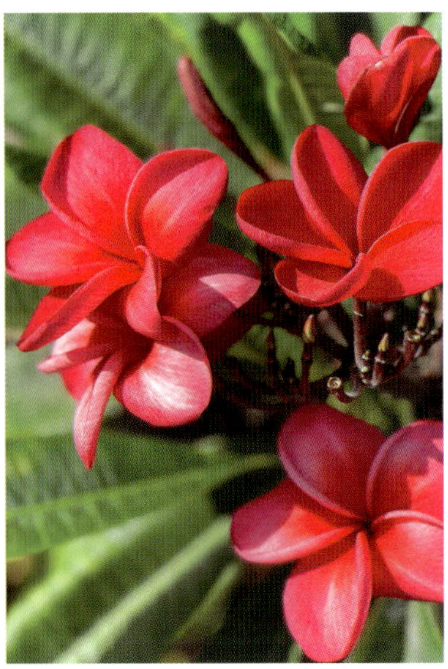

Figure 2.17 Fuchsia-colored plumeria. Maui, 2021, author's photo.

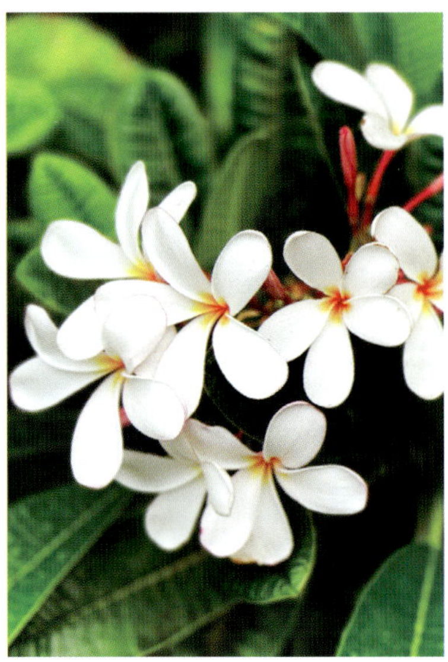

Figure 2.18 White plumeria. Maui, 2021, author's photo.

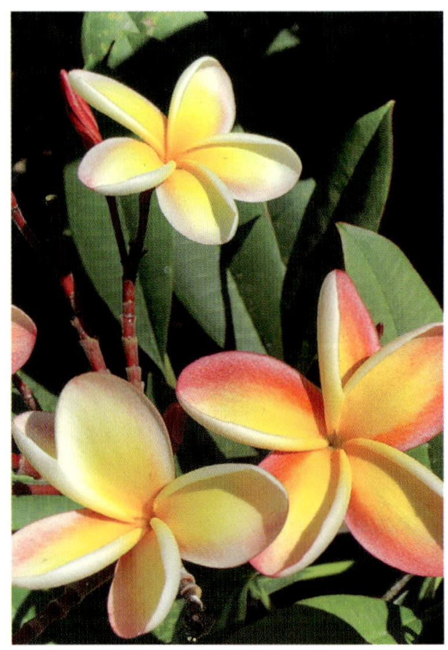

Figure 2.19 Rainbow plumeria. Maui, 2020, author's photo.

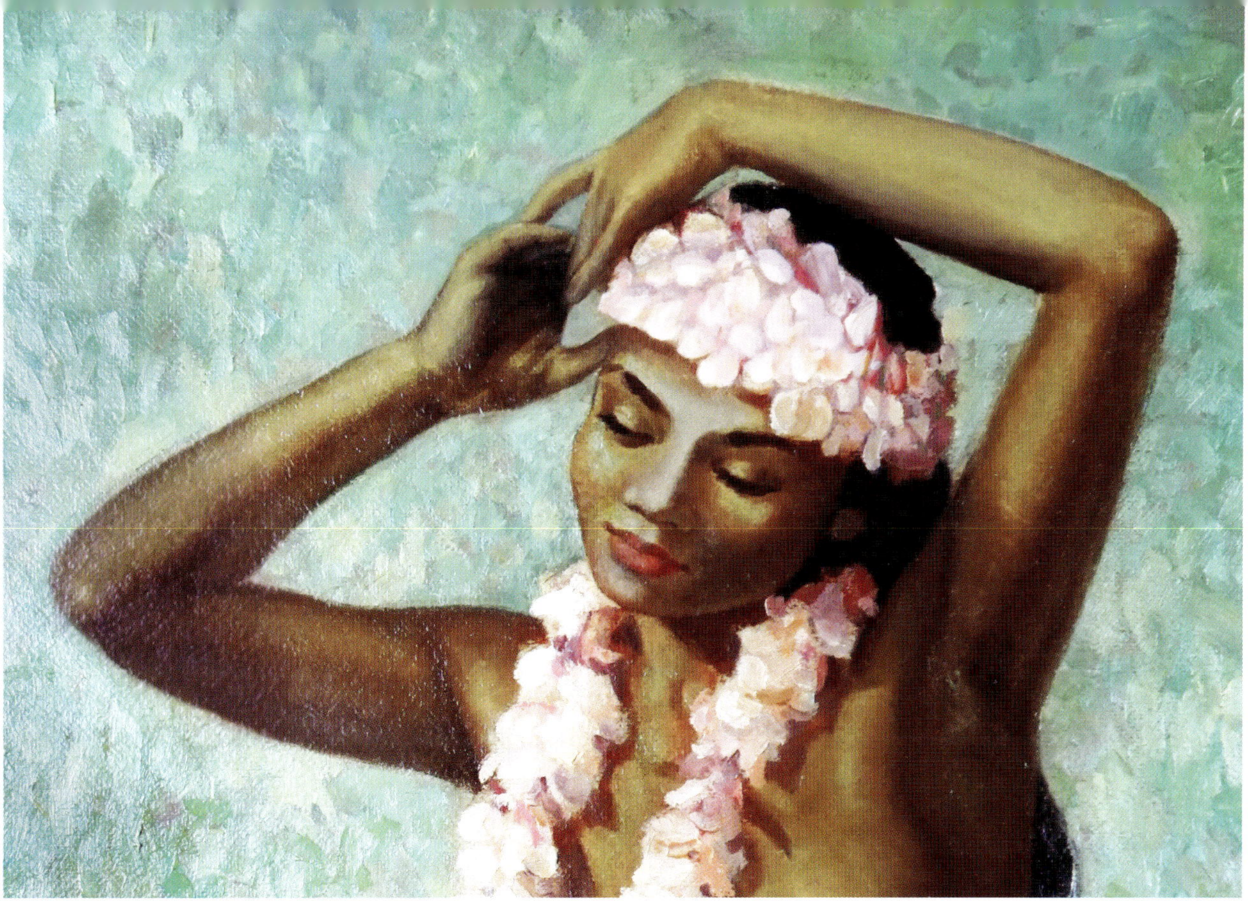

Figure 2.23 *Plumeria Lei & Lei Po'o*, Ralph Burke Tyree. 1970, oil on board, private collection.

Tyree includes the plumeria in many of his portraits of wahine, using the traditional flowers to draw attention to the model and indicate her South Pacific locale. These beautiful flowers bring a splash of color into otherwise often neutral compositions. Throughout his works, Tyree illustrates the many ways the plumeria is worn, including a singular flower over the ear (figure 3.9), a string of them as a lei (figures 3.10 and 3.12), and a magnificent crowning lei po'o (figures 2.1, 2.9, and 2.23).

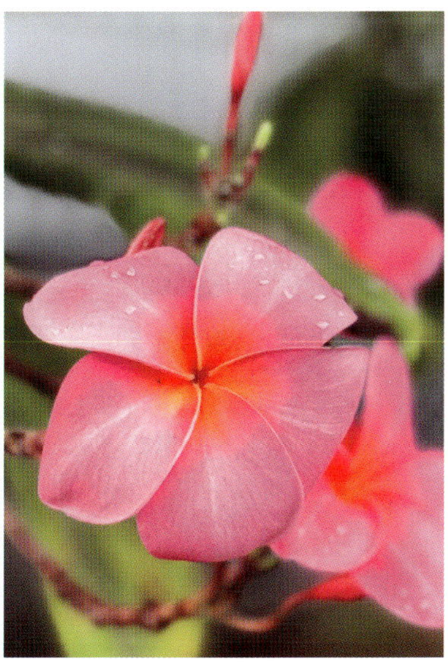

Figure 2.20 Pretty pink plumeria. Maui, 2021, author's photo.

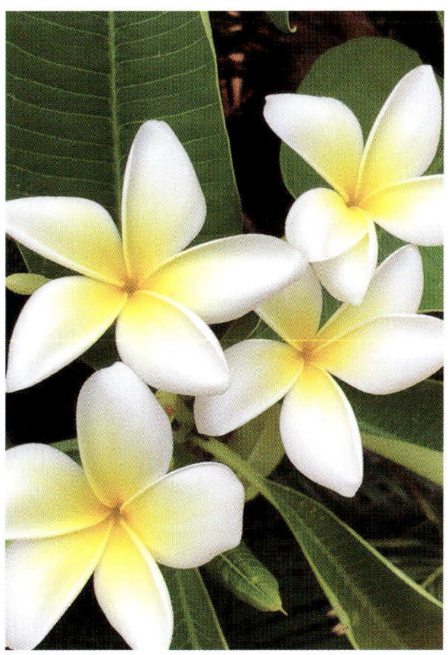

Figure 2.21 Yellow white plumeria. Maui, 2021, author's photo.

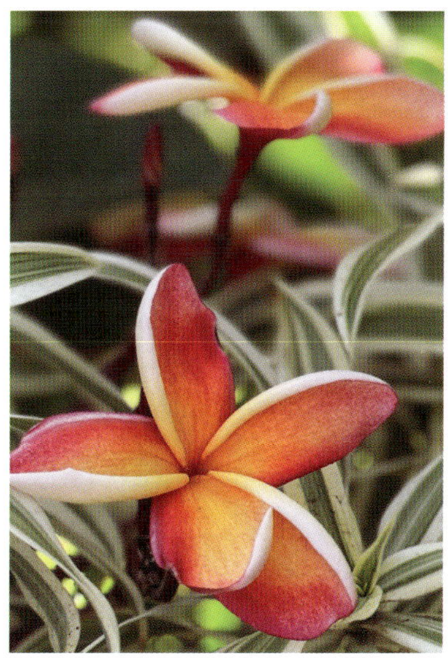

Figure 2.22 Rainbow plumeria. Maui, 2021, author's photo.

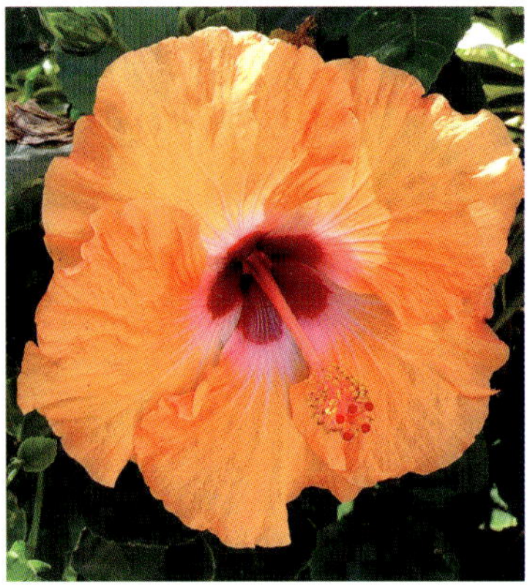

Figure 2.24 Orange hibiscus hybrid at its full eight-inch diameter. California, 2021, author's photo.

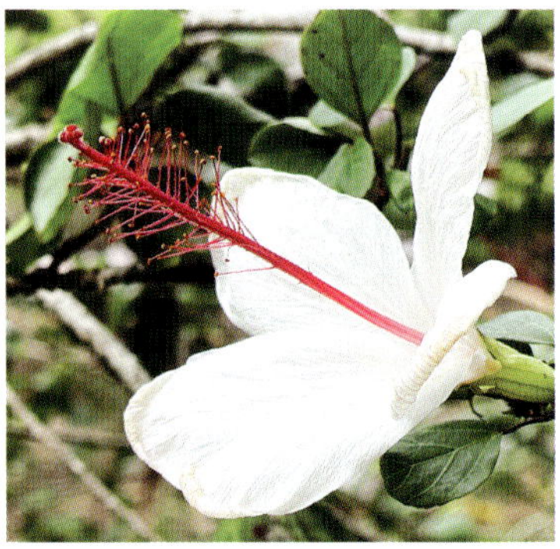

Figure 2.25 Native Hawaiian white hibiscus. Maui, 2021, author's photo.

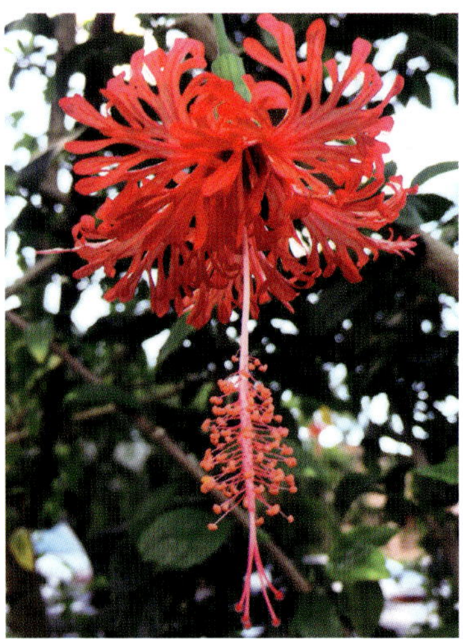

Figure 2.26 Hibiscus schizopetalus, which means split petal, discovered in Madagascar. This variety is now highly sought after. Maui, 2019, author's photo.

Hibiscus

Tyree's other preferred blossom was the **hibiscus** (*Hibiscus*), a genus of flowers in the *Malvaceae* family found throughout subtropical and tropical regions. Either a midsized bush or a small tree, these plants are recognizable for their large, showy flowers (figures 2.24–2.29). As a nyctinastic plant, the hibiscus has a circadian cycle, opening its flowers during the day and closing them at night. This evolutionary movement of the plant protects the flower from predators and saves its floral brilliance for daytime insects.

Including both annual and perennial varieties, there are several hundred natural species of hibiscus. These versatile flowers are most often used in landscaping, but also to make paper and rope, in tea, medicine, and jam, and even eaten on their own. With their increased popularity, commercial growers have developed hundreds of new, even larger species through hybridization. A personal favorite, my garden has a variety of these unique and spectacular flowers.

Tyree loved the hibiscus, often featuring it in his paintings. Like the plumeria, it is the perfect embellishment against his models' usually dark hair, maybe better due to its considerable size and dynamic jagged petals. He often includes a single hibiscus placed over the ear (figures 2.30, 3.1, and 3.8) or many blossoms strung together as a lei or a crowning lei po'o (figures 2.31, 3.3, and 3.11). In one case, he paints his model underneath a blooming hibiscus plant (figure 3.6), and in another, he uses it as an adornment for a Hawaiian Chief (figure 3.5). Tyree also paints the hibiscus on its own, as the subject (figure 3.7) and within a still life featuring tropical fruits (figure 12.2). These colorful and iconic flowers instantly transport the viewer to the tropical locales he sought to evoke.

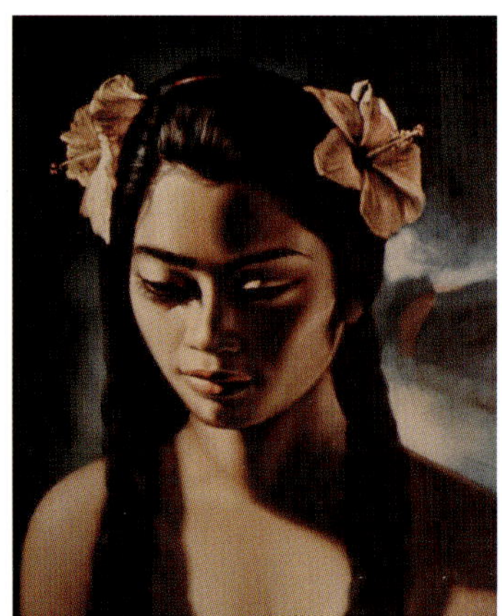

Figure 2.30 *Wahine With Braids and Hibiscus on Both Stides*, Ralph Burke Tyree. Circa 1964, oil on velvet, 16" x 12", private collection.

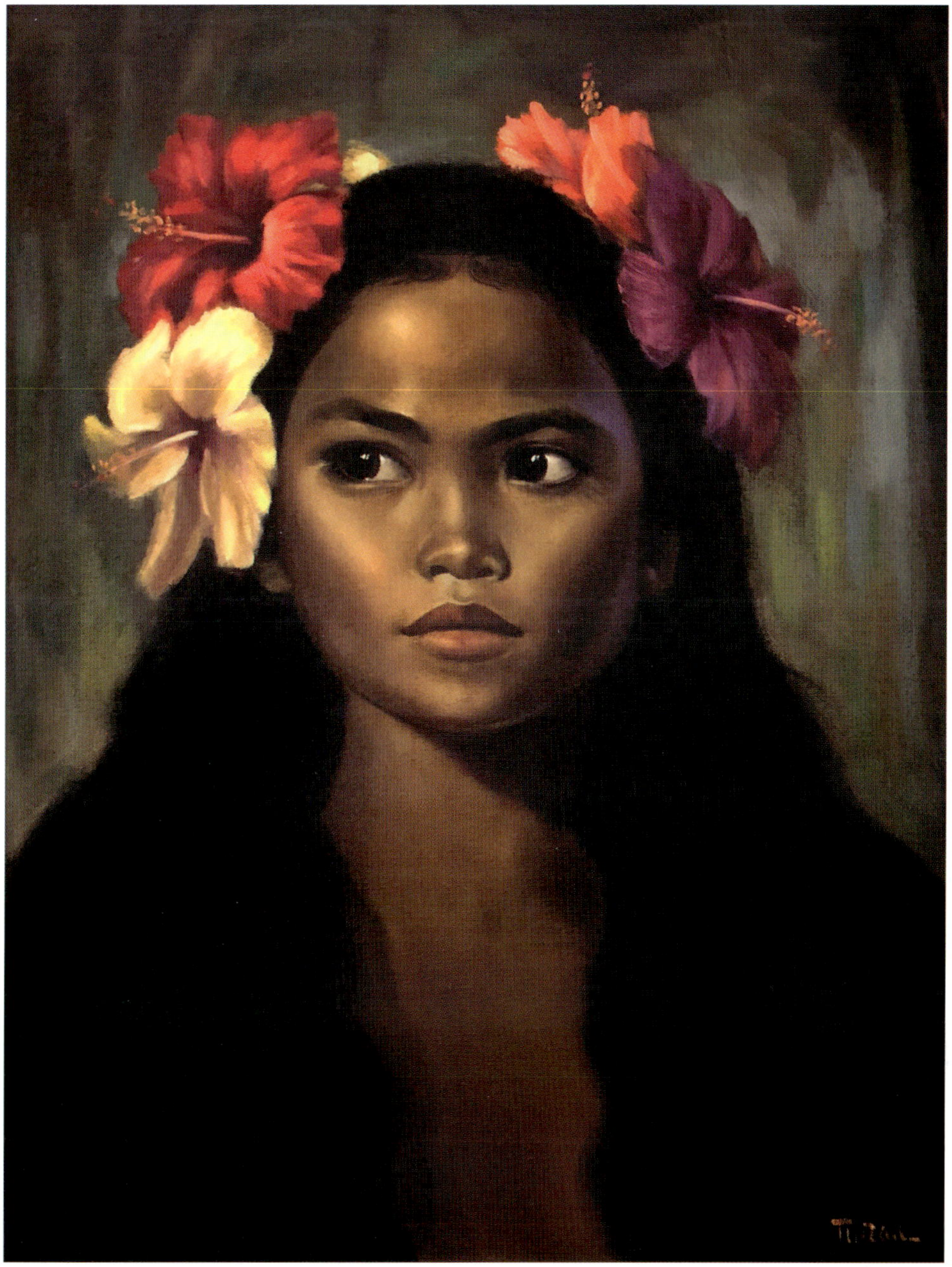

Figure 2.31 *Linea With Hibiscus Lei Po'o*, Ralph Burke Tyree. 1968, oil on board, 20" x 16", author's collection.

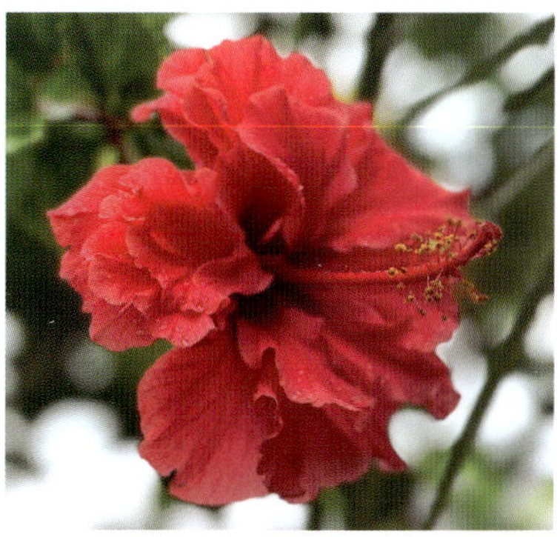

Figure 2.27 Double red hibiscus. California, 2021, author's photo.

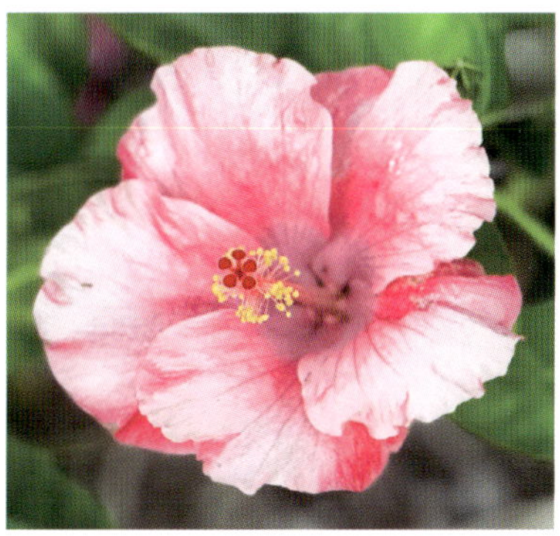

Figure 2.28 Pink hibiscus hybrid. California, 2021, author's photo.

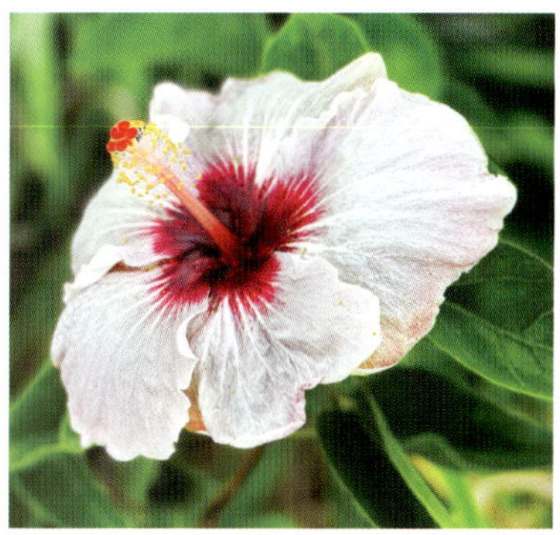

Figure 2.29 Large eight-inch white hibiscus hybrid. California, 2021, author's photo.

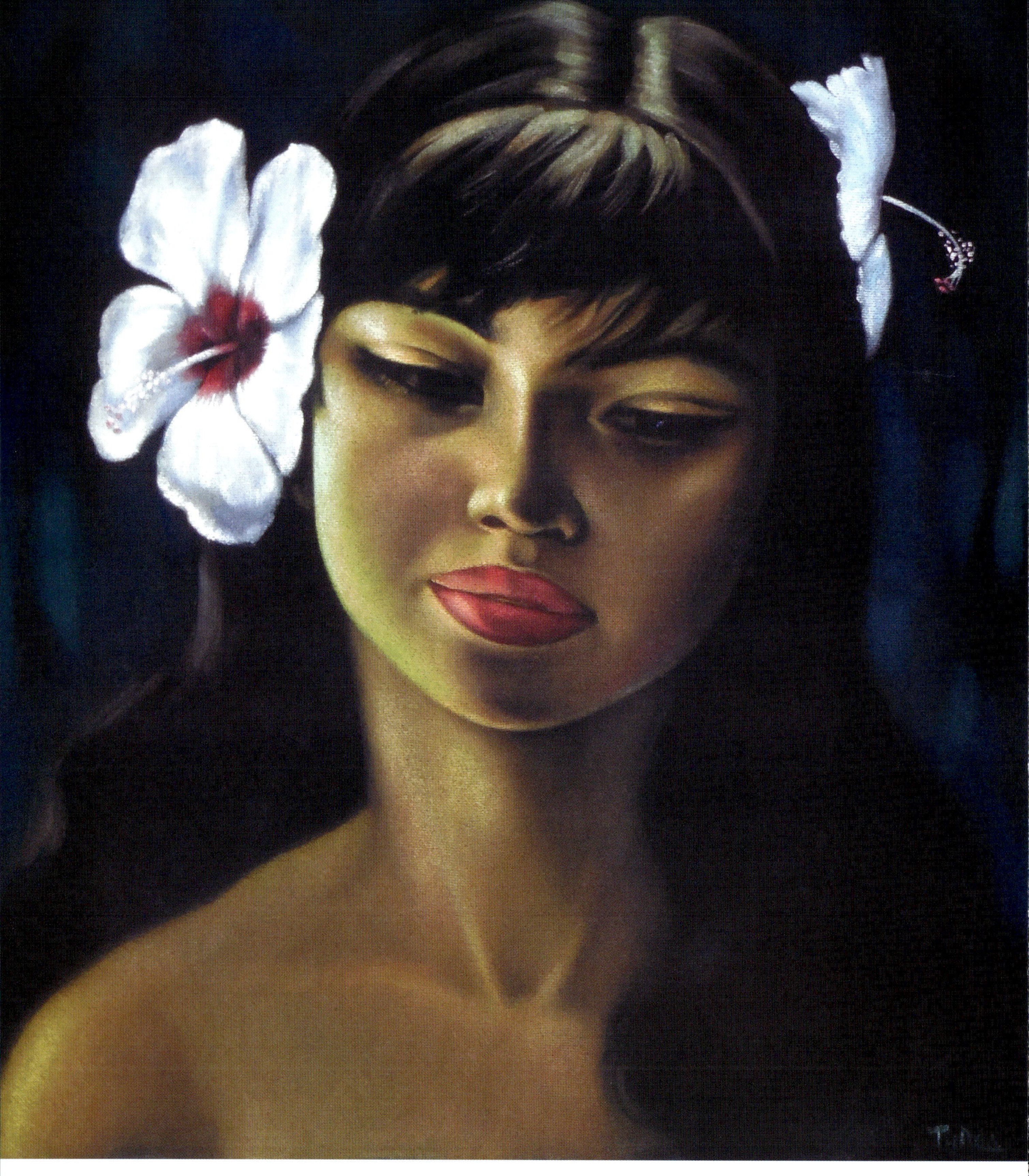

CHAPTER 3

Black Velvet, Genesis of a Genre

"Black velvet and that little boy's smile
Black velvet with that slow southern style
A new religion that'll bring ya to your knees
Black velvet if you please."

—WRITTEN BY CHRISTOPHER WARD AND DAVID TYSON, SUNG BY ALANNAH MILES, 1990

In early 1958, after more than six years away from the US mainland, the Tyree family left Hawaii and returned to Modesto. Tyree took a few odd jobs to make ends meet while continuing to paint professionally. Later that year, the increased stress from working multiple jobs, maintaining a painting career, and caring for a large family led him to a severe heart attack at age 36. At the time, cardiac infarct therapy merely consisted of advice to rest, lose weight, and improve one's diet. Tyree soon returned to work, undaunted and reinvigorated to continue painting, working hard, and following his desires wherever they took him. While recovering in 1960, Tyree began experimenting with oil paints on velvet instead of his usual board or canvas surface.

Black Velvet Origins

While the origins of velvet weaving are not clear, this lush fabric likely developed in China before being brought west by the Silk Road in the sixth century. As its popularity grew, thanks to its variety of textures and embossing capabilities, production centers sprung up across Asia, the Middle East, and Europe. An expensive process, creating this soft and luxurious fabric requires separating two pieces of woven material to form a pile about three millimeters high. Traditionally, silk is used, but rayon, cotton, linen, or mohair can also be used. Because of its unique texture and expense, silk velvet was initially almost exclusively used by royalty. This association with nobility continues into the present day.

Oil painting on velvet is a centuries-old technique, with some particularly exquisite examples held

Figure 3.1 *White Hibiscus in Apolima's Hair*, Ralph Burke Tyree. 1968, oil on velvet, 20" x 16", private collection.

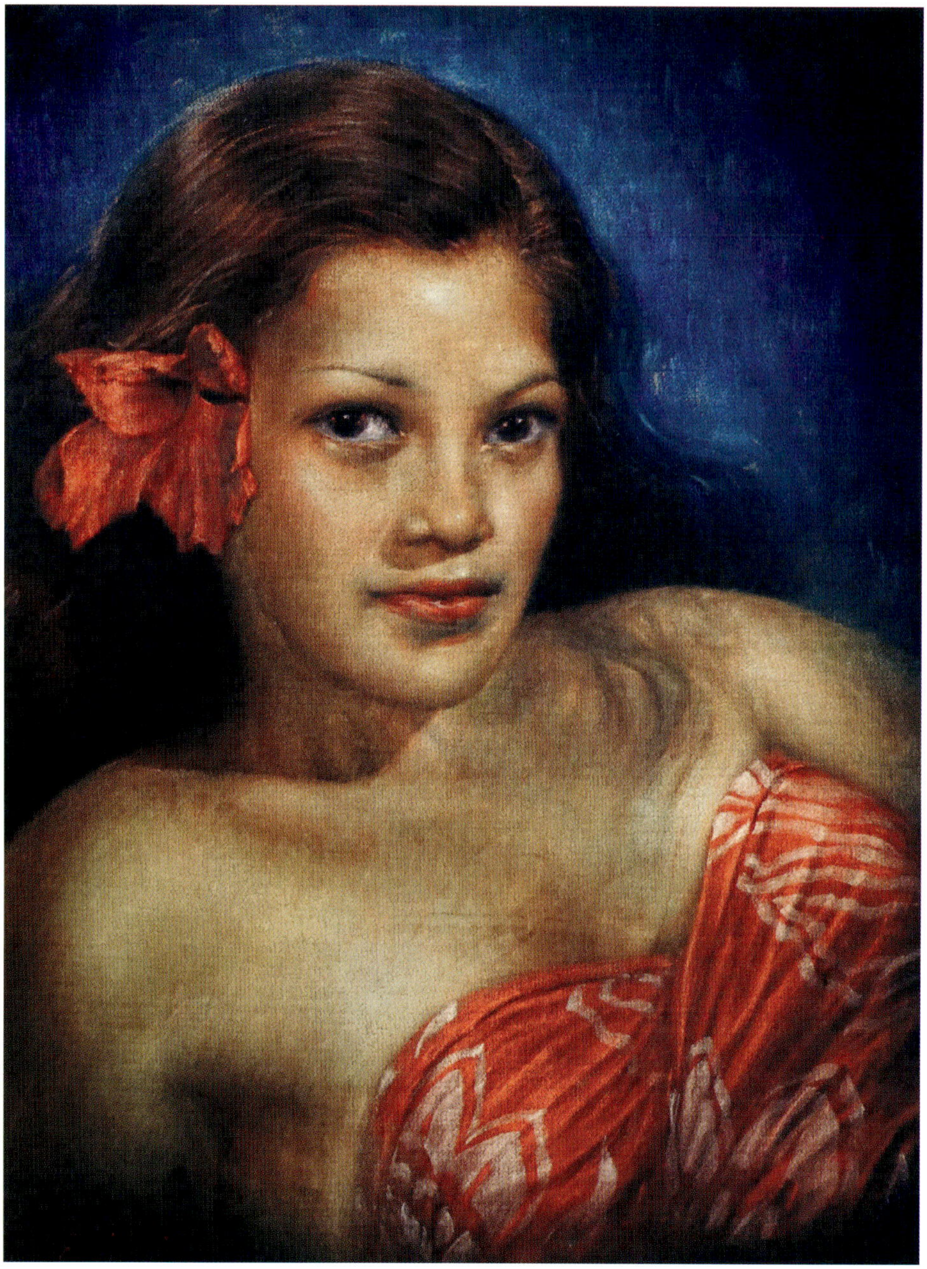

Figure 3.2 *Foxy Lady*, Edgar Leeteg. 1948, oil on velvet, 23″ x 17″. From ***Leeteg: Babes, Bars, Beaches, and Black Velvet Art***, CJ Cook, (San Diego: South Pacific Dreams, 2021).

in museums. Likely first developed and popularized in Kashmir by Russian Orthodox priests who painted religious icons, Marco Polo later observed velvet paintings of Hindu deities in India during the thirteenth century. Polo then brought the technique back to Europe, establishing it in the west. From there, Spanish conquistadors spread the technique to places like the Philippines and Mexico, where velvet painting on skirts and dresses became popular. Traditional oil painting on velvet is an exceedingly difficult and time-consuming skill to master. To paint on black velvet, the artist gradually applies light oils ground in a varnish emulsion, methodically coating the velvet from the depths of the backing to the surface of the pile. Canvas' flat surface cannot duplicate the three-dimensional aspect of painted velvet.

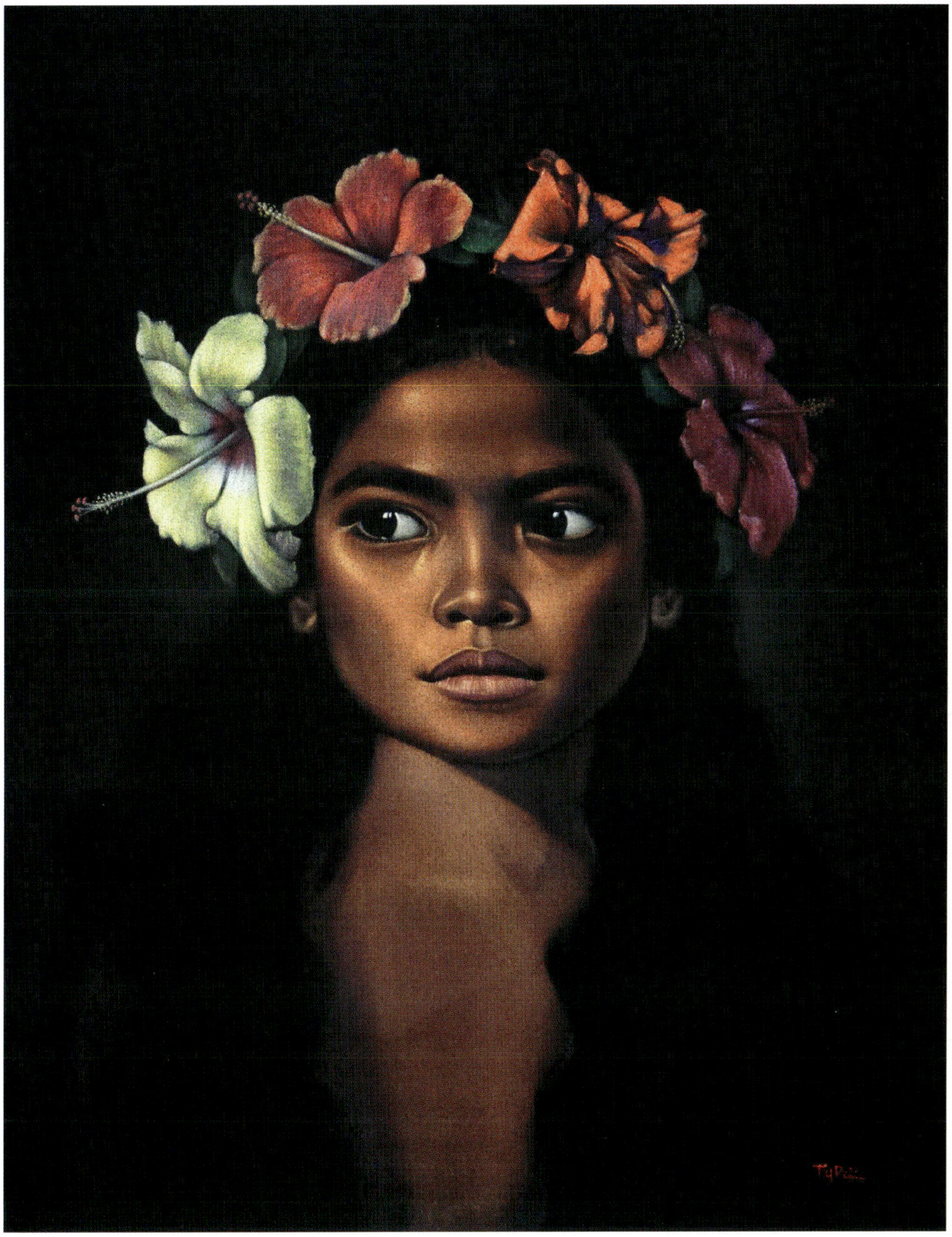

Figure 3.3 *Linea With Hibiscus Lei Po'o*, Ralph Burke Tyree. 1964, oil on velvet, 20" x 16", author's collection.

In 1960, Tyree developed his own unique method of painting on velvet and began selling these paintings. Using only the best French silk velvet, he gradually applied thin layers of oil paint, starting at the pile's base, methodically working toward the surface. The work was exacting, and any mistake could ruin the piece. With this change in medium, Tyree returned to painting portraits featuring his two favorite flowers, the hibiscus and plumeria, front and center.

Black velvet art was in demand in the Tiki bars and restaurants of Hawaii and California, thanks to Edgar Leeteg. His velvet paintings (figure. 3.2) were especially popular in these South Pacific-themed restaurants, and his success led many acolytes to attempt to cash in on this trend in the subsequent years. Originally a highway sign painter from Sacramento, California, Leeteg moved to the island of

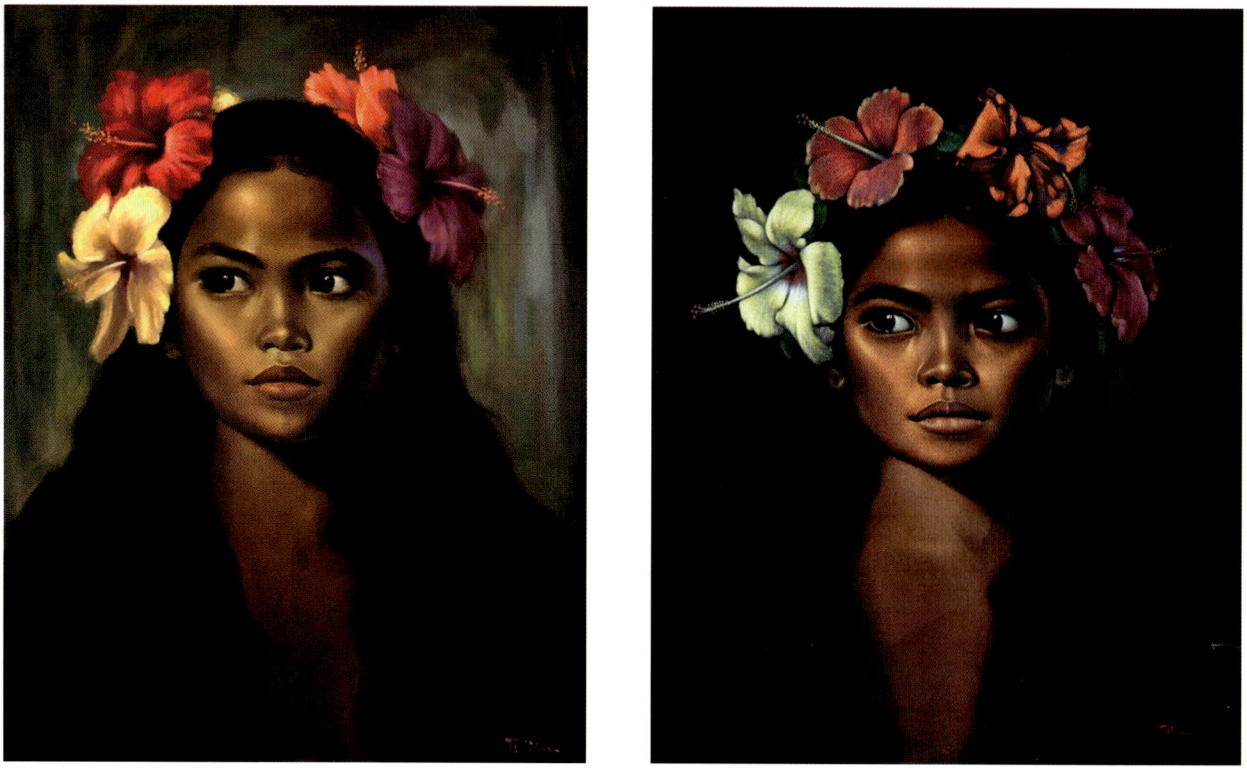

Figure 3.4 *Linea With Hibiscus Lei Po'o*, Ralph Burke Tyree. Oil on board *(left)* v. oil on velvet. What is your preference?

Tahiti in 1933. In Tahiti, he began painting on velvet, promoted himself, and partied, leading a wild lifestyle for the next 20 years. By 1950, he was labeled "Leeteg the Legend" by Author James Michener due to his status as one of the region's wealthiest artists and most infamous rascals. He remained in Tahiti until 1953, when he died suddenly in a motorcycle accident. Leeteg is the undisputed father of this genre of American velvet painting that continues into the present day.

While Tyree never met Leeteg, his use of black velvet inspired Tyree, who began painting portraits on this new surface, as Leeteg had so expertly done. Velvet allows the artist to reach new depths of light and shadow, especially when using chiaroscuro, a technique that utilizes contrast to create realistic volume (figure 3.3). Fifteenth-century Renaissance artists such as Leonardo da Vinci first popularized this advanced technique, which seventeenth-century Dutch artists like Rembrandt then promoted. While chiaroscuro is not medium-specific, Tyree preferred the depth and luminescence achieved when painting on the dark, textured surface of velvet instead of the pale, flat surface of canvas or board (figure. 3.4).

Black Velvet Paintings Become Kitsch

As the prices for velvet paintings rose, unsurprisingly, artists all over the planet began to use the medium. In the 1950s, cheaply made surface paintings on velvet gained popularity in Mexico and Southeast Asia as souvenirs for American tourists. By the 1970s, Ciudad Juárez, on the border of Mexico and the US, became the center of velvet painting.

The emergence of this technique, much simpler and less time-consuming than Tyree's, drastically altered the velvet painting industry. Not only was the method simplified, but cheaper materials, like

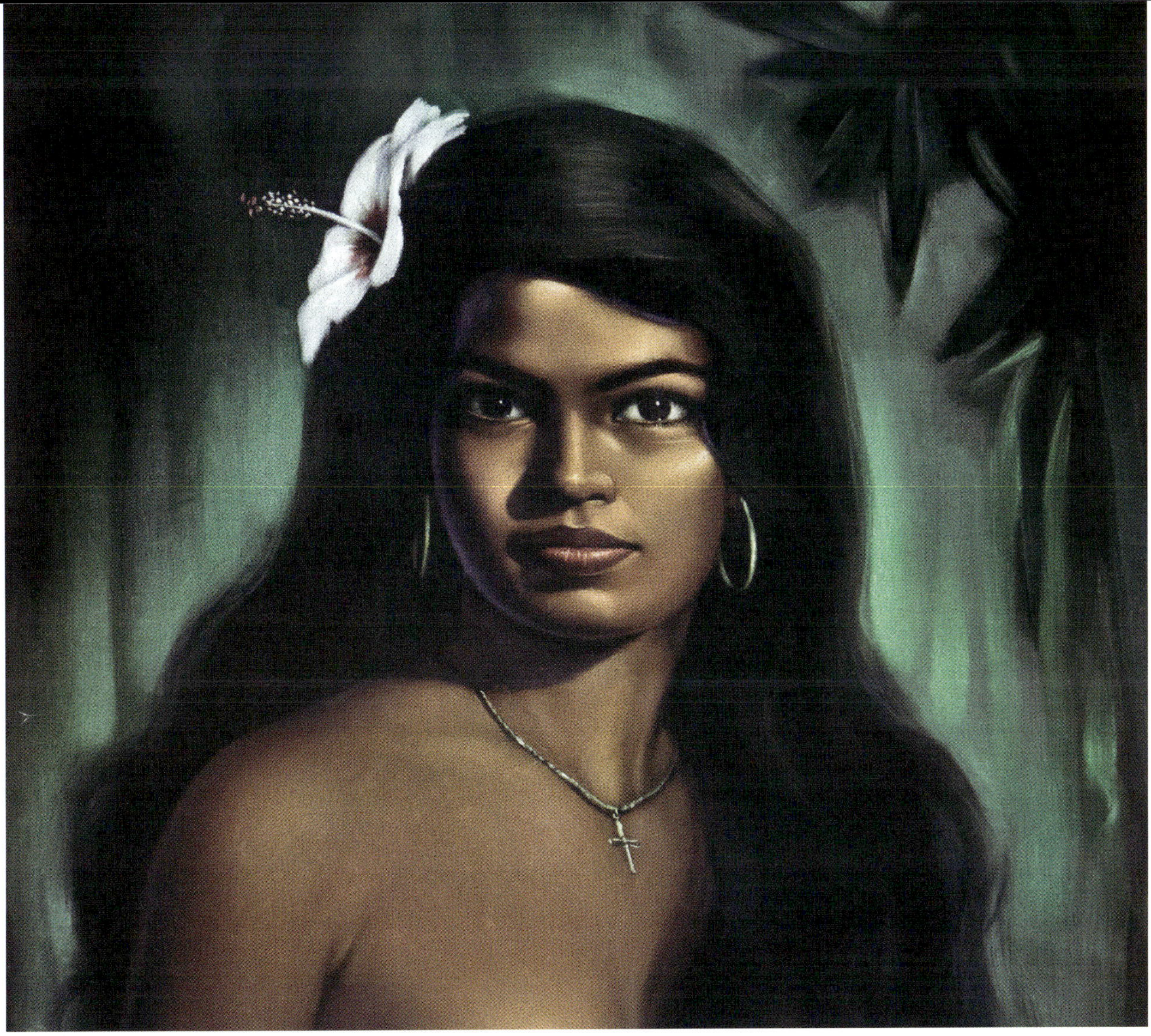

Figure 3.5 *White Hibiscus in Lote's Hair*, Ralph Burke Tyree. 1970, oil on velvet, 24" x 20", private collection.

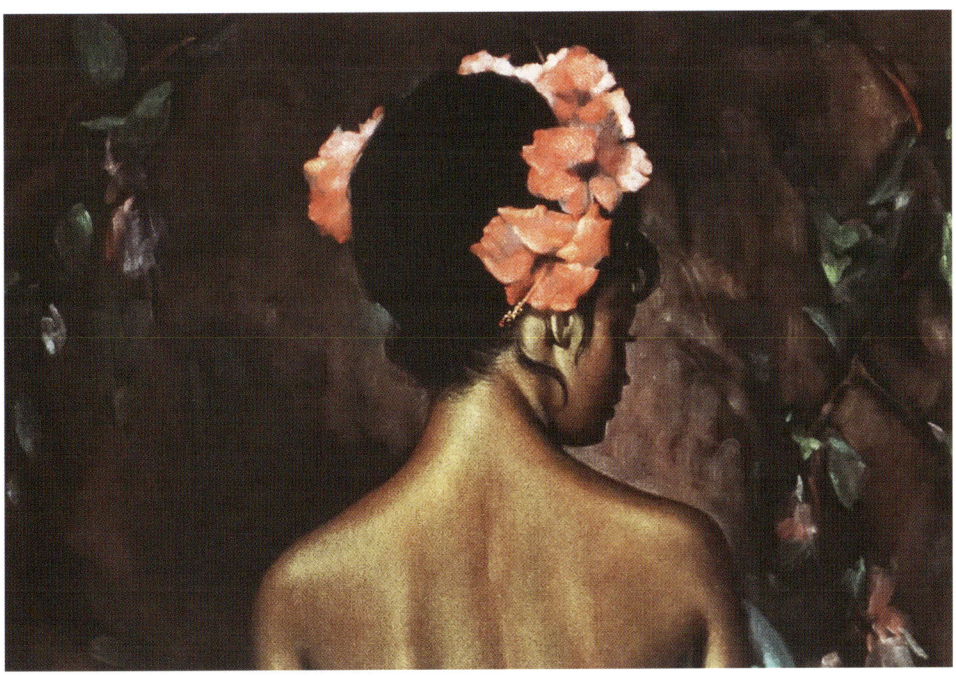

Figure 3.6 *Pink Hibiscus Lei Po'o*, Ralph Burke Tyree. 1968, oil on velvet, private collection.

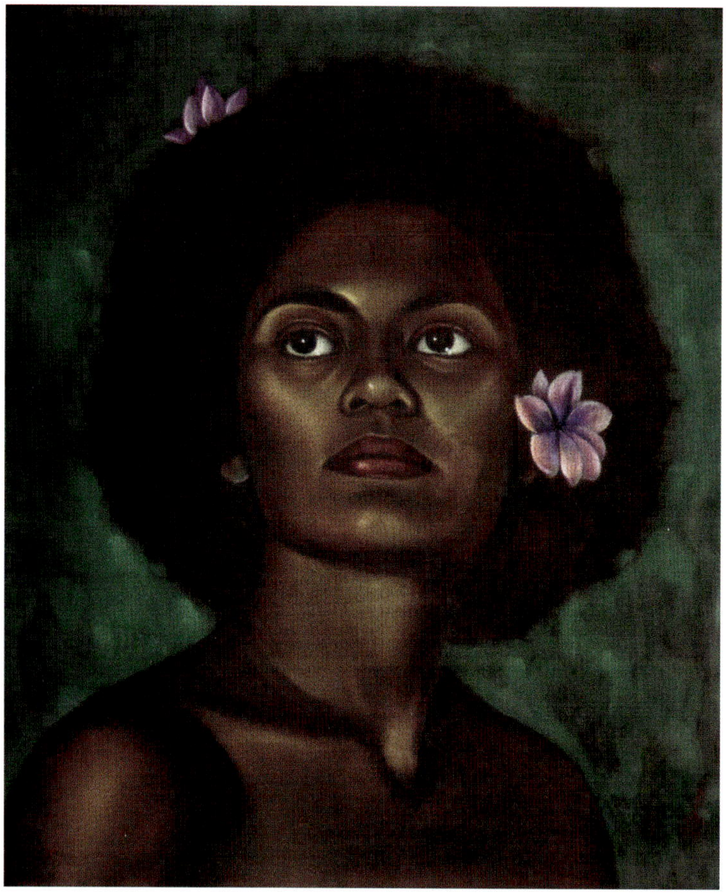

Figure 3.7 *Fijian Wahine with Pink Plumeria*, Ralph Burke Tyree. Note the plumeria over her right ear that implies she's available. 1965, oil on velvet, 20" x 16", private collection.

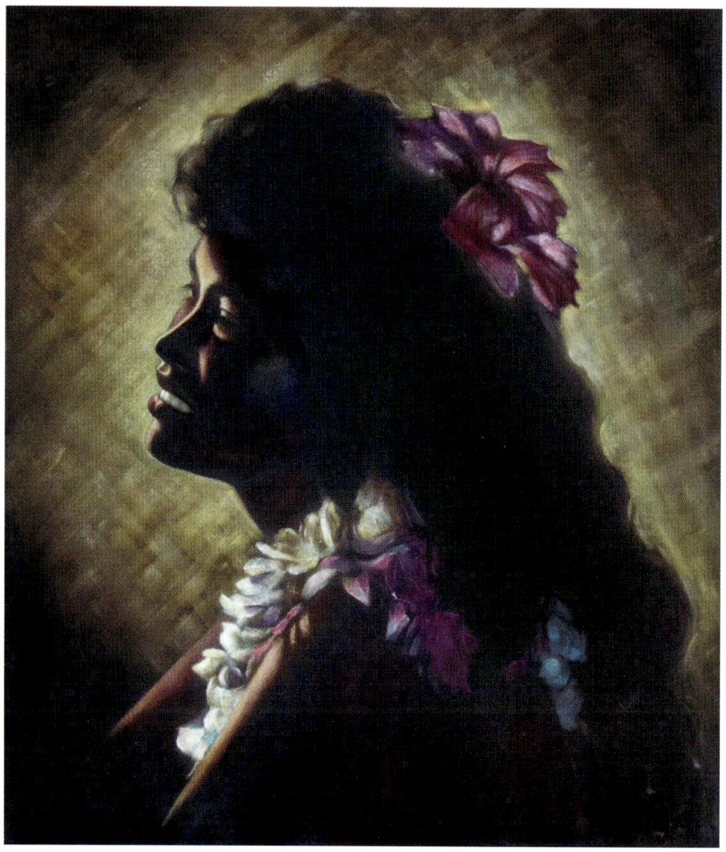

Figure 3.8 *Plumeria Lei*, Ralph Burke Tyree. 1970, oil on velvet, 16" x 20", private collection.

velveteen and acrylic paints, became favored. These paintings were now mass-produced and became seen as tacky, touristy kitsch instead of fine art. Popular subject matter, including celebrities, Jesus, and animals, including images of dogs playing poker, was repeated again and again. This influx rapidly chilled the market for fine art on black velvet and sent many professional artists, such as Cece Rodriguez, to other mediums. Undaunted, Tyree persisted in painting on velvet, explaining,

> "Basically, there are two types of velvet paintings: surface painted and in-depth painted. Surface-painted velvets are standard, usually done in Mexico or the Orient. These pieces are painted heavily in oils or acrylics and are not permanent."

Tyree's unique mastery of the material allowed him to create incredibly realistic textures and brilliantly deep colors and shadows. His careful technique maintained the velvet's softness, leading him to refer to his paintings as "touchable art, provided your fingers are not greasy." He summarized thusly,

Black velvet is a medium that invites touch.
There is a sensuality to silk velvet—a look and a feel—that mediums like canvas and board lack and cannot recreate.

More Hawaiian Adventure

In 1964, after over six years on the mainland and the birth of their son, Jeff, Tyree moved his family back to Maui where he and Margo's seventh child and sixth son, Marc, was born. By this point, Tyree was principally working with oil paints on velvet. Inspired by Maui's beauty, he painted portraits en plein air of the Hawaiian people in their native surroundings. Other times, Tyree would photograph his models in lush natural settings and finish the painting back in his studio. He always worked on multiple pieces at once, allowing one layer of paint to dry while working on another piece, a critical step when painting the velvet nap, essential to create depth. He often had four or five paintings going simultaneously due to the time-consuming nature of drying oil paints in humid Hawaii (figures 3.5–3.11).

The Tyree family briefly moved back to Turlock, California, after a year and a half on the island. He sold his work along the US Pacific Coast to the increasingly popular Tiki bars and restaurants outfitted with fishnets, glass floats, Tikis, and the ubiquitous paintings of native girls on black velvet. His primary client was the famous Tiki restaurateur Hop Louie, who owned nightclubs and venues across California. Tyree's residency in California was mutually beneficial for the two, allowing Louie to buy many pieces from the emerging master of velvet art.

Only several years later, in 1968, Tyree, craving inspiration, moved back to Hawaii and established a new residence on the Big Island, where Margo and the younger children joined him. At this time, he began featuring flowers in his paintings as the primary subject while also continuing to use them as hair accouterments in his portraits of wahine. This would be his last move to the tropical islands he loved so much before his final move back to California in 1972.

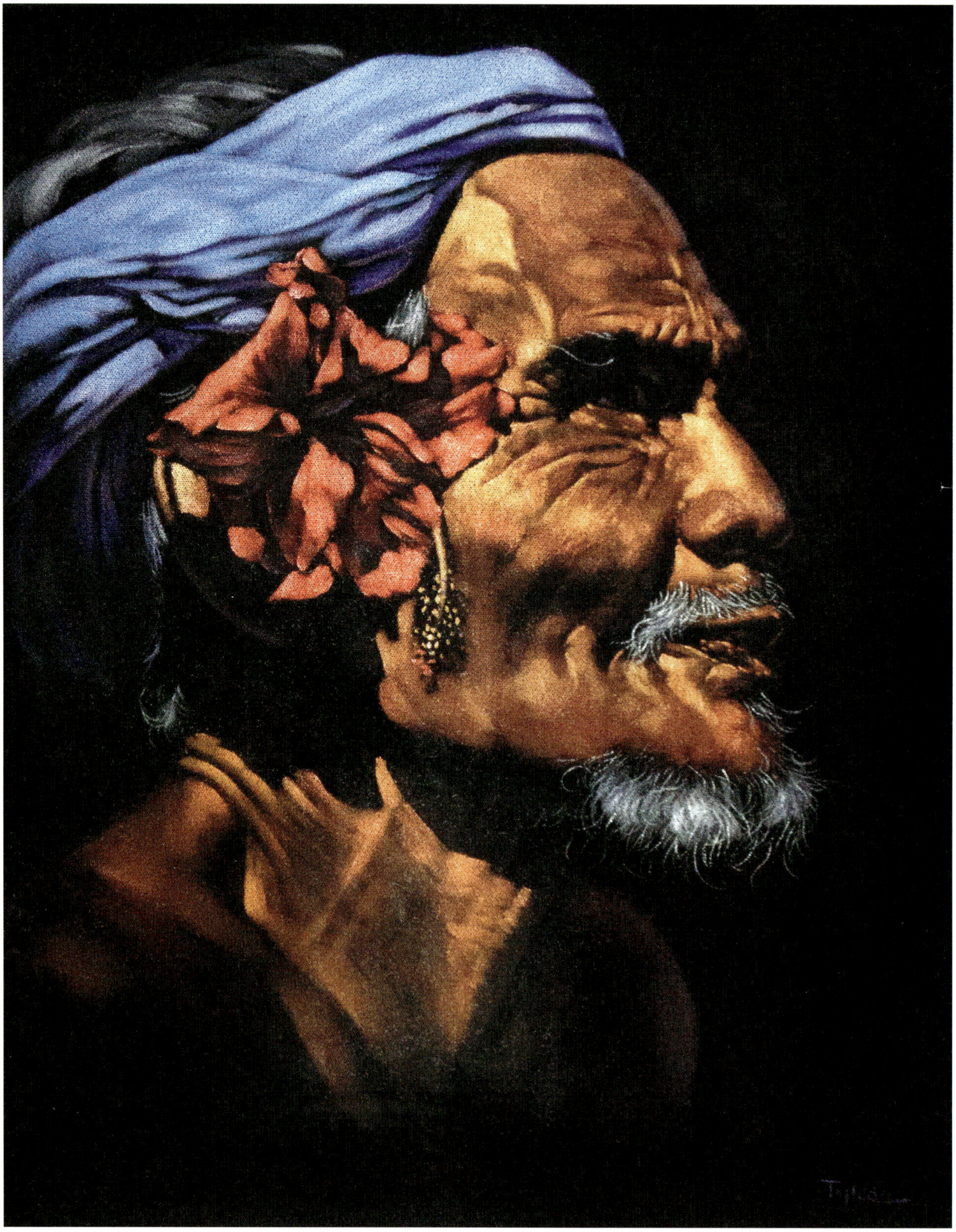

Figure 3.9 *Red Hibiscus and Old Hawaiian Chief*, Ralph Burke Tyree. 1971, oil on velvet, 20" x 16", author's collection.

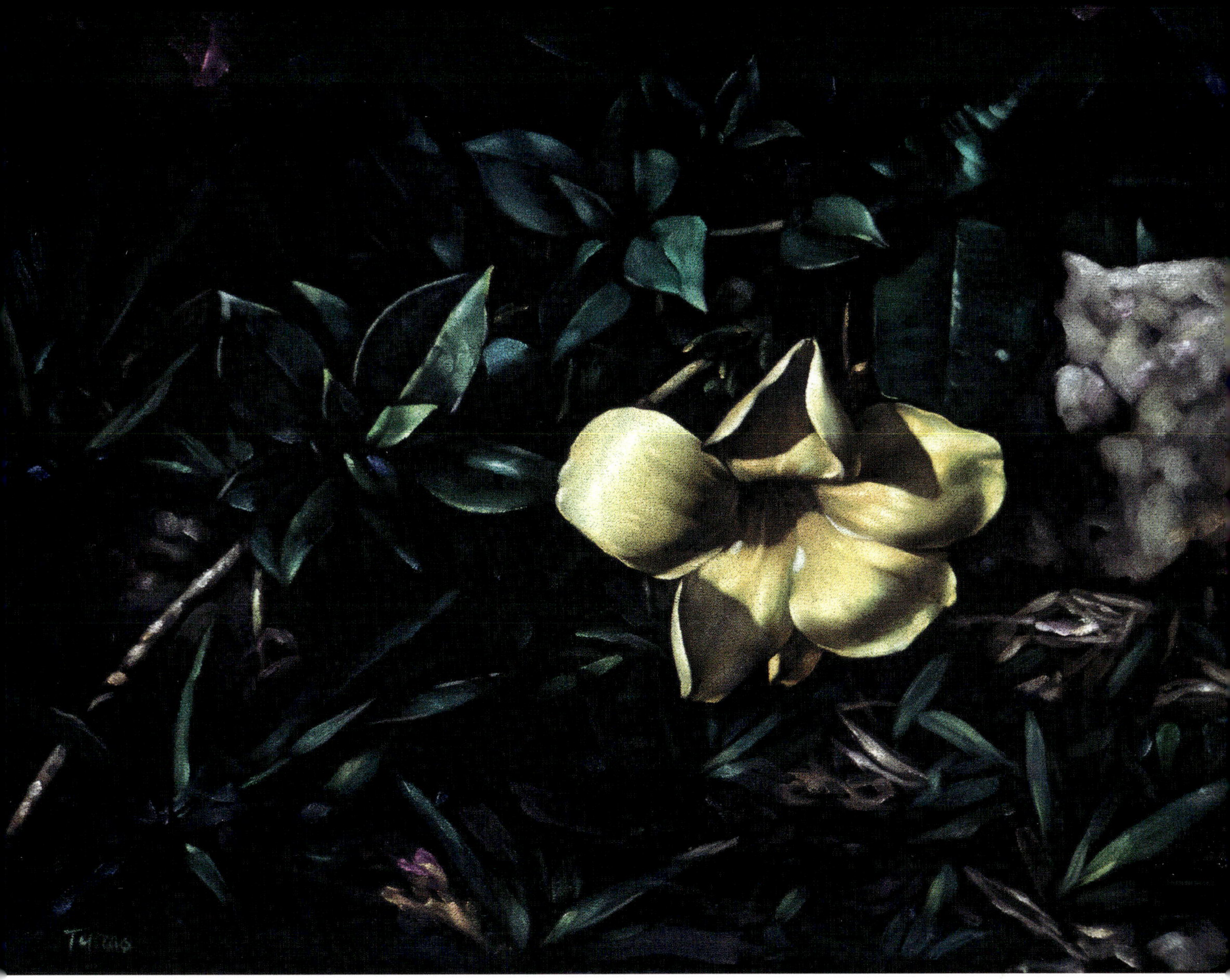

Figure 3.10 *Hibiscus*, Ralph Burke Tyree. 1970, oil on velvet, 16″ x 20″, private collection.

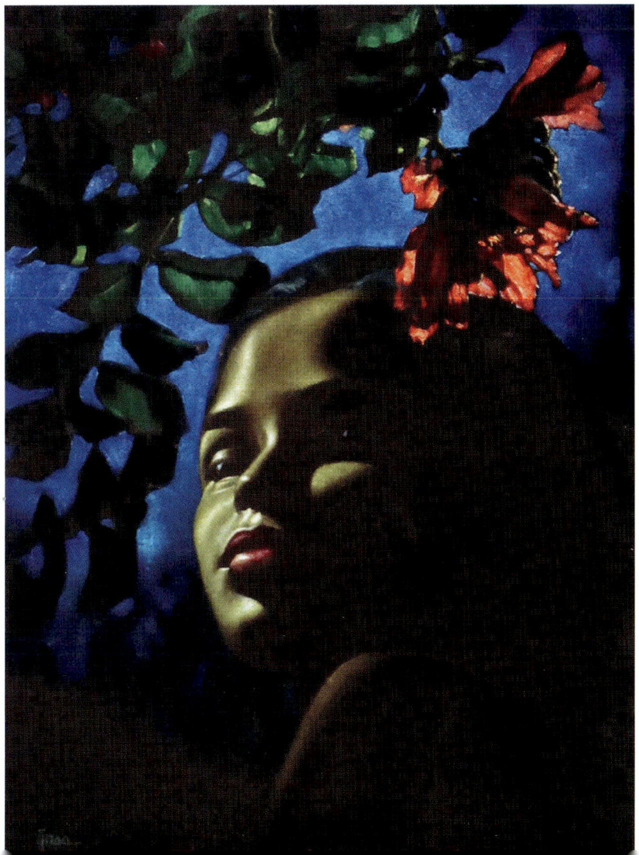

Figure 3.11 *Kalia Under Red Hibiscus Bush*, Ralph Burke Tyree. 1970, oil on velvet, 20″ x 16″, private collection.

Quince Flower

The **quince** (*Cydonia oblonga*) is a fruiting shrub native to mountain woodlands of Western Asia from the *Rosaceae* family (figure 3.12). With thorny branches and delicate red, orange, white, or pink flowers, this frost-hardy plant produces a yellowish-green fruit (figure 3.13). Hard and tart, the pear-shaped fruit is most often used to make jams, pastes, and alcoholic beverages. A quince shrub can grow up to 26 feet in height and 20 feet wide. Because of its lovely blooms, it is often grown ornamentally, with branches of flowers used in floral arrangements. Quince thrived in the ancient world and is specifically associated with Mesopotamia and Cydonia.

When painting the quince, Tyree focused on one of its beautiful flowers, as seen in figure 3.14. The vibrant pink, yellow, and green of the plant's bloom and leaves stand out against the brown of the quince's branches. Figure 3.15 provides a closer inspection, and the viewer finds a bee sitting on the flower collecting pollen. The bee is shown as a tiny part of a larger composition, much like it is within its hive and like its hive is within the larger ecosystem.

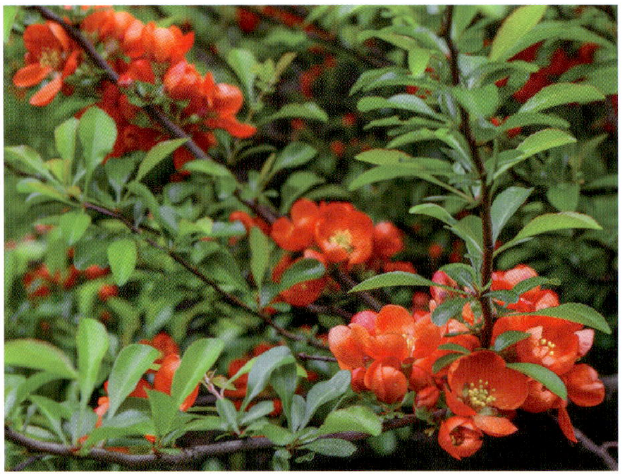

Figure 3.12 Quince flowers from Japan. Photo by Yoksel Zok on Unsplash.

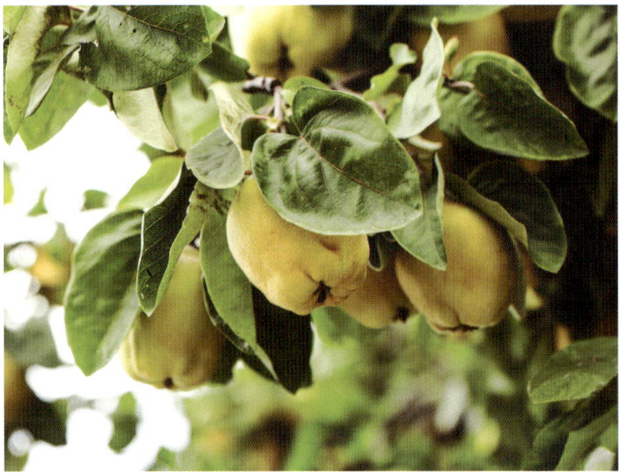

Figure 3.13 Quince leaves and fruit. Utilisateur:Spone.

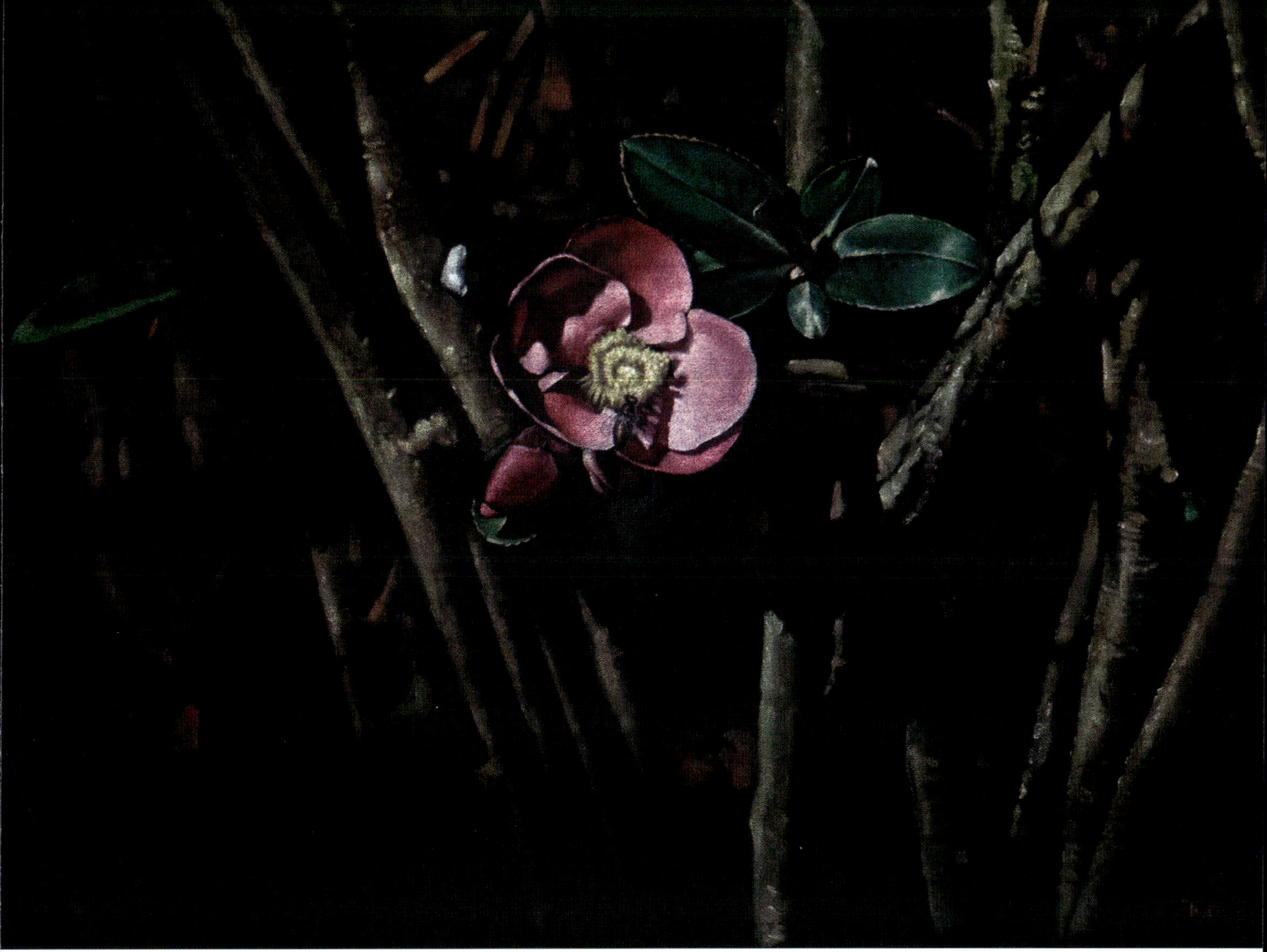

Figure 3.14 *Quince Flower*, Ralph Burke Tyree. 1967, Oil on velvet, 16" x 20", private collection.

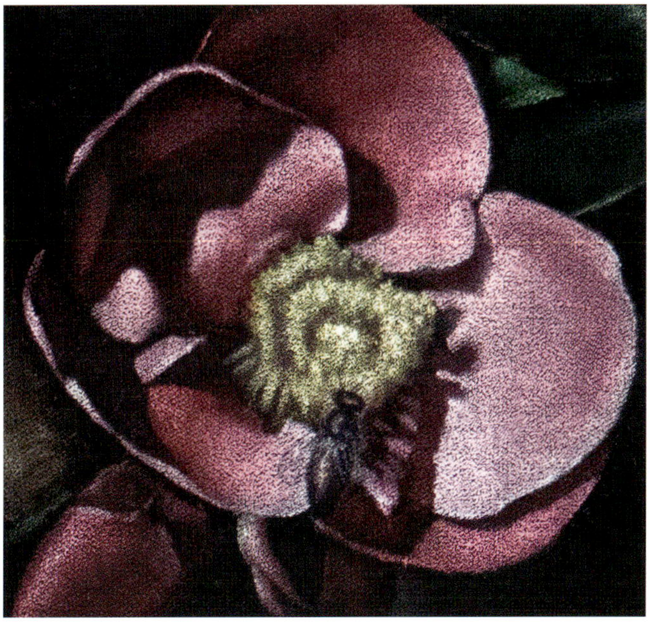

Figure 3.15 Close-up of *Quince Flower* to show honey bee.

Passion Fruit (Liliko'i)

The **passion fruit** (*Passiflora edulis*), known as liliko'i in Hawaiian, is a member of the *Passiflora* genus of about 550 species. It is recognizable for its large, colorful, exotic flowers (figure 3.16) and oval-shaped fruits that contain edible juice and seeds (figure 3.17–20). Sixteenth-century Spanish missionaries in South America used this popular flower to explain the story of Jesus Christ's crucifixion while trying to convert the natives to Christianity. The missionaries called it *flor de las Cinco llagas*, meaning flower of the five wounds, in reference to the five wounds of Christ and the five anthers of the flower. Through this, it was used to tell of the passion of Christ and eventually became known as passion fruit and passion flower.

Native to Brazil, Paraguay, and Argentina, this vine is now commercially cultivated across the tropics and subtropics for its fruit and juice and represents a significant portion of local horticulture industries (figures 3.19 and 3.20). It has become an invasive species in Hawaii and many tropical islands as it spreads and proliferates in the moist climes, quickly overwhelming native flora. Popular and ubiquitous, the passion flower is Paraguay's national flower and has at least two songs in its honor "Liliko'i" (2006) by Paula Fuga and "Passionfruit" (2017) by Drake.

Tyree painted a passion flower in oil paints on velvet, surrounding the beautiful white bloom with green foliage (figure 3.21). The flower is the clear focal point, depicted with intricate detail and directly facing the viewer. After the eye lands on the flower, it moves across the light-dappled leaves to the large passion fruit that hangs in the upper right corner of the composition. Tyree accomplishes this through his brilliant use of light and shadow.

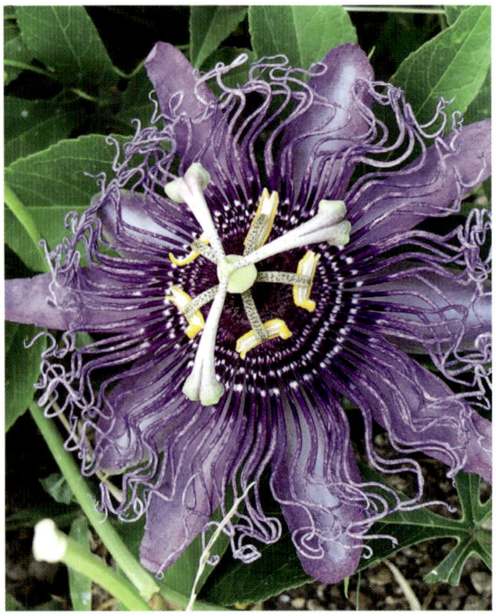

Figure 3.16 Purple passion flower, California. 2019, author's photo.

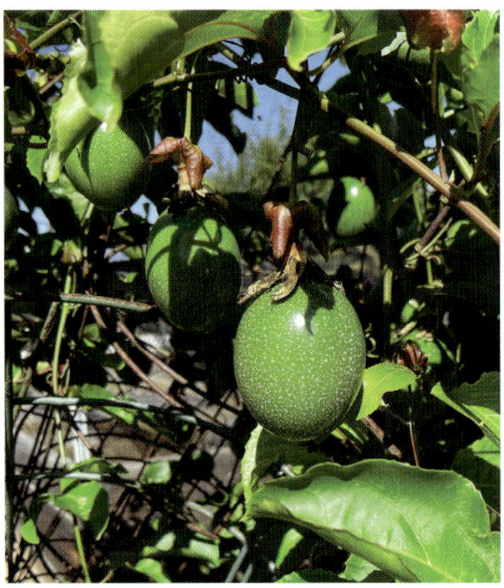

Figure 3.17 Passion fruit, California. 2021, author's photo.

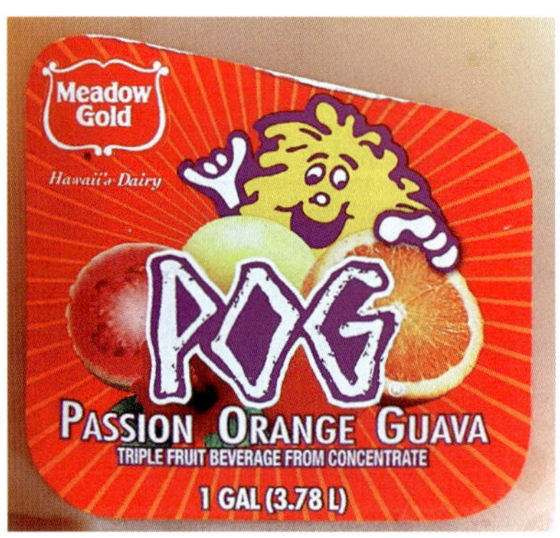

Figure 3.18 POG, a mix of passion fruit, orange, and guava juice, popular in Hawaii. 2022, author's photo.

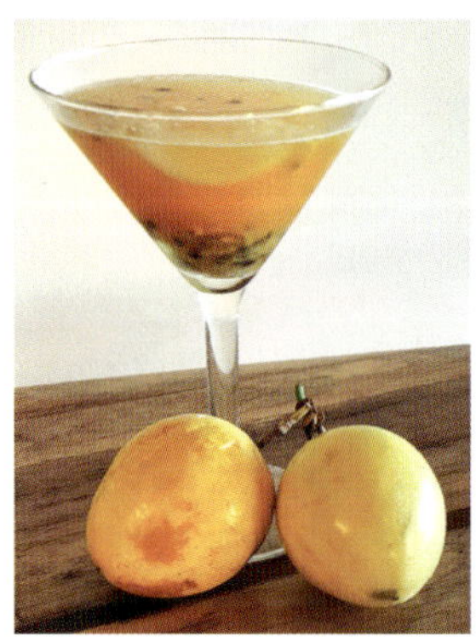

Figure 3.19 Passion fruit martini, Maui. 2021, author's photo.

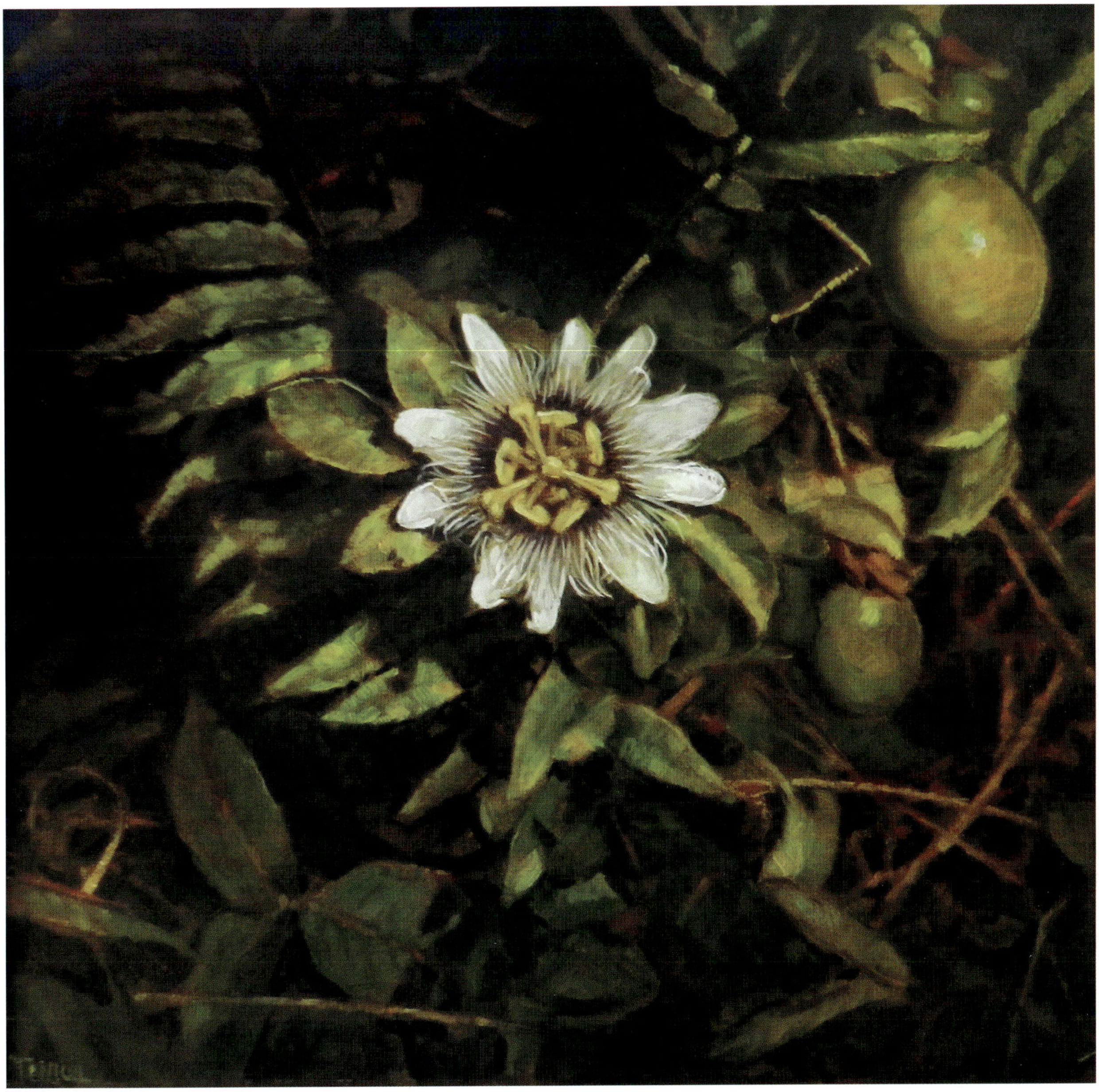

Figure 3.21 *Passion Flower and Fruit*, Ralph Burke Tyree. 1967, oil on velvet, 16" x 20", private collection.

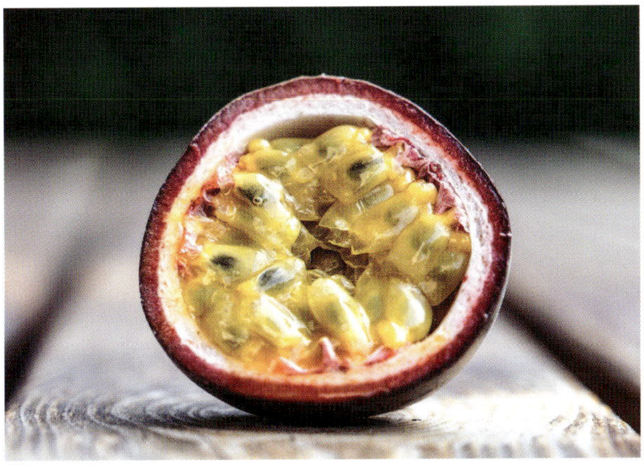

Figure 3.20 Passion fruit, Michael Kucharski on Unsplash.

Magnolia

Magnolia (*Magnolia*) is a large genus of over 200 species in the *Magnoliaceae* family native to Asia, the Americas, and the West Indies (figure 3.22). Named in honor of seventeenth-century French botanist Pierre Magnol, the family consists of both evergreen and deciduous trees. Up to 95 million years old, this family may include the most primitive flowering plants on Earth. The perfume from a couple of magnolia trees' large, sweet-smelling flowers can permeate the air during their springtime bloom. A popular symbol, it is the official flower of a handful of different cities, regions, and countries and is often associated with the Southern United States. While most often used in horticulture, certain varieties are used in recipes, in traditional medicines, and for timber.

Tyree featured the magnolia tree in paintings several times, both as the primary subject and as a floral backdrop for a model. He painted a close-up of a magnolia branch in full bloom in 1968 (figure 3.23). In this piece, he uses light and shadow exquisitely, bathing the tree in light from the lower right corner and casting bluish shadows across the deep green leaves and bright white petals. With this lighting, the atmospheric background of blue, green, and purple evokes a dramatic, ethereal quality. Four years later, he painted a sea of magnolia flowers behind a wahine using a similar color palette (figure 3.24). The background is painted in an impressionistic style and evokes a similar ethereal quality.

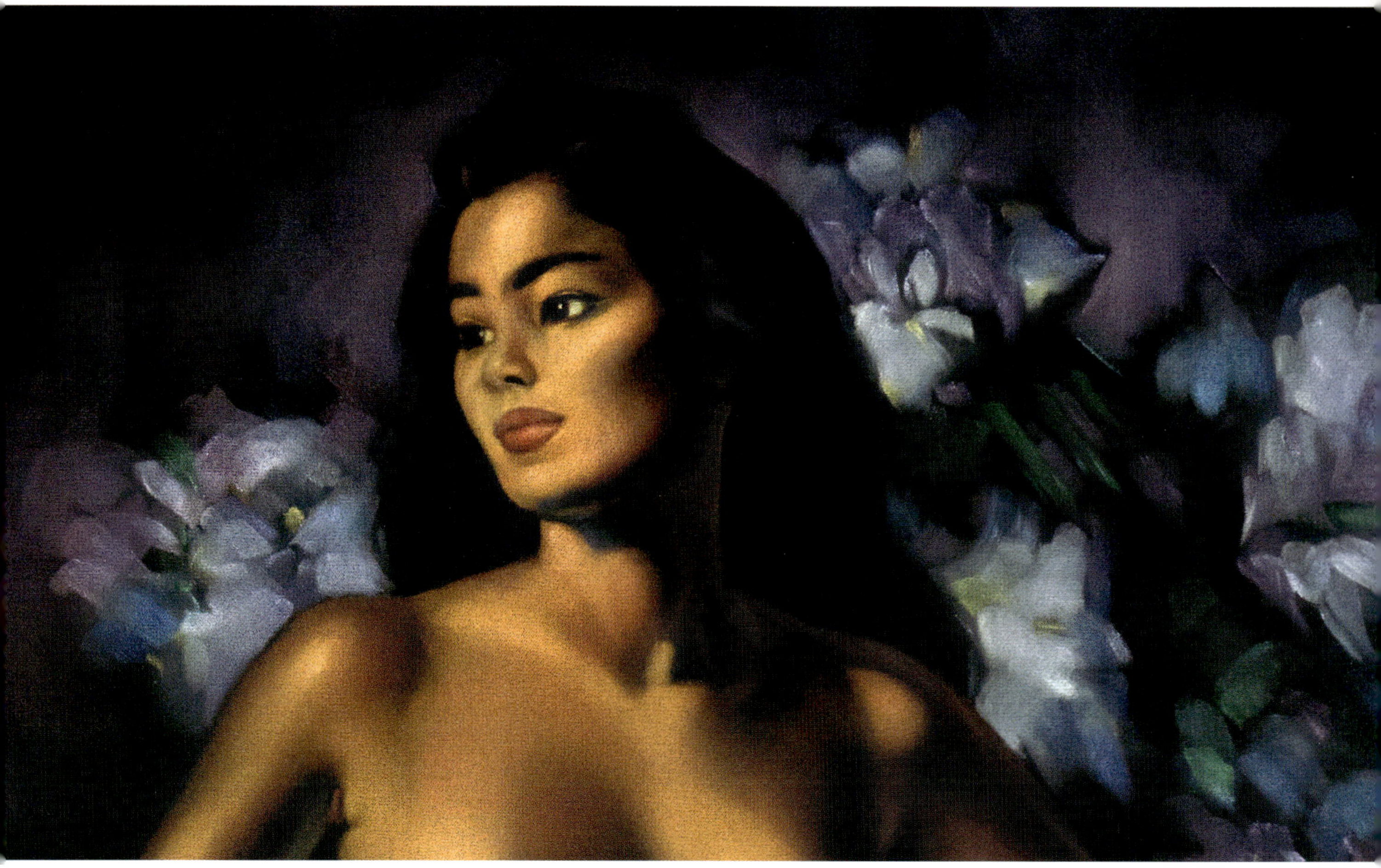

Figure 3.24 *Angelic Beauty in a Sea of Magnolia*, Ralph Burke Tyree. 1972, oil on velvet, private collection.

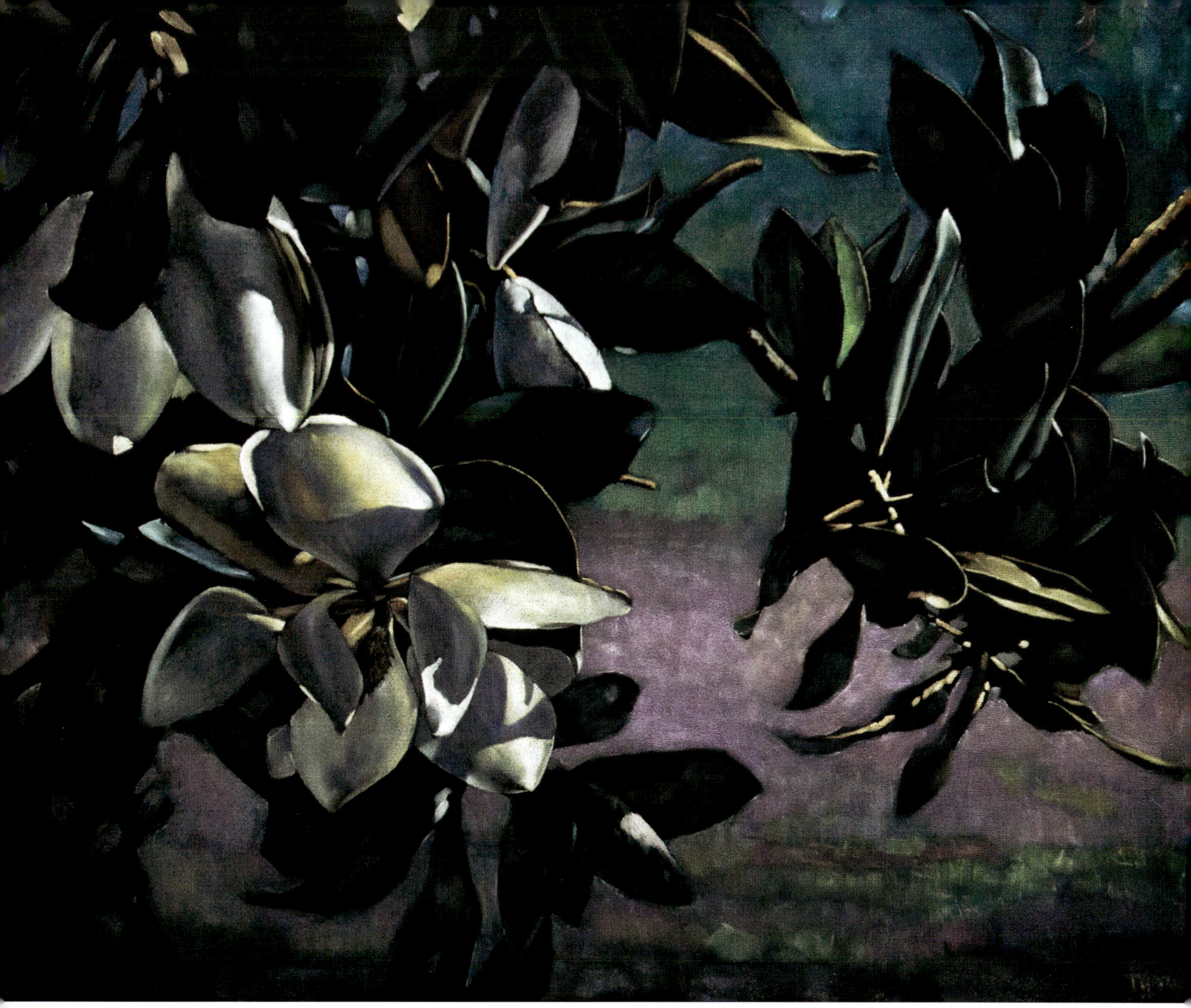

Figure 3.23 *Magnolia*, Ralph Burke Tyree. 1968, oil on velvet, 24" x 28", private collection.

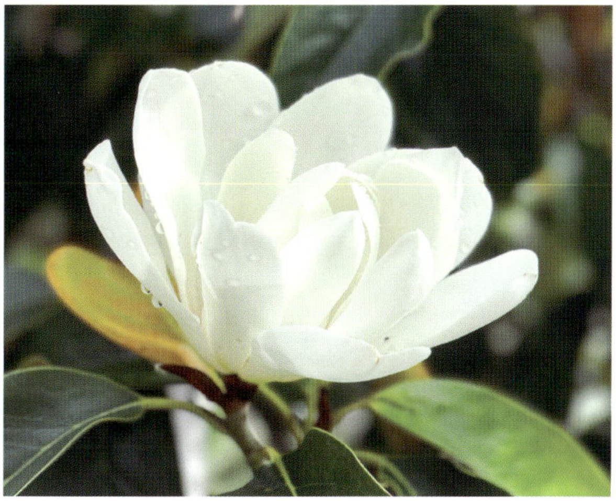

Figure 3.22 Early-blooming and very fragrant magnolia, California. 2022, author's photo.

Cup of Gold

Cup of gold (*Solandra maxima*), also called golden chalice vine or Hawaiian lily, is a species of flowering vine in the *Solanaceae* family (figure 3.25). Aptly named, the plant has eye-catching, massive yellow flowers that emit a pleasant coconut scent at night. Even the plant's glossy leaves are striking. Thanks to its thick stems, this robust vine can grow up to 40 feet long. Every part of the plant is toxic and can be deadly if ingested. This vigorous creeper is endemic to Mexico, Central America, and Colombia and has been introduced to other, usually dry, tropical environments. In the Hawaiian Islands, it thrives as an invasive species, climbing over and smothering native trees to take over large areas.

Painting the cup of gold flower at least twice in the same year, Tyree seems to have been taken by these alluring flowers. Figure 3.26, done in oil paints on velvet, shows the flower front and center in vibrant contrast to the darkness surrounding it. The golden flower seems to glow as its perfume permeates the nighttime air and its leaves melt into the blackness of the velvet. Figure 3.27, painted on board, depicts a cup of gold over the right ear of a wahine. While she is the subject of the portrait, she is placed in a three-quarter view and looks off into the distance beyond the viewer. This turns the large, yellow bloom to face the viewer directly. With its pale color, the eye goes straight to the flower before moving to the woman's face.

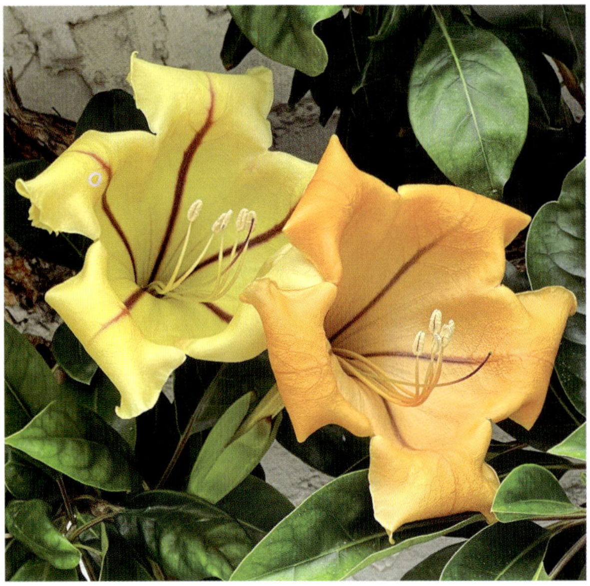

Figure 3.25 Cup of gold, Los Angeles. 2023, author's photo.

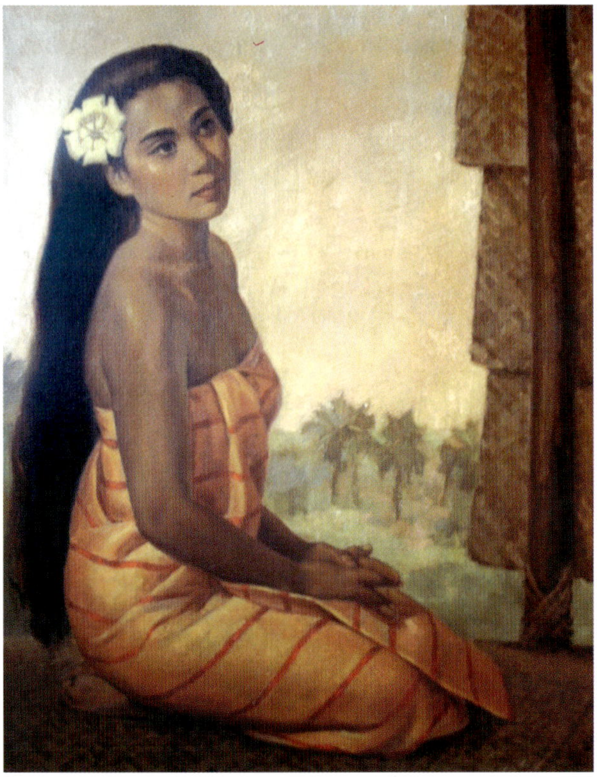

Figure 3.27 *Cup of Gold Over the Right Ear of a Wahine*, Ralph Burke Tyree. 1976, oil on board, 20" x 16", private collection.

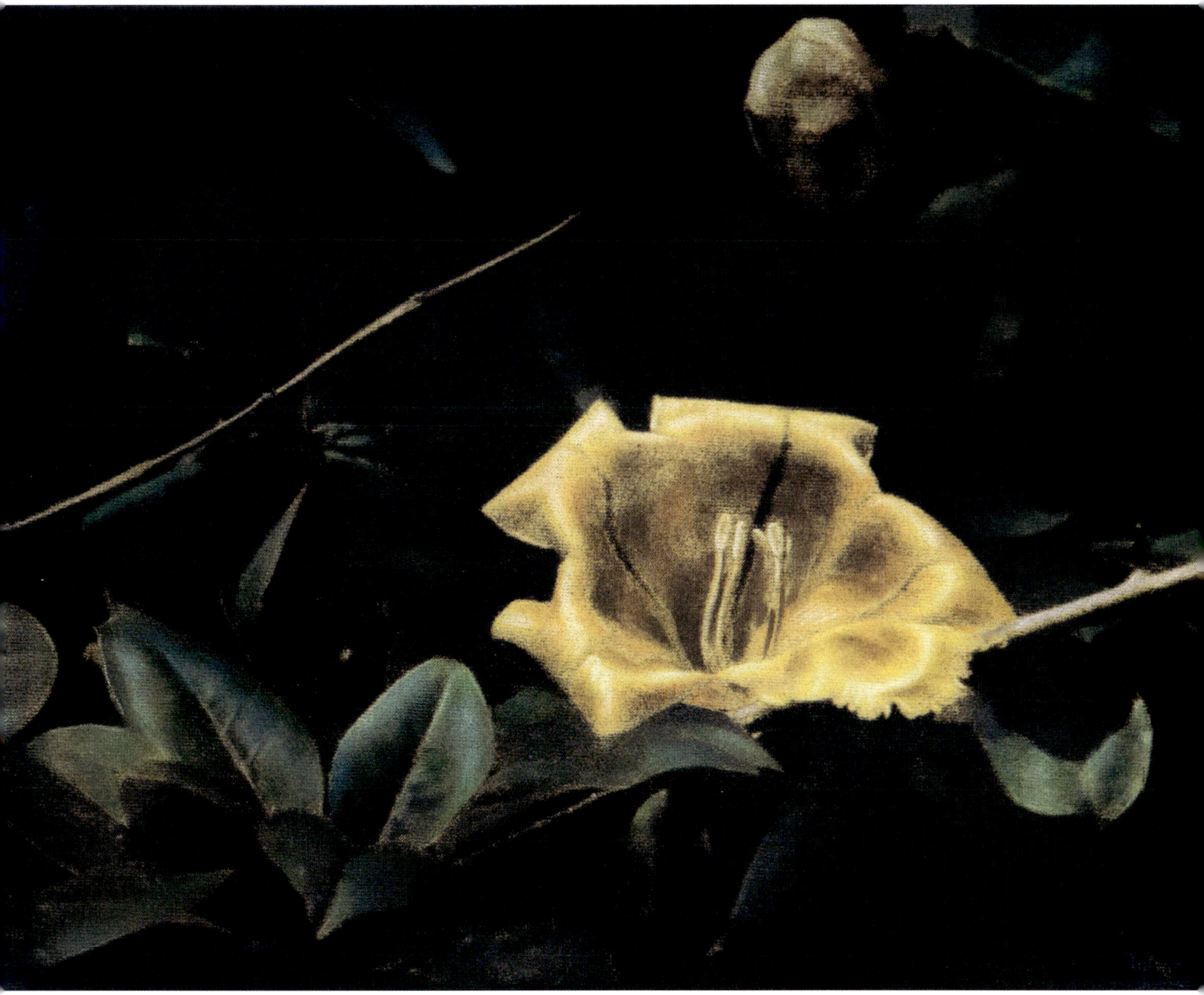

Figure 3.26 *Cup of Gold*, Ralph Burke Tyree. 1976, oil on velvet, 20" x 16", private collection.

Orchid

The **orchid** (*Orchidaceae*) from the *Asparagales* order is one of the most wide-ranging families of flowers, both in terms of variety and distribution (figure 3.28). With about 28,000 species worldwide, this family of colorful flowers is one of the most significant and numerous on the planet. All orchids have petal symmetry and a lip, called a labellum, for pollinators to land on in place of a petal, as seen in Haeckel's lithograph (figure 3.29) and in figures 3.30–3.32. These charming flowers have inspired a massive horticulture following, with organizations spanning the globe focused on their cultivation and collection. Since the nineteenth century, over 100,000 new hybrids and cultivars have been created. Like the magnolia, the orchid is a popular choice as the official flower for a handful of cities and nations.

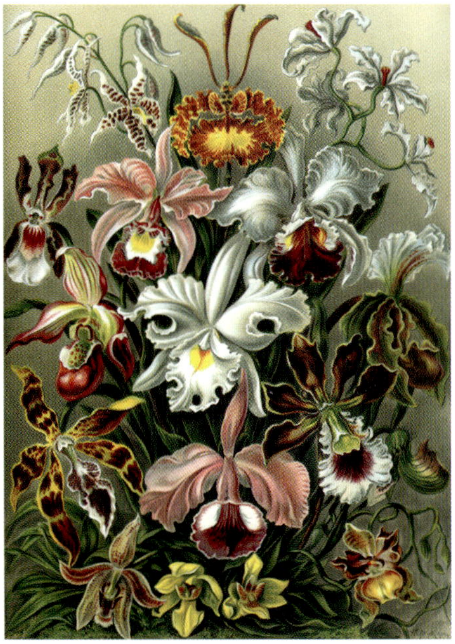

Figure 3.29 Plate 74: Orchidae, showing different varieties of orchids, Ernst Haeckel. 1904, lithographic color plate, from Kunstformen der Natur.

One variety, the vanilla orchid (*vanilla*), produces fruit pods used to make vanilla flavoring (figure 3.33 and 3.34). Three species of vanilla are grown globally for this purpose. I had the opportunity to visit a plantation in Tahiti and see the process first-hand. However, Madagascar and Indonesia are the largest producers. Growing vanilla is labor-intensive because the flowers are manually pollinated, making it the second most expensive spice after saffron. More than a cooking additive, it is also used to create scents for perfume and candles.

Tyree used the orchid in his paintings as part of a beautiful floral headpiece (figure 3.35). He used the purple dendrobium in the floral headpiece of a stunning oil-on-velvet portrait. The vibrant flowers sit in a wahine's hair, and braids of panex leaves cascade down past her shoulders. She faces the viewer straight-on with her face brightly lit as bold shadows fall across her face.

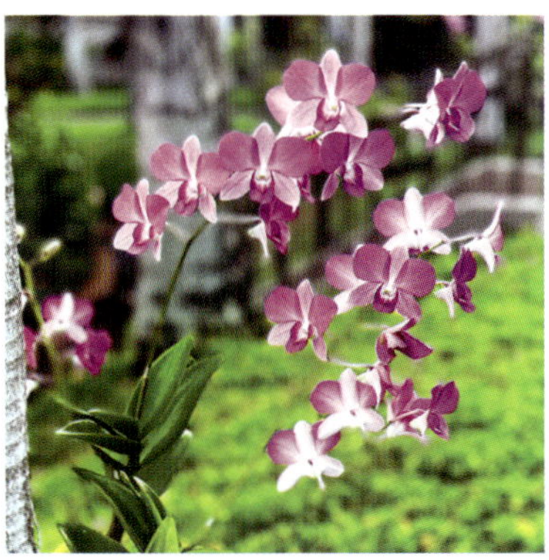

Figure 3.28 Orchid as an epiphyte growing on a coconut palm, Maui. 2021, author's photo.

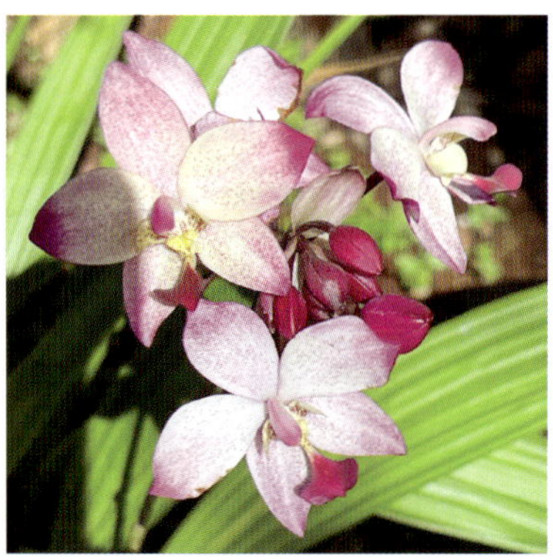

Figure 3.30 A small terrestrial orchid with petal symmetry and purple labellum, Maui. 2021, author's photo.

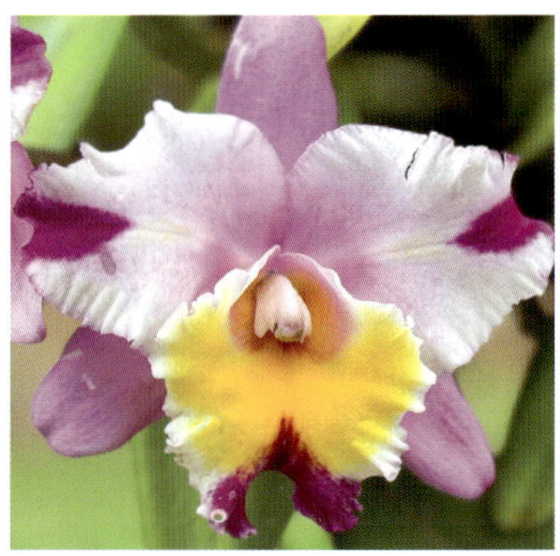

Figure 3.31 Cattleya Orchid, Maui. 2021, author's photo.

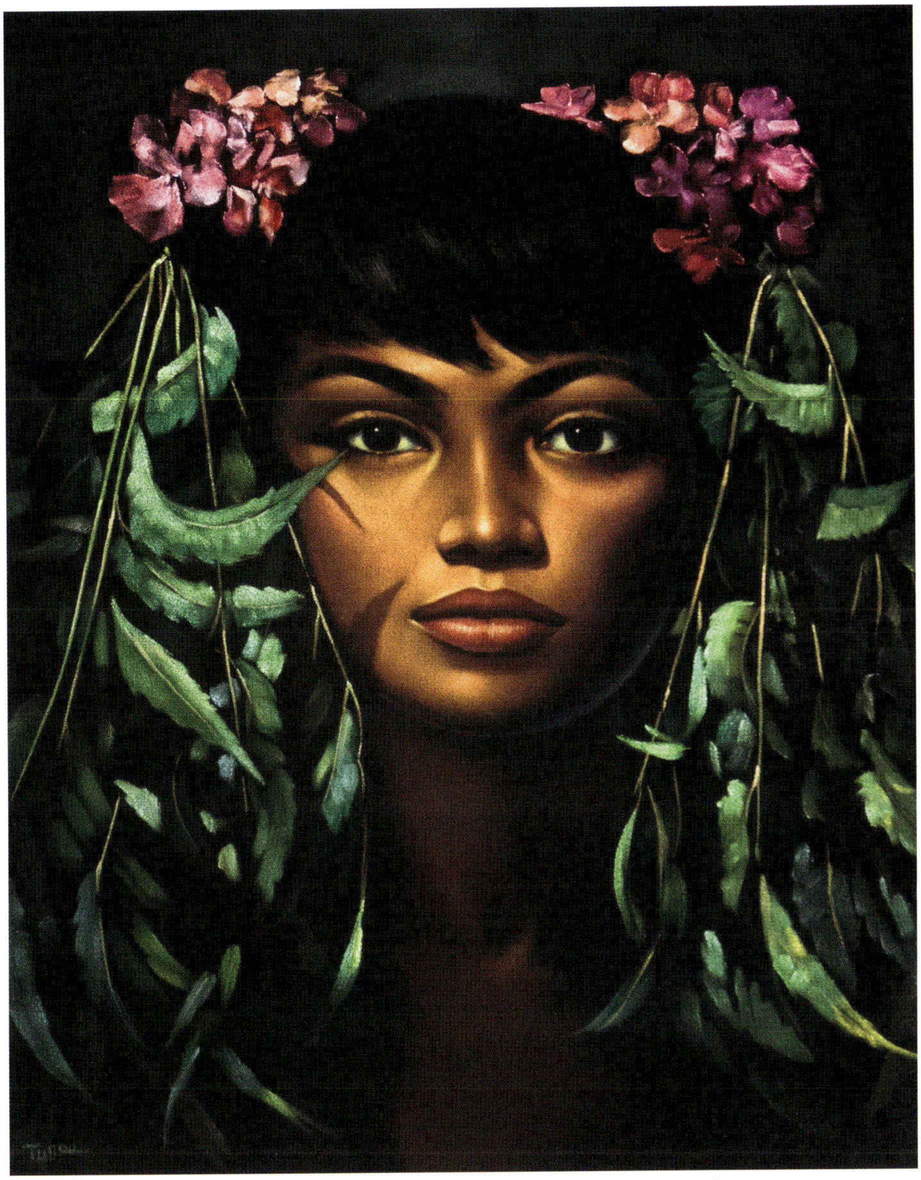

Figure 3.35 *Purple Dendrobiums and Panex leaves in Kulio's Hair*, Ralph Burke Tyree. 1968, oil on velvet, 20" x 16", private collection.

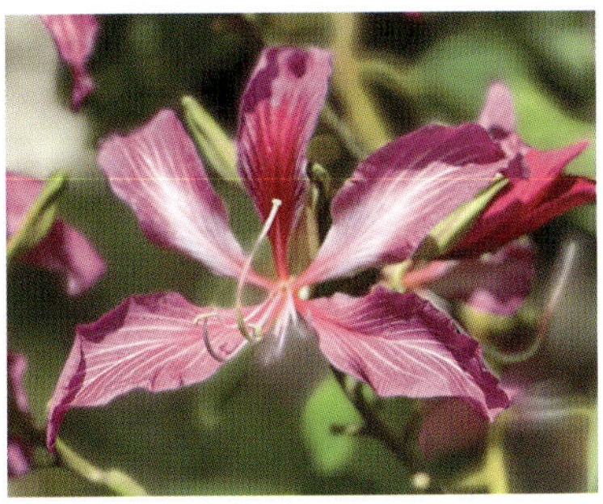

Figure 3.32 Flower from the Hong Kong orchid tree. Not actually an orchid because, while it has petal symmetry, it doesn't have a labellum. Maui, 2021, author's photo.

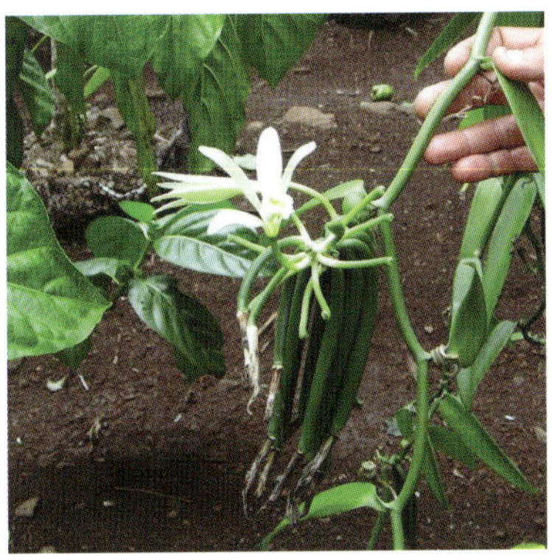

Figure 3.33 Vanilla orchid, Huahine, Tahiti. 2006, author's photo.

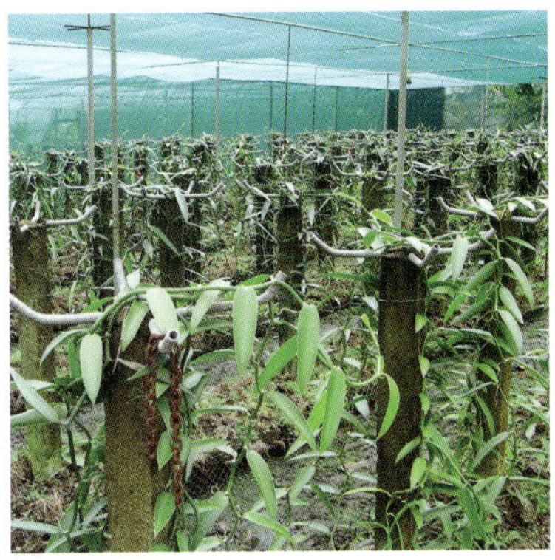

Figure 3.34 Vanilla growing in a shade shelter, Huahine, Tahiti. 2006, author's photo.

Bromeliad

Much smaller than the massive orchid family, the **bromeliad** (*Bromeliaceae*) family in the order *Poales* consists of around 3,500 diverse species. Mostly local to the tropical Americas, their foliage ranges from fine needles to broad, flat leaves in a rosette pattern, as illustrated in figure 3.36. This family's many species include well-known plants like Spanish moss (figure 3.37) and the terrestrial pineapple (figures 3.38 and 3.39), the only bromeliad cultivated commercially for food. Many bromeliads store water in their overlapping leaves, creating a habitat for small invertebrates. Most reproduce by sending out offshoots, called pups, from the mother plant's base.

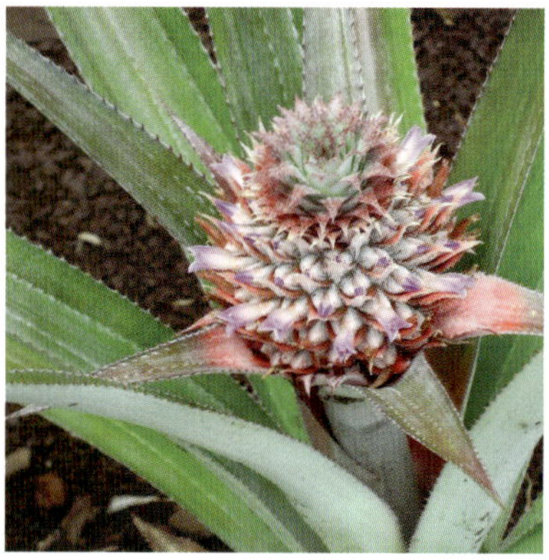

Figure 3.36 Anas bromeliad huahine, Tahiti. 2006, author's photo.

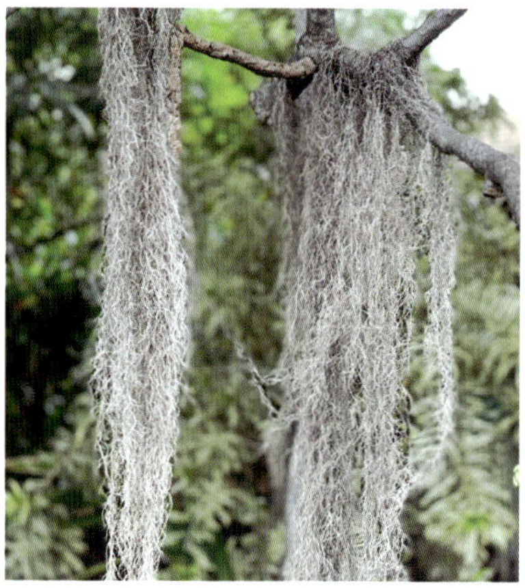

Figure 3.37 Spanish moss, Maui. 2021, author's photo.

Tyree first painted a bromeliad when he included pineapples in his oil on canvas painting of a Portuguese lady (figure 3.40). He created this work as a gift for Margo's oldest sister, Mabel, highlighting his wife's Portuguese roots. The traditionally dressed woman sits on a rock in front of a bright blue sky filled with fluffy clouds next to several pineapples. In this work, the pineapples add color and interest to the composition as they sit to the right, with their dynamic leaves mirroring the woman's form on the left.

Later, as seen in figure 3.41, Tyree painted a wahine in the jungle gazing at the viewer past a bromeliad on a branch. As the woman and bromeliad duel for the viewer's attention, Tyree draws the eye to the plant by painting it in sharp focus and highlighting its spikey leaves with soft light. The beautiful woman is slightly out of focus, seeming to emerge from the jungle behind her. Despite this, her piercing stare from the background cannot be ignored and ultimately overshadows the bromeliad.

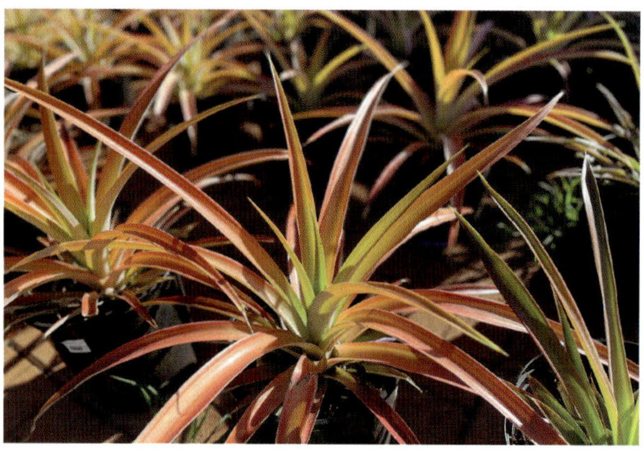

Figure 3.38 Young pineapple plants, Maui. 2021, author's photo.

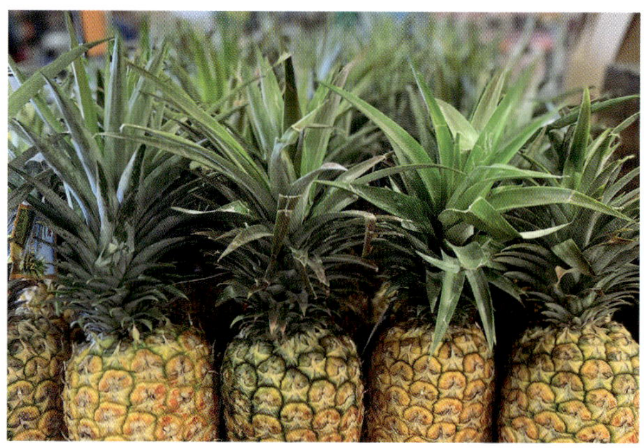

Figure 3.39 Pineapples waiting for a friendly home, Maui. 2021, author's photo.

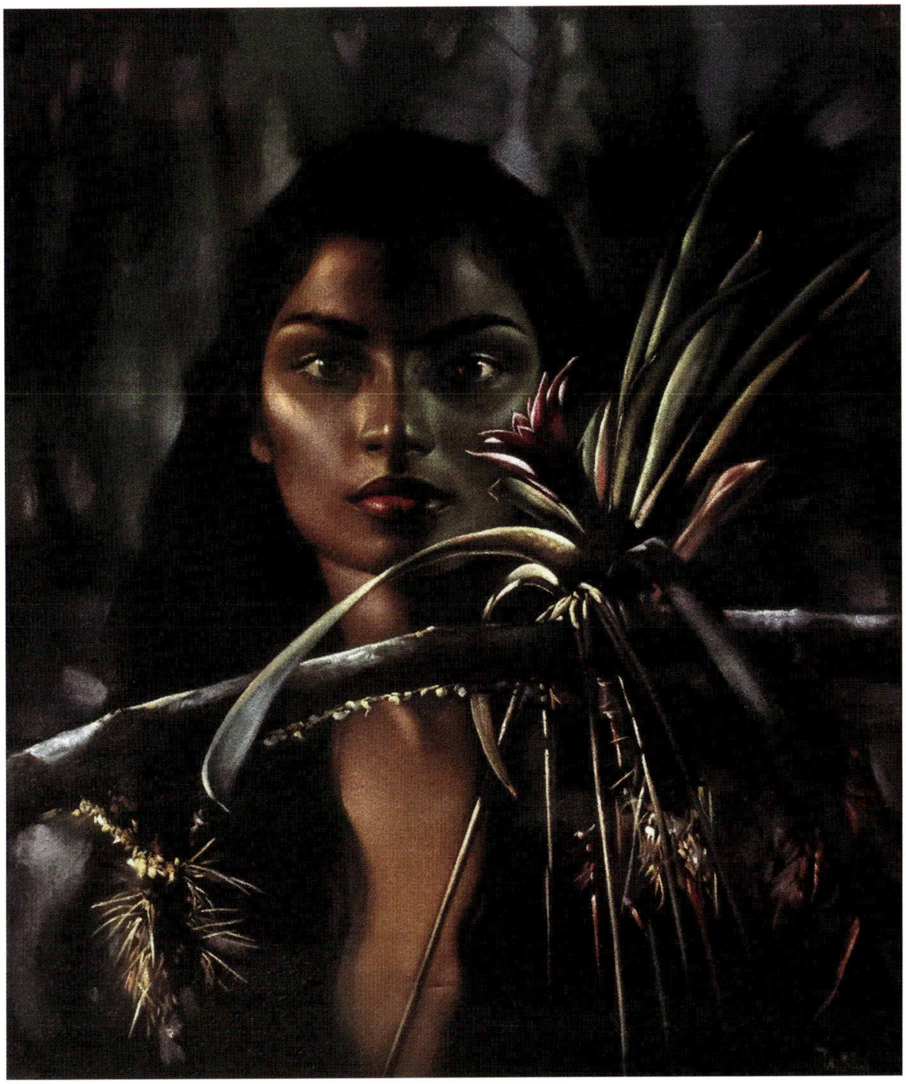

Figure 3.41 *Bromeliad and the Wahine*, Ralph Burke Tyree. Oil on velvet, 24" x 20", private collection.

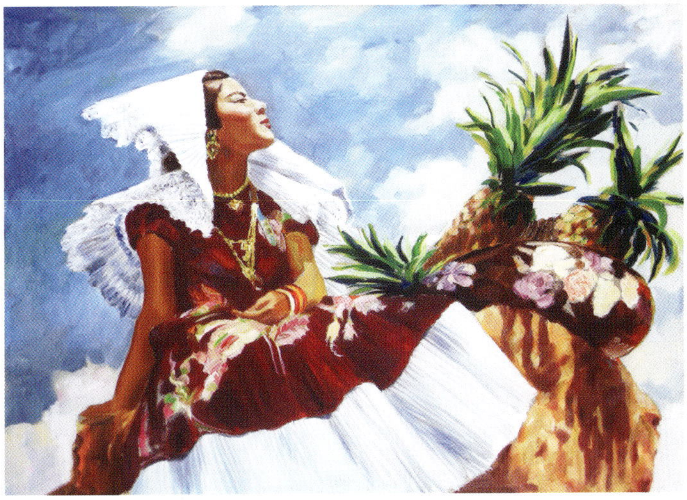

Figure 3.40 *Portuguese Lady and Pineapples*, Ralph Burke Tyree. 1946, oil on canvas, 19" x 26", private collection.

Banana

The **banana** (*Musa*) is a genus from the *Musaceae* family most known for its elongated, edible fruit (figure 3.42). One of the largest flowering plants, most varieties are tall and tree-like with huge leaves. The fruit comes in many sizes and colors, including yellow, green, and purple. First domesticated in Papua New Guinea, they were spread worldwide by sea trade as early as 2,000 BC (figure 3.43).

Today, these fruits serve as a staple in many cultures' diets. Additionally, the flowers and trunk are used in South and Southeast Asian cuisine, the leaves are used to hold food, and the fiber is used to make textiles and paper. Broadly cultivated bananas descend from two species split into two groups—cooking and dessert. Bananas used for cooking are often called plantains in Europe and North America. However, plantains are only one variety of the starchier and less sweet cooking bananas. Artists often depict the banana as a classic tropical fruit delight, similar to the coconut palm.

When painting a banana plant, Tyree chose to paint its blossoms instead of its instantly recognizable fruit (figure 3.44). The reddish cone-shaped blooms hang off the banana's leafy branch, almost entirely in the tree's shade. The leaves extend upward, out of the composition, placing the flowers against the turquoise sky. Mostly hidden by dark shadows, a bright light that instantly draws the eye hits the largest flower. Further highlighting the blossoms, the branch above and the dark skyline of trees below frame them. Working with oil paints on velvet, Tyree again demonstrates his mastery of light and shadow.

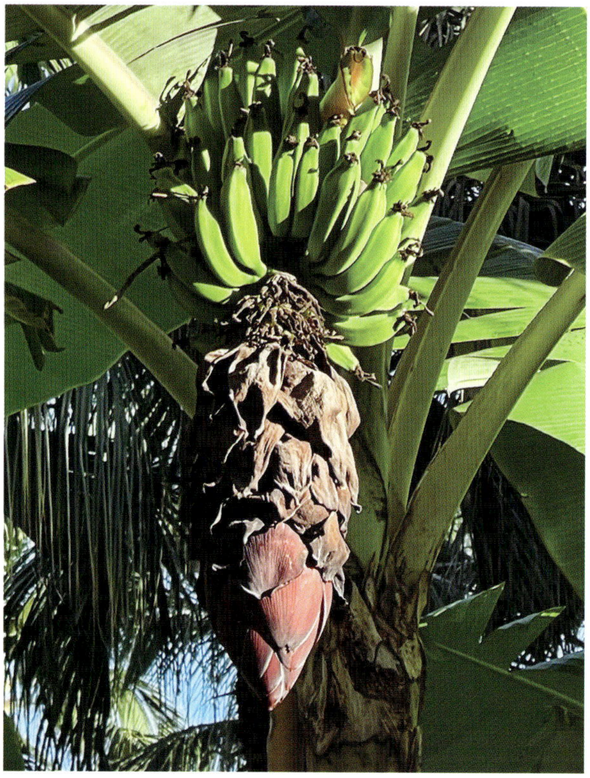

Figure 3.42 Large banana tree with both the purple flower and developing bananas above. Kauai, 2022, author's photo.

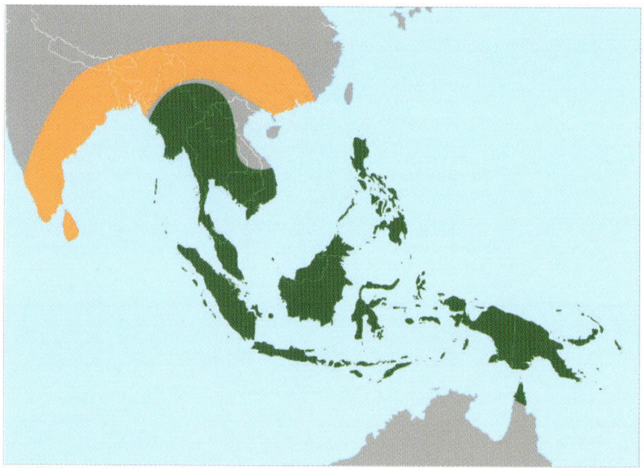

Figure 3.43 Original native ranges of the two ancestors of edible bananas. Papua New Guinea. 2021, Obsidian Soul.

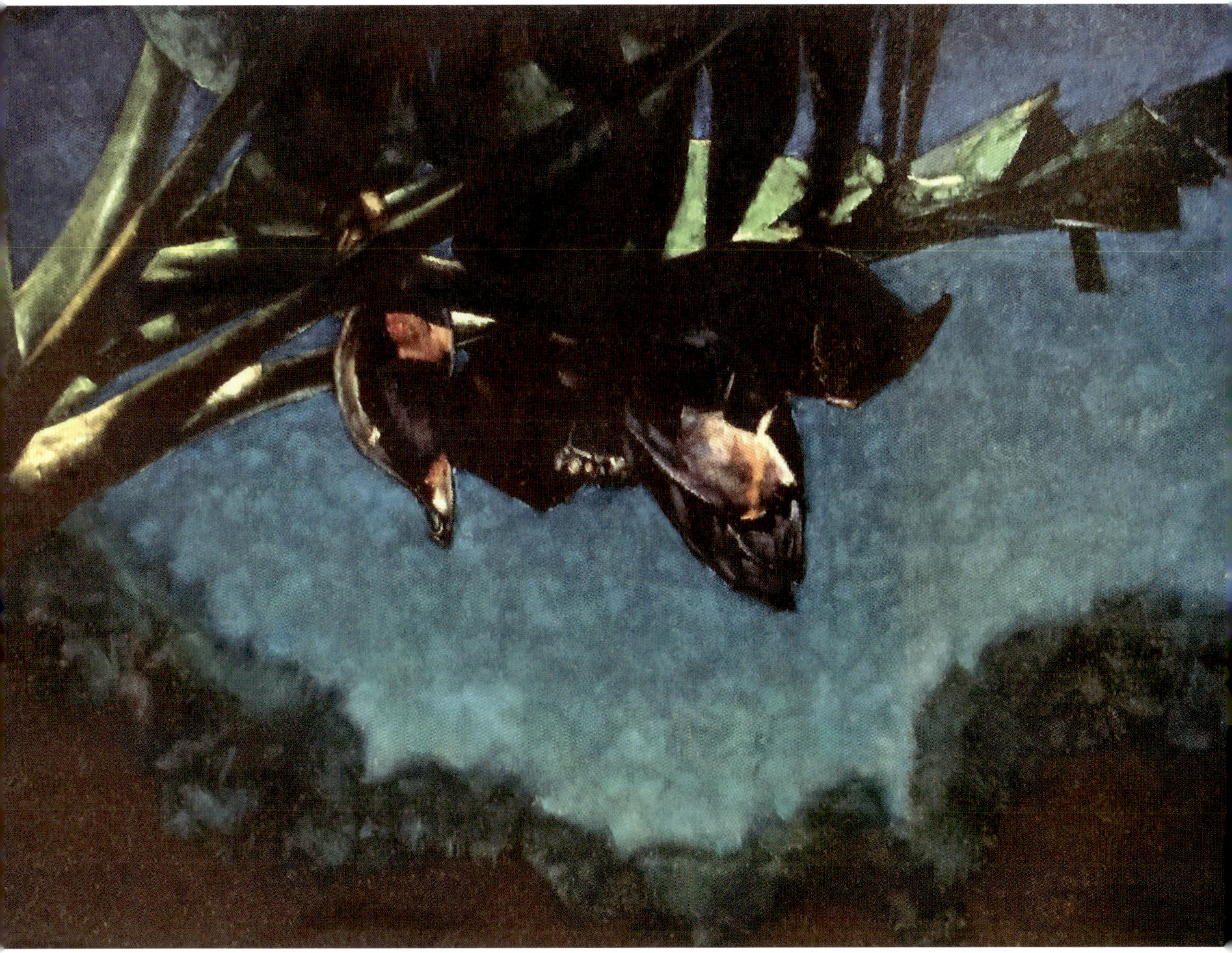

Figure 3.44 *Banana Blossoms*, Ralph Burke Tyree. 1975, oil on velvet, 16″ x 20″, private collection.

Palm Tree

The **coconut palm** (*Cocos nucifera*) is the prototypical island palm from the *Arecaceae* family of palms (figure 3.45). While humans have distributed this tree across the globe, it naturally occurs in coastal areas with sandy, salty soil and high humidity. With a skinny trunk and long fronds, it can be up to 100 feet tall. It flourishes on many beaches, even while contending against strong ocean breezes, thanks to its fibrous root system that primarily remains close to the surface. This swaying beach tree produces coconuts, a fruit that has long sustained its tropical cohabitants. The rest of the tree, including its husks, leaves, trunk, and even roots, is also utilized by humans. Fittingly, its name in Malay translates to "the tree of a thousand uses," and in the Philippines, it is called the "tree of life." Not only is it ubiquitous across tropical cultures, but it also has considerable religious or mythical significance. In the Hindu tradition, the coconut signifies abundance and plays a vital role in many rituals and festivals. Multiple origin myths exist across the Pacific in which the coconut plays a central role.

Tyree featured coconut palms in his works on several occasions. While stationed in Samoa, he painted a scene of a fishing village (figure 3.46). A woman walks along the waterfront in the foreground with palms behind her, cutting across the frame. Huts dot the coastline in the background amongst densely packed palms and vegetation. Here, the palms set the tropical location.

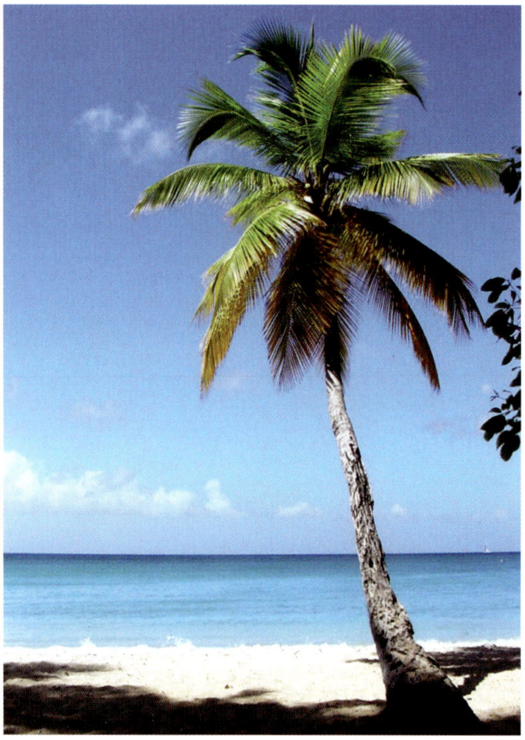

Figure 3.45 Coconut Palm. Maarten-van-deheuvel-Siu on Unsplash.

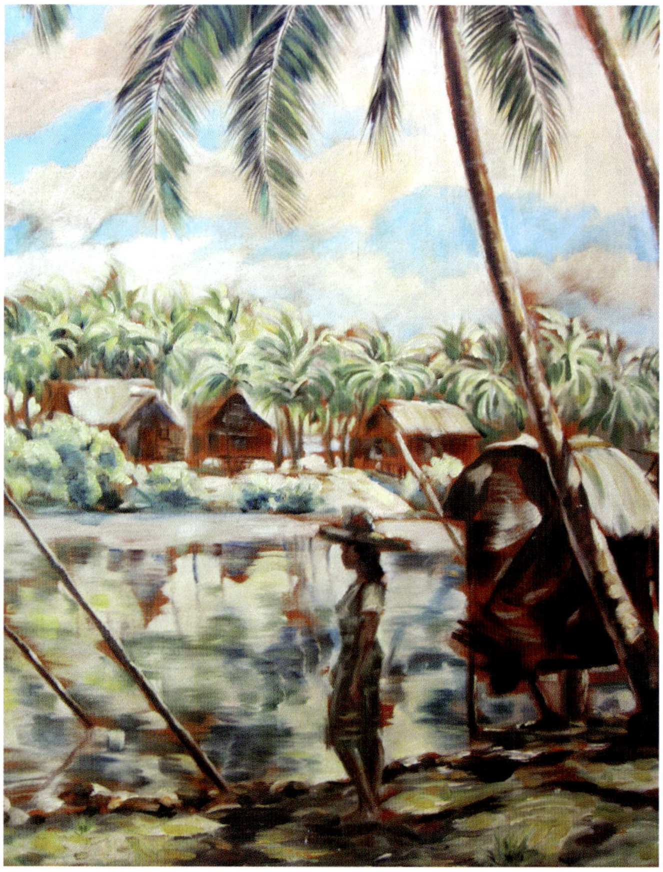

Figure 3.46 *Native Fishing Village Featuring Coconut Palms*, Ralph Burke Tyree. Samoa, 1944, oil on canvas, 20" x 16", private collection.

Pandanus

Pandanus (*Pandanaceae*), commonly known as pandan or screw palm, is a genus of palm-like trees in the *Pandanales* order (figure 3.47). They are not closely related to the palm, despite their similarity in appearance and co-presence on tropical and subtropical isles. Instead of the palm's single, thin trunk, the pandanus has a stout trunk with many supportive stilt roots around the base, which help the tree stabilize in sandy soil. The variety in Micronesia and Hawaii has leaves that are ideal for weaving and are often used to make mats, hats, baskets, and thatching. Throughout Asia, the pandanus has made its way into cooking, with the leaves sometimes referred to as the "vanilla of Asia."

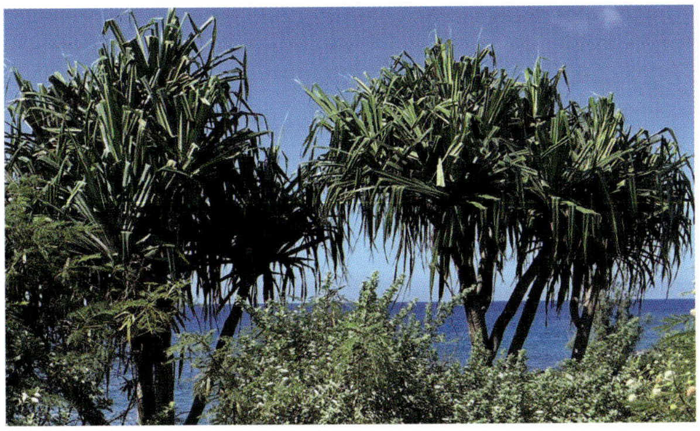

Figure 3.47 Pandanus, Kauai. 2022, author's photo.

Tyree painted a wahine under a pandanus, shown in figure 3.48, using the tree to make her tropical locale evident at first glance. Working with oil paints on board, the tree's long, pointed leaves frame the woman within the larger composition. As she walks through the jungle, she stops and places her hand on a tree trunk, turning to look at the viewer. This engagement invites the viewer into the piece and, thus, into the island paradise.

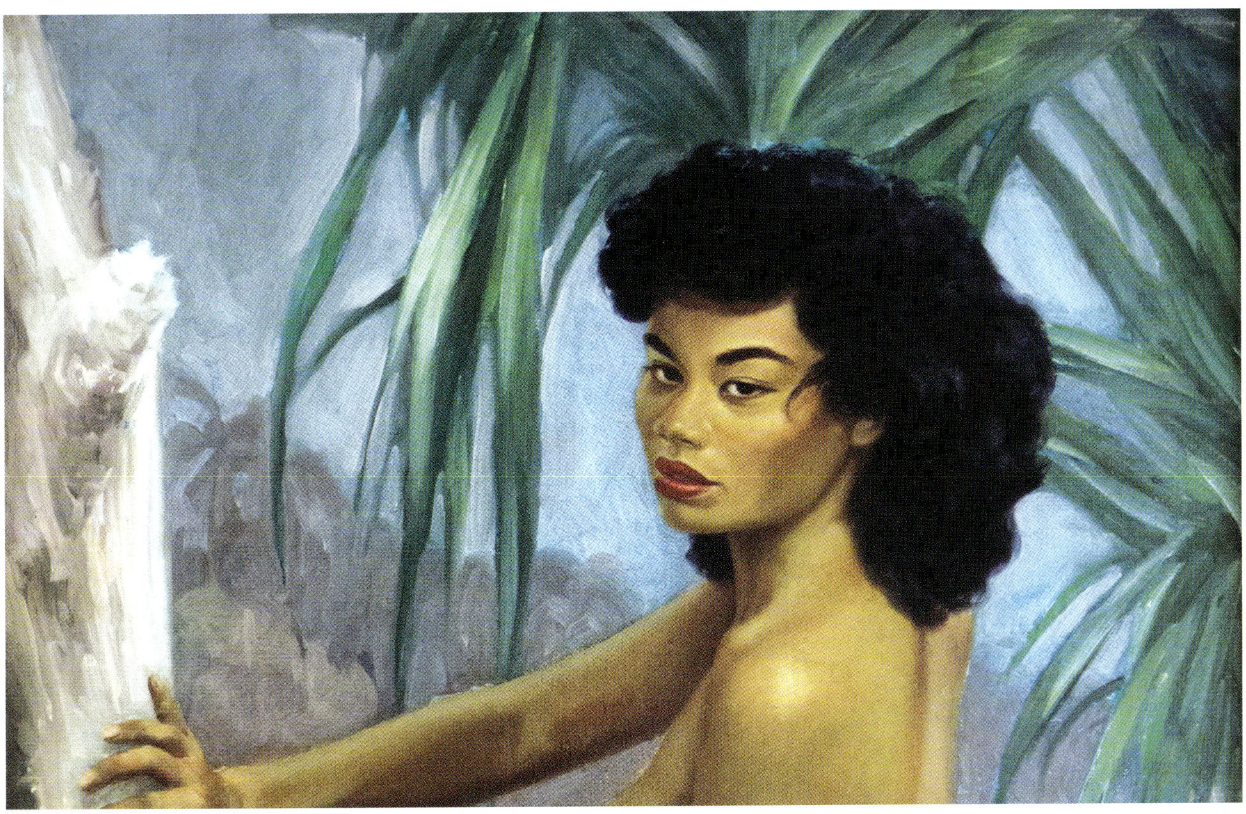

Figure 3.48 *Wahine Under the Pandanus Tree*, Ralph Burke Tyree. 1955, oil on board, private collection.

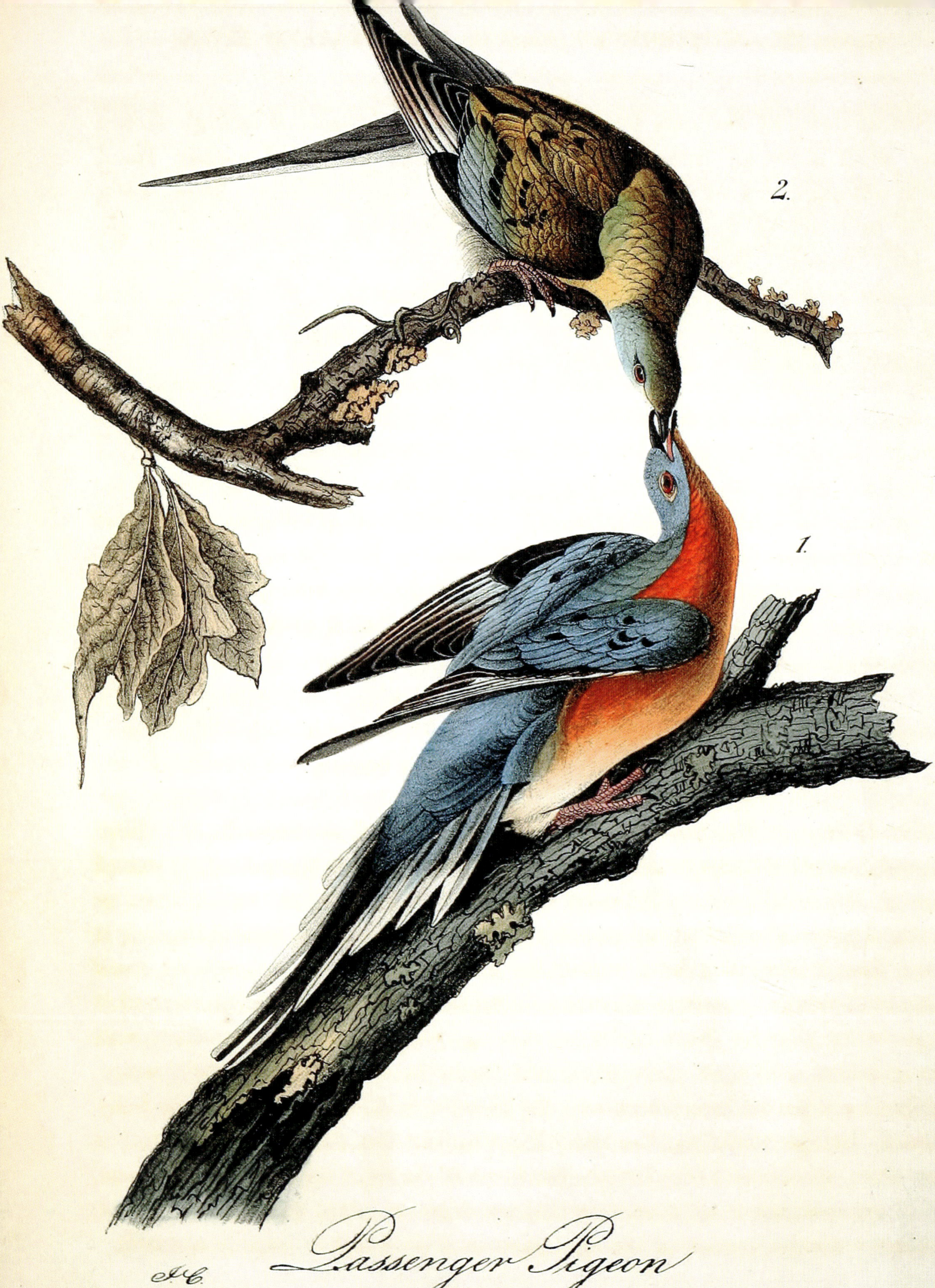

Passenger Pigeon

1. Male. 2. Female.

CHAPTER 4

Threatened, Endangered, & Extinct

"All the world's a stage,
And all the men and women merely players.
They have their exits and their entrances."

—WILLIAM SHAKESPEARE

Tyree continued oil painting on black velvet and board for the rest of his career, but in 1970 he drastically changed subject matter. After decades of primarily creating portraits of people, usually wāhine, he shifted to featuring animals as the subjects of his paintings. He had always felt a deep love for animals, and this love inspired a new and noble goal—painting endangered beasts. On this, he remarked, "Generally, I paint [animals] that have been around a long time. . . , and if we're very fortunate, they'll be around tomorrow."

"The texture of velvet is closer to the texture of natural things than canvas. Birds and animals are soft to the touch, and so is velvet."

During Tyree's final stay in the tropics, he took up residency on the Big Island of Hawaii and began painting local creatures like barn owls and frogs. Over the next decade, he depicted animals from around the entire planet. This included species common to North America, like squirrels, blue jays, and bobcats, as well as exotic animals from far-away places, like cotton-top tamarins from South America, caracals from Africa, and red pandas from Asia.

What is an Endangered Animal?

According to the National Geographic Society, loss of habitat and decreased genetic variation are the two biggest reasons animals become threatened by extinction.

Loss of habitat can occur naturally or because of human activity. An example of natural habitat loss is the asteroid strike that led to the extinction of dinosaurs. When the asteroid struck Earth about 65

Figure 4.1 *Passenger Pigeon Pair*, John James Audubon. 1827, watercolor, plate LXII from *The Birds of America*, London: 1827–1838, courtesy of the National Gallery of Art.

million years ago, the planet experienced a sudden drop in temperature. With this, dinosaurs lost their habitat and quickly went extinct. On the other hand, humans have cleared thousands of acres of jungle in the Amazon rainforest for farming and development, destroying the habitats of many local species and endangering them. This also has a broader impact on the entire atmosphere as photosynthesis decreases, lessening the amount of carbon dioxide removed from the atmosphere and converted into oxygen. Without this conversion, carbon dioxide accumulates as a layer in the atmosphere, heating the planet. As the world's largest rainforest, the Amazon is critical for this process.

In California, booming development has encroached on the mountain lion's natural habitat. While this has caused a notable decrease in the lion's numbers, remote mountainous areas remain in the Sierra Nevada and Rocky Mountain ranges, allowing the species to avoid becoming endangered. This encroachment has also caused an increase in encounters between humans and mountain lions. In these close encounters, the cat usually loses as humans turn to firepower to protect children, pets, or livestock.

The other primary danger to a species' survival is **loss of genetic diversity** within the population. Generally, the greater number of individual members in a species, the greater its genetic variation. Inbreeding, particularly with closely related animals such as siblings or first cousins, results in the increased manifestation of recessive genes and, in smaller populations, genetic defects. This brings greater disease susceptibility, lower birth numbers, and earlier deaths. Like habitat loss, this can happen naturally or due to human activity. During the last Ice Age, cheetahs were isolated and forced to inbreed over an extended period, which decreased the species' genetic diversity. As a result, today's cheetahs are less adaptable, less likely to survive to maturity, and more challenging to breed in captivity than other big cats. Overhunting and overfishing often create the same problem, reducing populations and, thus, lessening a species' genetic diversity and slowly causing the population to implode.

IUCN and Its Red List of Threatened Species

Since its establishment in 1964, the IUCN's Red List of Threatened Species has become a complete information source on the world's species' extinction risk. Using precise criteria relevant to all species and all regions of the world, the IUCN evaluates the extinction risk of species and subspecies. This crucial tool helps to identify species that require targeted recovery efforts and locations that need protection. Due to the Red List's strong scientific base, it is widely recognized as the most authoritative guide on the health status of our world's plants and animals. Ultimately, the IUCN Red List helps to guide and inform future conservation and funding priorities.

IUCN members have comprehensively assessed over 150,000 species, recording, and evaluating information like range, population size, habitat, ecology, use, threats, and conservation actions. Regularly, the IUCN assesses newly recognized species and re-assesses the status of existing species to ensure information is current. Based on criteria such as rate of decline, population size, geographic distribution, and degree of fragmentation, species are placed into one of nine threat categories. In order, these threat categories are extinct (EX), extinct in the wild (EW), critically endangered (CR), endangered (EN), vulnerable (VU), near threatened (NT), least concern (LC), data deficient (DD), and not evaluated (NE) (figure 4.2).

Unfortunately, extinction threatens over 40,000 species on the Red List. As shown in figure 4.3, this includes 41% of amphibians, 37% of sharks and rays, 34% of conifers, 33% of reef-building corals, 26% of mammals, and 13% of birds. In the face of an increasing number of threatened species, the IUCN is working to halt or reverse the Earth's current decline in biodiversity. It aims to add another 20,000 species to the Red List, improving coverage and providing a better base for conservation and policy decisions. Through this, it will become a more complete 'Barometer of Life.'

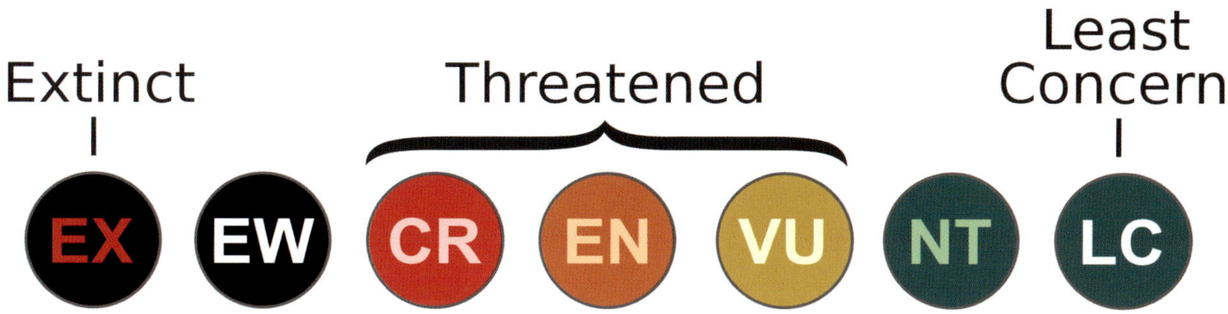

Figure 4.2 IUCN scale of Red List Classifications
EX=extinct EW=extinct in the wild
CR=critically endangered EN=endangered
VU=vulnerable NT=near threatened
LC=least concern, courtesy of IUCN

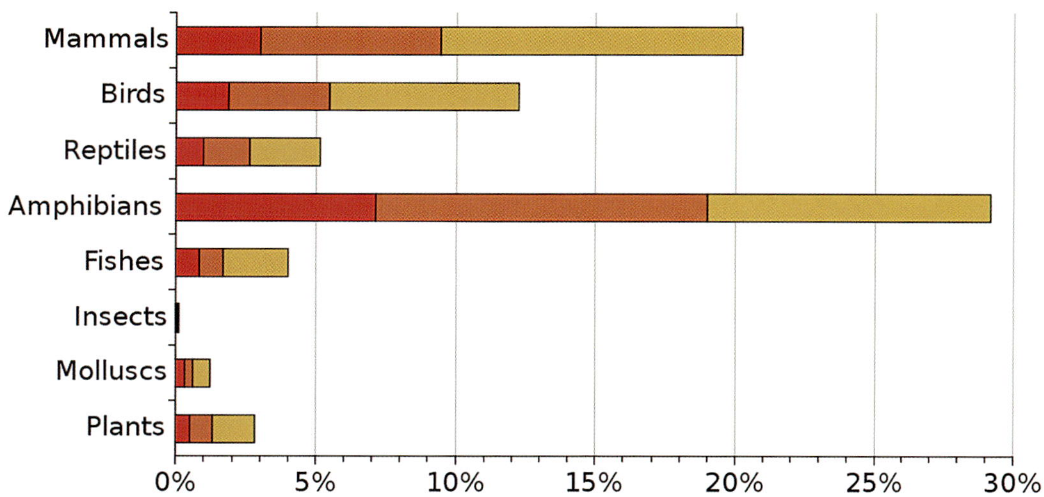

Figure 4.3 Percentages of different species listed as critically endangered, endangered, and vulnerable on the 2007 IUCN Red List, courtesy of IUCN Red List.

What is CITES?

The Convention on International Trade in Endangered Species of Wild Fauna and Flora (CITES) is a multilateral treaty that protects endangered plants and animals from international trade. Adopted in 1975, CITES controls the international importation, exportation, and transportation of endangered species, animals, plants, and products containing their parts, like fur, ivory, horns, or scales. Today, 184 countries have signed onto this world agreement, and it protects more than 38,000 species.

Passenger Pigeon Goes Extinct

The **passenger pigeon** (*Ectopistes migratorius*) is a wild pigeon from the *Columbidae* family, which was once the most abundant bird in North America and now provides a notable case study of how a prolific and widespread species can become extinct. In the 1800s, when John J. Audubon painted his famous pair of passenger pigeons (figure 4.1), thousands of flocks of these now-extinct birds darkened skies across Canada and the US (figure 4.4). Native Americans had long hunted this pigeon, but the practice dramatically escalated when the Europeans, and their guns, arrived. Pigeon meat was commercialized, and planned hunts of this seemingly plentiful bird became common (figure 4.5).

In addition to hunting, loss of habitat also contributed to the waning and subsequent extinction of the passenger pigeon. The influx of European settlers also brought widespread deforestation, destroying much of the pigeon's habitat. Between hunting and habitat loss, the bird's population dropped, and inbreeding accelerated. With this, came a rapid decline in the passenger pigeon's numbers during the late nineteenth century and, in 1901, hunters shot the last wild passenger pigeons.

Subsequent breeding of captive passenger pigeons was largely unsuccessful, and all the birds held by zoos gradually succumbed. The Cincinnati Zoo housed the final surviving passenger pigeon, named Martha for First Lady Martha Washington. Unfortunately, all attempts to breed her were unsuccessful, and with her death on September 1, 1914, the passenger pigeon became extinct. Never before had a species' exact extinction date been so well documented. Sad as this is, it illustrates the importance of population diversity and habitat preservation to protect threatened species.

Today, there is some hope that the passenger pigeon will return thanks to modern technological advances in cloning. Martha's body was frozen in a block of ice and shipped to the Smithsonian Institution in Washington, DC. She is now housed at the Museum of Natural History (figure 4.6). Soon, it may be possible to create clones using DNA from remains like Martha's that are held in museums, but only time will tell.

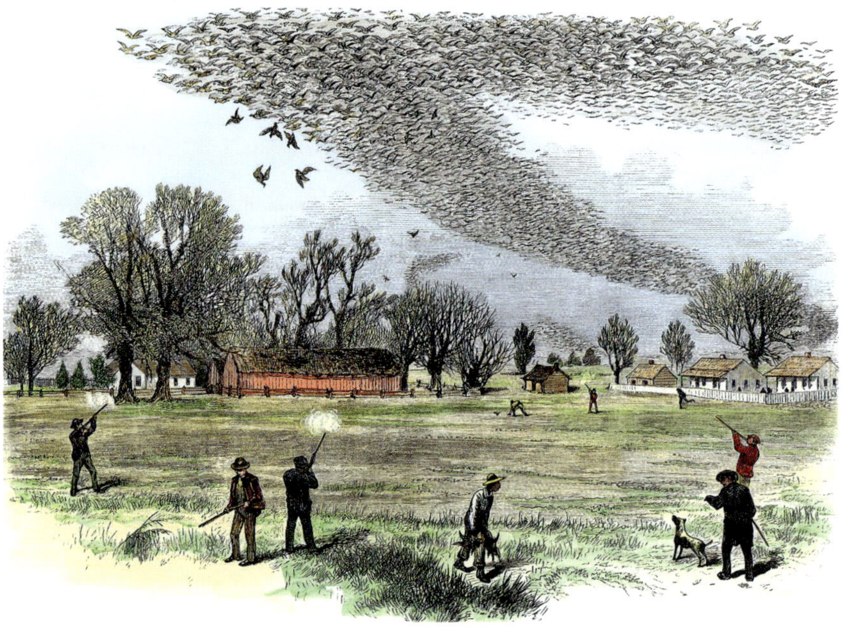

Figure 4.4 *Passenger pigeon flock being hunted in Louisiana,* Smith Bennett. 1875, woodcut based on sketch, from the *Illustrated Shooting and Dramatic News,* 3 July 1875.

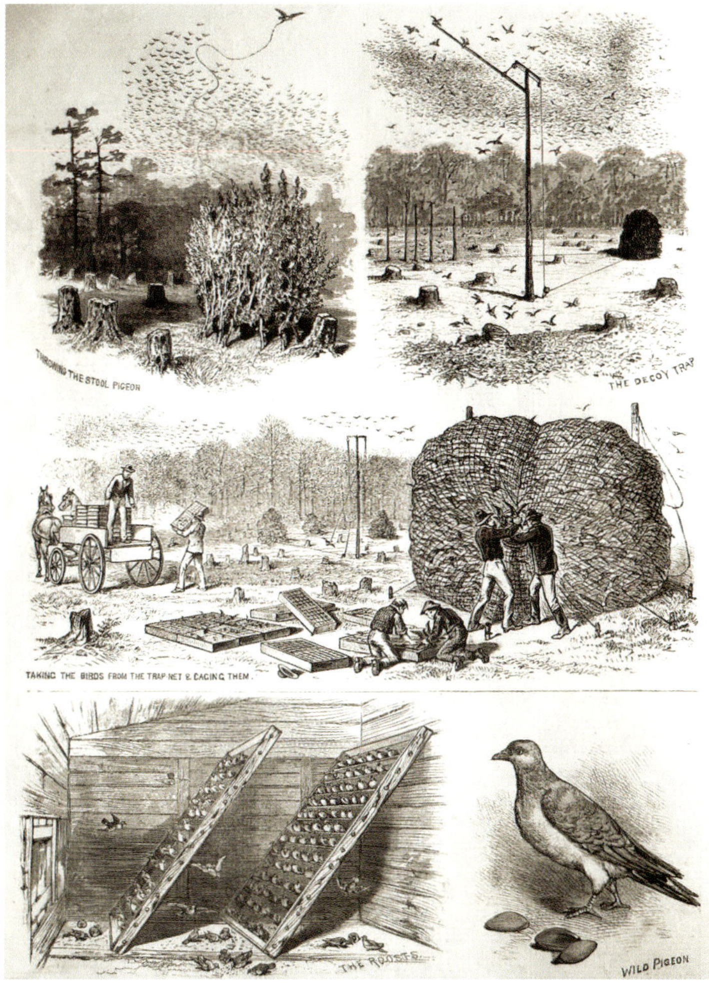

Figure 4.5
Methods of capturing passenger pigeons for shooting contests, staff artist. 1881, from *Frank Leslie's Illustrated News* (vol. LII, no. 1344, pp. 299–300).

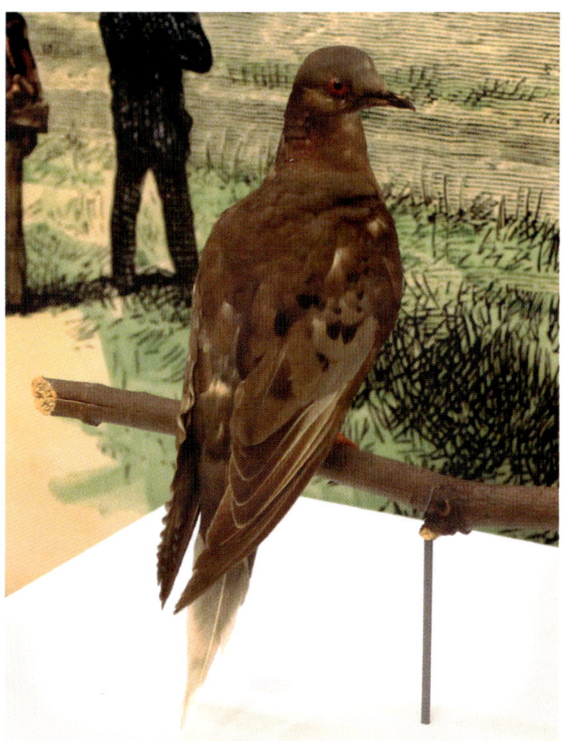

Figure 4.6 Martha, the last passenger pigeon, mounted in a display case in the National Museum of Natural History. 2015.

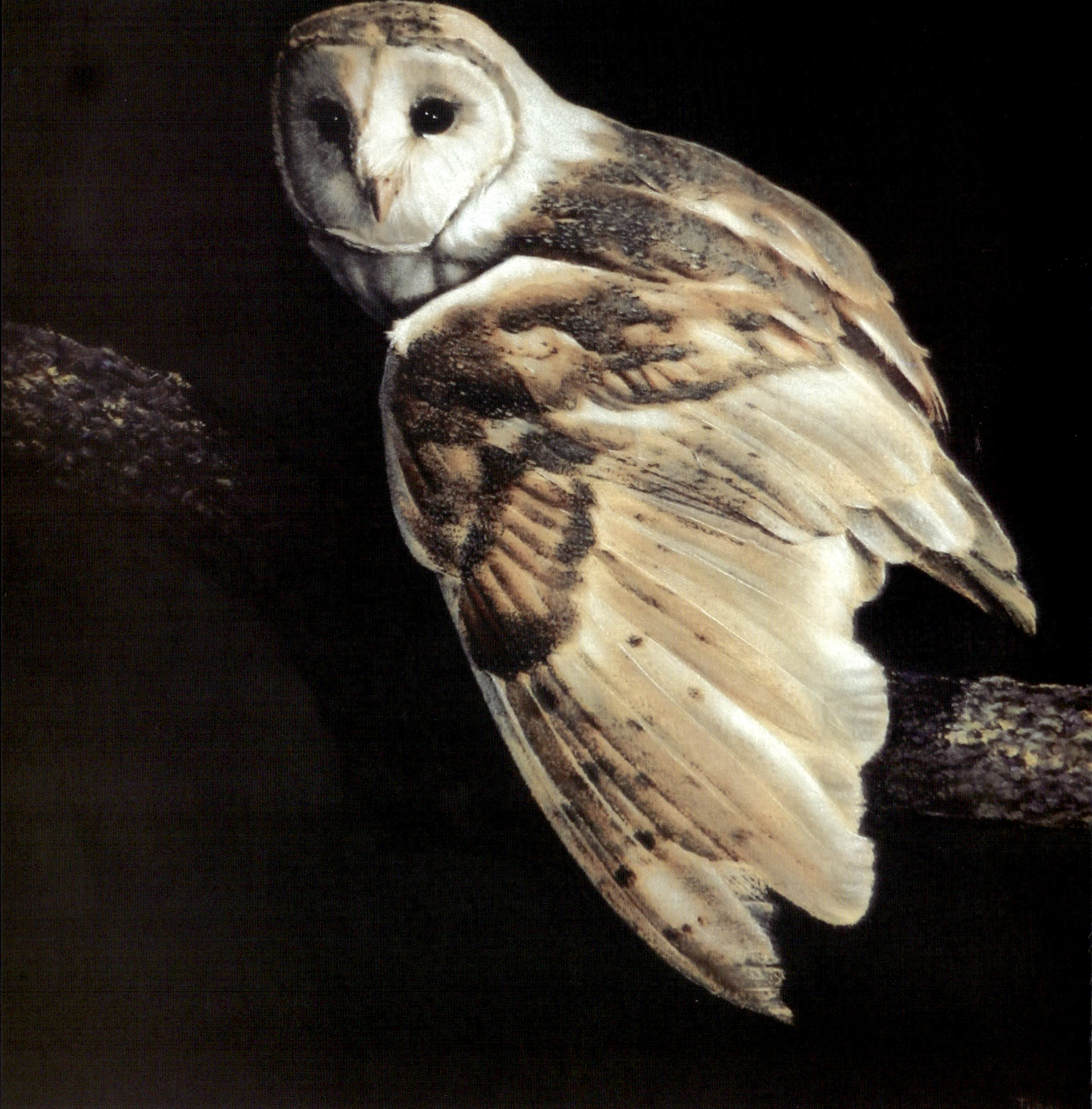

Figure 5.1 *Barn Owl*, Ralph Burke Tyree. Oil on velvet, private collection.

CHAPTER 5

Birds of a Feather

"As the eagle was killed by the arrow winged with his own feather, so the hand of the world is wounded by its own skill."

—HELEN KELLER

Tyree loved birds of all kinds, from aggressive loggerhead shrikes to giant condors and cranes, and he began to paint these feathery friends often. Of all the birds, the owl was always his favorite. When Tyree started looking to animals for subject matter, many iconic raptors, like the bald eagle and the California Condor, were on the brink of extinction. This was primarily due to the widespread use of the pesticide Dichlorodiphenyltrichloroethane (DDT) over the first half of the twentieth century. Using his brushes and oil paints, Tyree passionately depicted these elegant and endangered birds at this dire point in their existence as conservation efforts came to the forefront.

SMALL BIRDS

The Woodpecker—Tap, Tap, Tap

The **northern flicker** (*Colaptes auratus*) is a woodpecker native to North and Central America, Cuba, and the Cayman Islands from the *Picidae* family, which also includes piculets, wrynecks, and sapsuckers (figure 5.2). Known by many names, including the common flicker, yellowhammer, yarrup, heigh-ho, and gawker bird, it is notable for its high-pitched call that sounds like a laugh. This medium-sized bird has a speckled brown, white, and black pattern and is identifiable by the dark patch on its breast and the red patch on the back of its head. Like most woodpeckers, the northern flicker lives in forests and woodlands, where it uses its sharp beak to carve out nesting cavities in tree trunks. Unlike other woodpeckers, it does not hunt for bugs in the tree bark, instead finding insects on the ground or mid-air. While the northern flicker is not at risk, other woodpeckers native to the Americas, like the ivory-billed woodpecker and the imperial woodpecker, are critically endangered due to habitat loss as humans clear land for agriculture.

Woodpeckers are generally viewed as exciting and entertaining birds, even working their way into popular media throughout history. They are given religious or spiritual symbology with their inclusion in mythologies, including Greek, Roman, Norse, Celtic, and Native American. More recently, this fascinating bird joined the *Looney Tunes* crew with the creation of the animated character Woody Woodpecker.

Despite a generally positive relationship with humans, conflict between woodpeckers and humans sometimes occurs because of the holes these birds create in trees and wooden structures, as demonstrated by my local acorn woodpecker (figures 5.3 and 5.4). Despite this, their foraging is ultimately beneficial insect control. Specifically, northern flickers are predators of the European corn borer, a moth that threatens crops and costs the US government over one billion dollars annually.

In figure 5.5, Tyree depicts a northern flicker on velvet perched on a tree trunk just below its carved-out home. The bird appears watchful and wary of its home and babies. With beautifully dappled plumage and a red patch on its head, the flicker perches at the center of the composition. The velvet surface allows the artist to give the bird's mostly light-colored feathers a natural softness and remarkable detail. The bird stands in contrast with the dark background and the rough tree bark, allowing the flicker to hold all the attention.

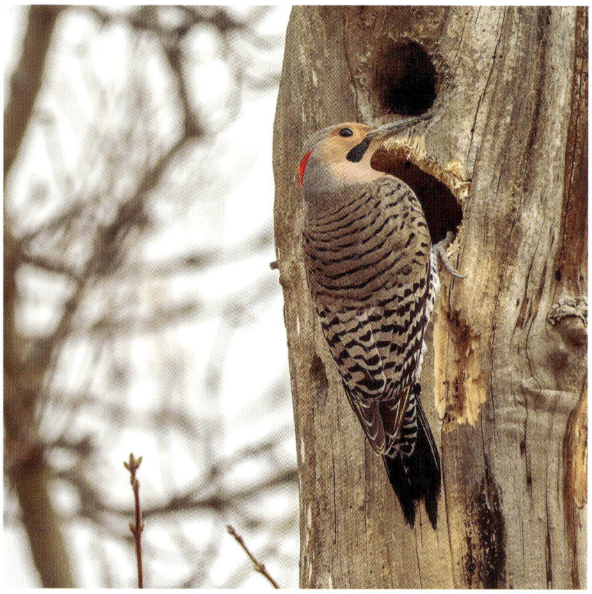

Figure 5.2 Northern flicker woodpecker. Patrice Bouchard on Unsplash.

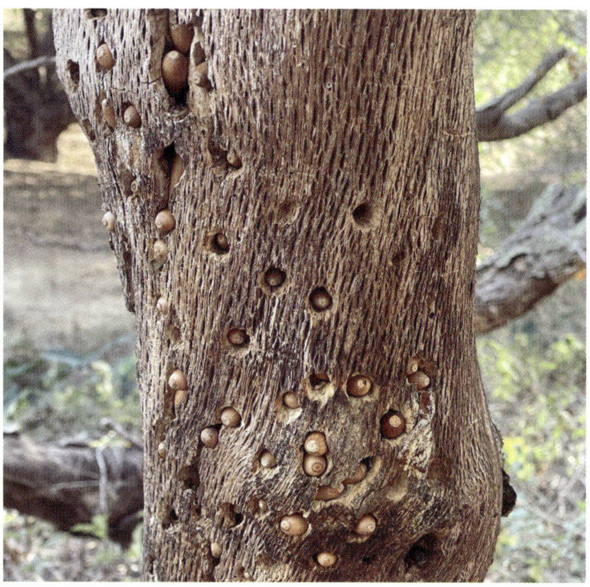

Figure 5.3 Acorn woodpecker's acorn storage depot in an old oak tree. 2022, author's photo.

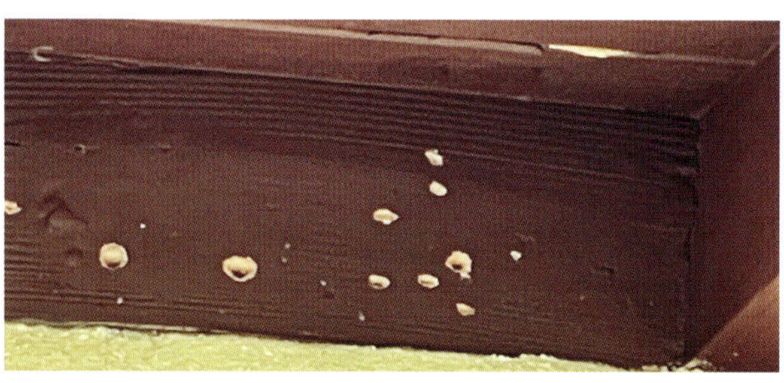

Figure 5.4 Acorn woodpecker holes under my home's eve, a noisy daily occurrence. 2022, author's photo.

Figure 5.5 *Flicker Woodpecker*, Ralph Burke Tyree. 1975, oil on velvet, 20" x 16", private collection.

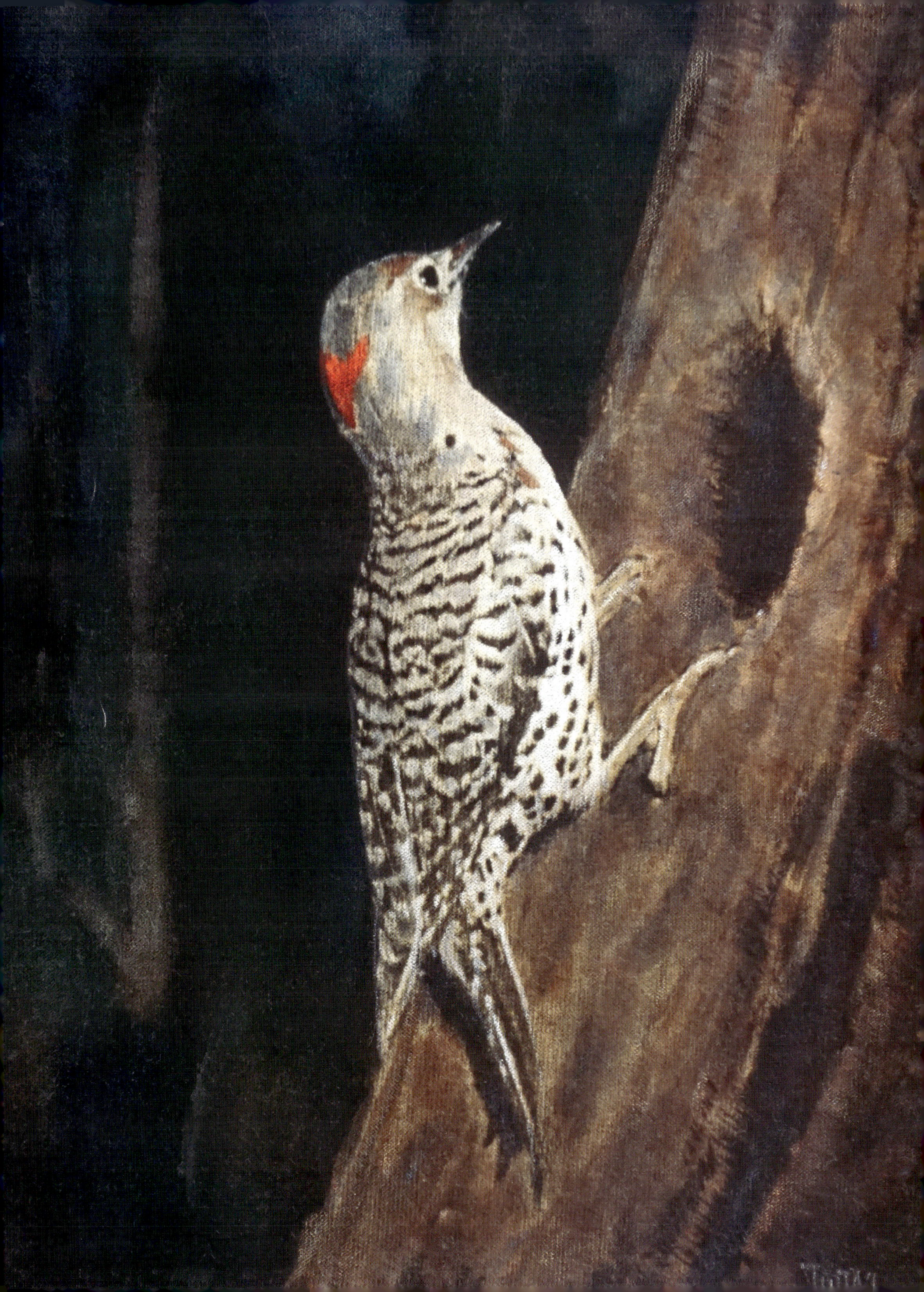

Sapsuckers

Like woodpeckers, **sapsuckers** (*Sphyrapicus*) are a genus in the *Picidae* family. As their name implies, they feed primarily on tree sap, drilling into living trees to create holes that the sap oozes out of, making it easy to sip up (figure 5.6). This practice attracts the woodpecker's other favorite food, insects, which are quickly gobbled up or taken back to their nests to feed their young. Sapsuckers are often considered pests due to the many holes they punch in trees, damaging and, sometimes, killing them. The four species of these slender black, white, red, and occasionally yellow birds are native across North America and Central America. According to Urban Dictionary, a yellow-bellied sapsucker is a coward who won't leave their bunker to come out and fight. The IUCN does not consider any of these species threatened.

Tyree painted a parent sapsucker returning to one of its young in its nest (figure 5.7). The young bird pops its red-crested head out of a small hollow in a multicolored tree as the parent perches just below, tilting its head to feed the baby. Painted in shades of red, yellow, and blue—the three primary colors—the tree trunk fills most of the frame. The most basic concept of color theory revolves around the primary colors and the array of colors created by mixing them. In this piece, Tyree experiments with color mixing as he works with these colors. He uses red, yellow, and blue to create a kaleidoscopic depiction of a naturally brown object, seemingly winking at the rule that mixing all three primary colors creates brown. Even though the tree in the painting is not brown, our brain logically knows the bark of a tree is brown and mentally blends the colors.

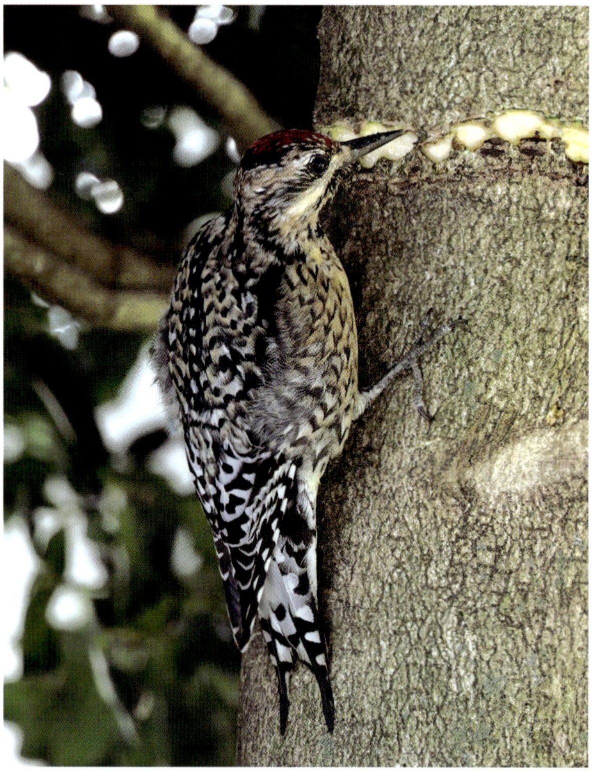

Figure 5.6 Yellow-bellied sapsucker girdling a holly tree. Andrew McCartney, 2021.

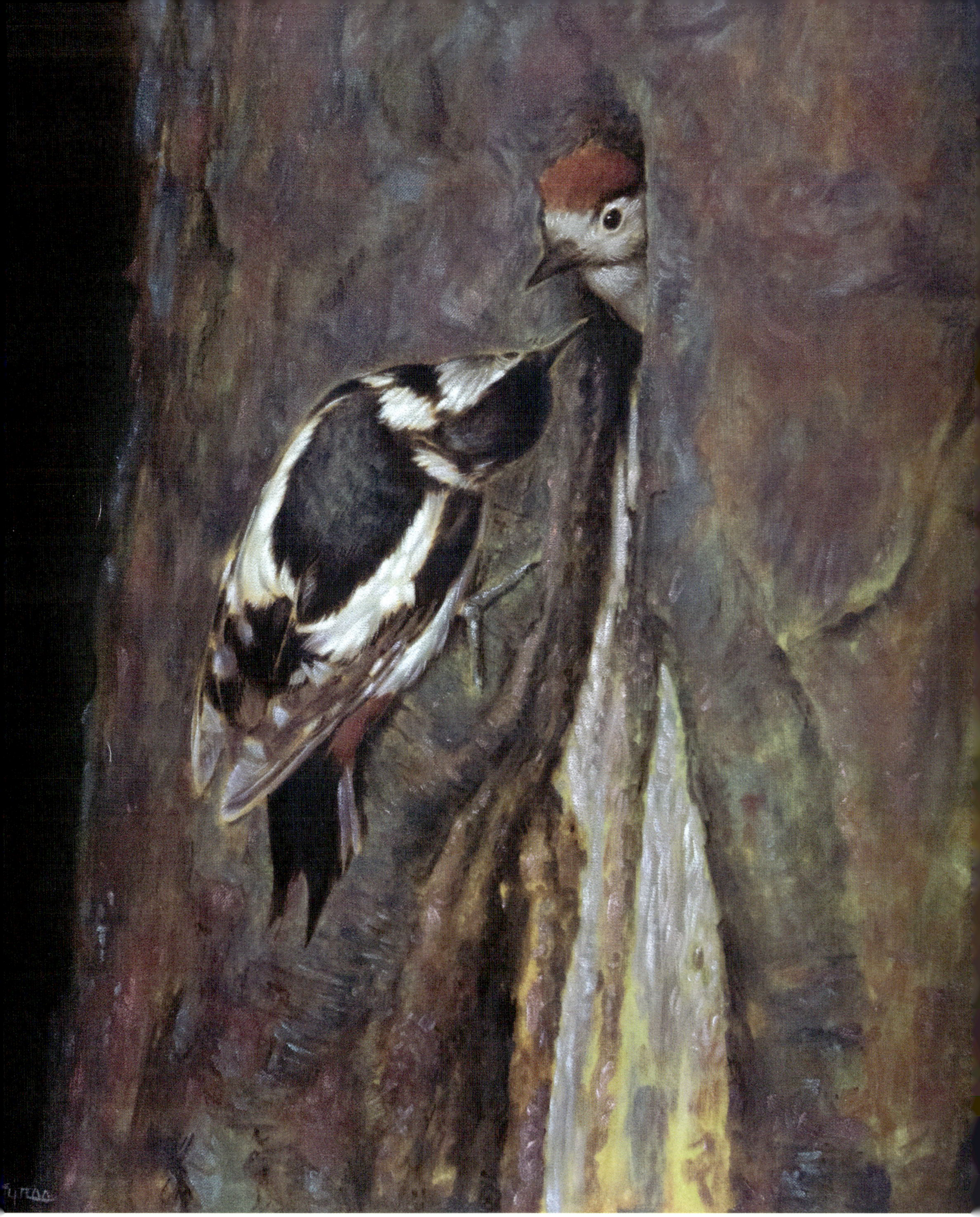

Figure 5.7 *Sapsucker Woodpecker*, Ralph Burke Tyree. Oil on board, 16″ x 12″, Tyree family.

Blue Jay—Clever and Curious

The **blue jay** (*Cyanocitta cristata*) of the *Corvidae* family is, as its name would suggest, a bright blue bird (figure 5.8). It is predominately blue with a white chest, a black collar, graphic white and black markings, and a prominent blue crest. The other half of the blue jay's name comes from its noisy, talkative nature. Although the blue jay is infamous for raiding other birds' nests to kill their young, this is far less common than often thought. In actuality, it mainly feeds on seeds, acorns, nuts, grains, fruit, and small invertebrates. Daring and combative, it is territorial and will scream if a threat enters its domain and even chase away larger predators like hawks, owls, and humans. It is also curious and clever, with an affinity for reflective objects and a demonstrated ability to use tools, at least while in captivity.

This migratory bird lives in various forests and residential habitats across North America, its native range. With the blue jay's prevalence across the US and Canada, it is not at risk for endangerment. This prevalence has also led to its inclusion in Native American and African American lore and, more recently, its use as a mascot for sports teams like the Toronto Blue Jays.

In figure 5.9, Tyree places the blue jay on an abstract blue-purple background to heighten the observer's awareness of the bird's dramatic color. The blue jay's coloring is entirely on display as it rests on a small, leafless branch. With its back turned, the bird swivels its head, looking at the viewer with its crest bristled, suggesting it has been startled by the approach. This positioning also serves to show off the blue jay's beautiful plumage.

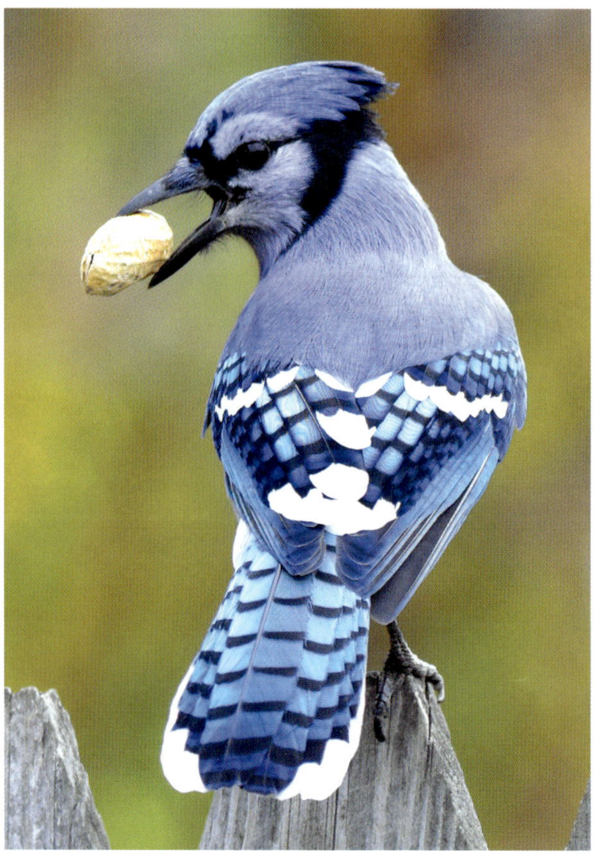

Figure 5.8 Blue jay with peanut. Jack Bulmer on Unsplash.

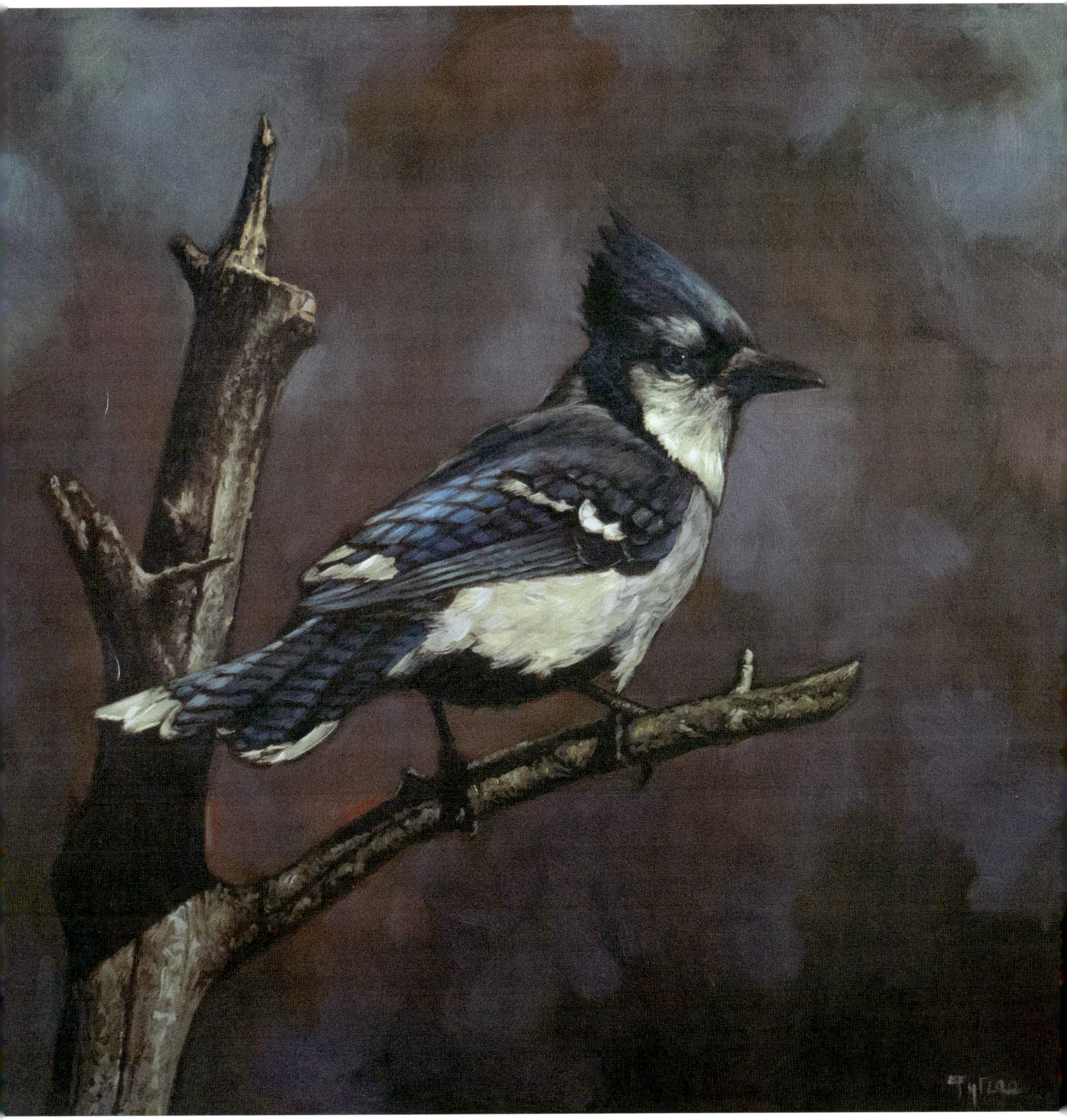

Figure 5.9 *Blue Jay*, Ralph Burke Tyree. 1975, oil on board, 16" x 16", author's collection.

Loggerhead Shrike—A Tough Little Guy

The **loggerhead shrike** (*Lanius ludovicianus*), also called the butcherbird because of its predatory tendencies, is the only member of the *Laniidae* family of shrikes native to North America (figure 5.10). The name "loggerhead" refers to the large size of its head in contrast to its body. This small bird weighs about two ounces and is only eight to ten inches long. It has grey plumage with a whitish breast, a black mask around the eyes, and black wings spotted with a white patch. The loggerhead shrike's song is a mix of harsh guttural warbles and squeaky whistles. Despite its small size and weak talons, this songbird behaves like a raptor, stalking the target before diving from an overhead perch or impaling it on a thorn or other sharp object (figure 5.11). Insects are its primary prey, but it regularly takes down larger prey like amphibians, reptiles, birds, and rodents.

Once found widely across North America, the loggerhead shrike's numbers have significantly declined since the 1960s (figure 5.12). As a result, the IUCN Red List now categorizes this unique songbird as near threatened. Population decline is likely due to a mix of habitat loss, pesticide use, and human interference. Unfortunately, two subspecies, the eastern loggerhead shrike and the San Clemente Island shrike, have been particularly impacted and are critically endangered, with less than 100 individuals remaining in the wild.

Tyree's velvet painting of four cute baby loggerhead shrikes on a branch belies this honorary raptor's macabre hunting techniques (figure 5.13). His expert use of the velvet portrays the soft down of the fledglings. Sitting in a row with their backs turned, and heads in profile, the little grey, black, and white birds create a pattern while still exhibiting individual characteristics. The branch they sit on blends into the dark, indistinct background, making the birds the sole focus of the painting.

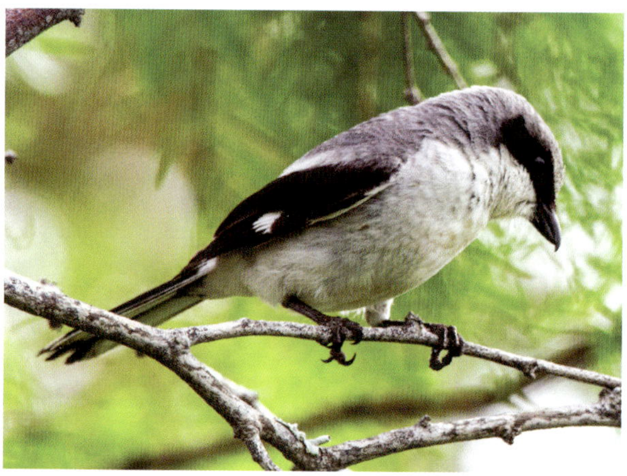

Figure 5.10 Loggerhead Shrike.
Joshua J. Cotten on Unsplash.

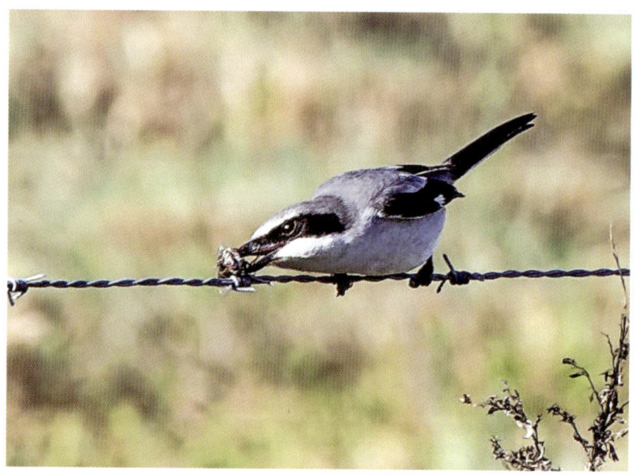

Figure 5.11 Loggerhead shrike with its impaled dinner.
Andy Morffew, Itchen Abbas, Hampshire, UK.

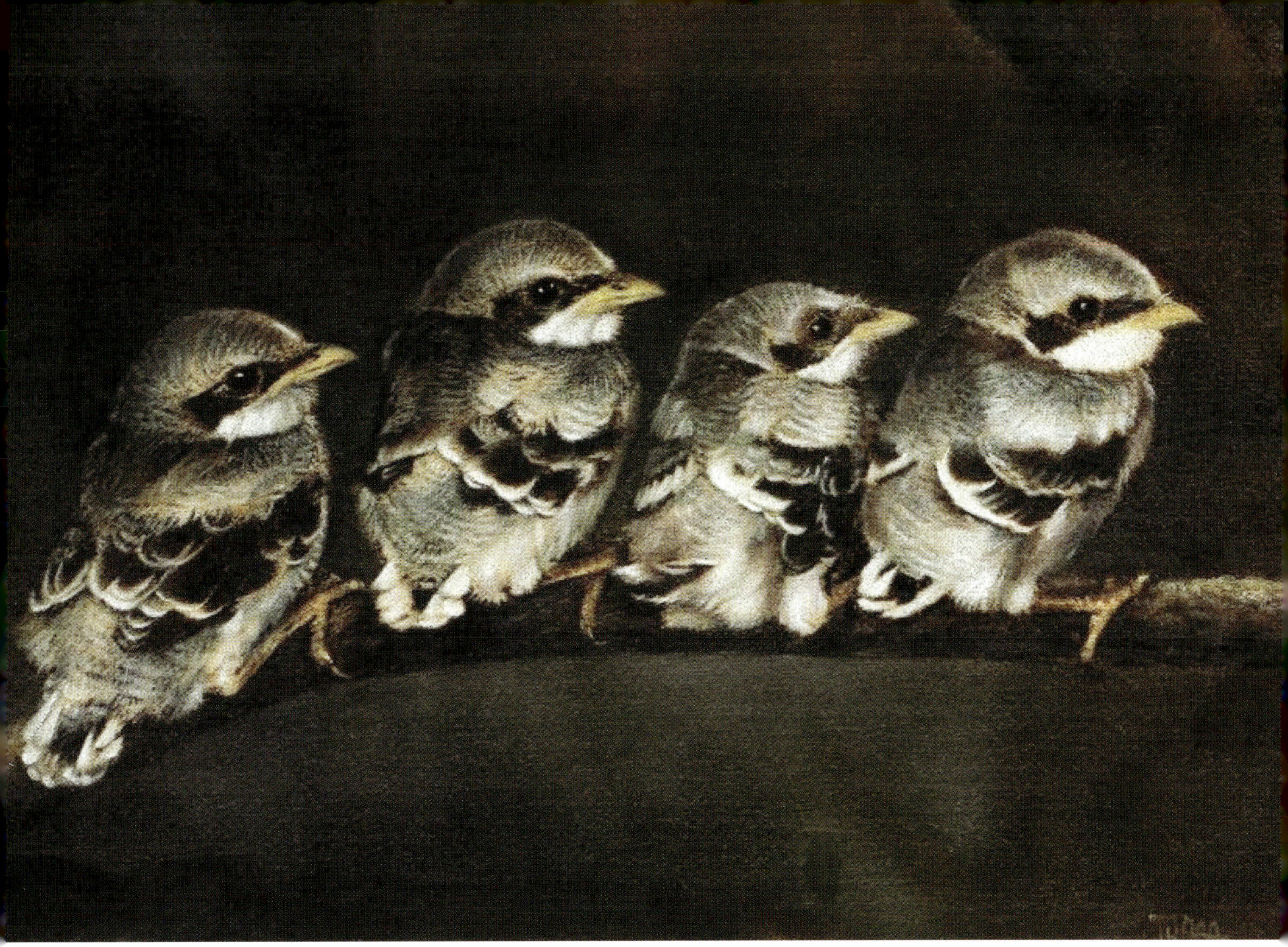

Figure 5.13 *Loggerhead Shrike Babies*, Ralph Burke Tyree. 1977, oil on velvet, 16" x 20", private collection.

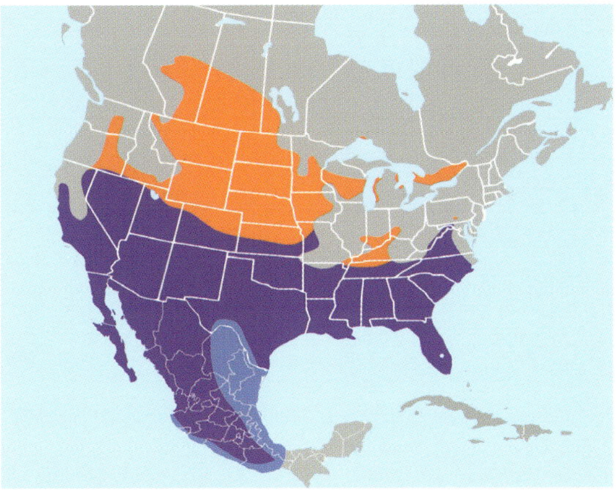

Figure 5.12 Loggerhead shrike distribution.
Cephas, 2018.

Orange=breeding
Purple=year-round
Blue=non-breeding

BIG BIRDS

Cormorant—Master Diver

The **pelagic cormorant** (*Urile pelagicus*), also called Baird's cormorant or the violet-green cormorant, is a black cormorant with a metallic sheen (figure 5.14). It is part of the *Phalacrocoracidae* family of cormorants, consisting of about 40 species of large water birds with wingspans up to three feet and long downward-hooked bills. The pelagic cormorant is one of the smaller varieties, with relatively short wings that are useful for hunting underwater but require a long water take-off like a seaplane. Like all cormorants, the pelagic cormorant is a great swimmer and diver, thanks to its webbed feet. Impressively, this variety can chase down prey as deep as 140 feet underwater, typically small fish and shrimp. On land, cormorants are clumsy and comical, waddling about like penguins. They form large colonies and nest in trees or cliffs along shorelines worldwide, with the pelagic cormorant found along the coasts of the Pacific Ocean and throughout associated coastal waterways.

All cormorants produce waterproofing secretions from their uropygial glands, which gives their feathers an oily appearance. Sometimes called the preen gland, it is located dorsally at the base of the tail. When preening, the bird distributes the oil around its feathers. Thus, after a water hunt, these birds often preen before drying their feathers in their classic spread-winged pose, as seen in figure 5.15.

Figure 5.16 shows Tyree's oil on board painting of a solo pelagic cormorant standing on a rock. The bird fills the frame, its oily waterproofing illustrated by its gleaming black-green feathers. By placing the bird against a red background, green's complimentary color, Tyree emphasizes the green tint of its feathers. Again, like in his painting of a sapsucker and its nestling, he utilizes fundamental principles of color theory to great effect.

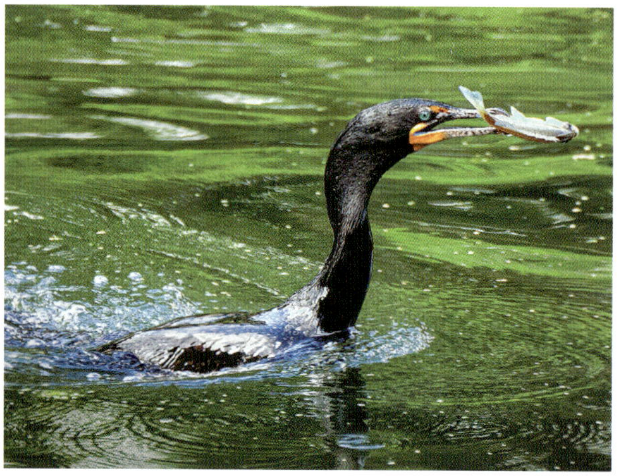

Figure 5.14 Cormorant with dinner.
Robert Thiemann on Unsplash.

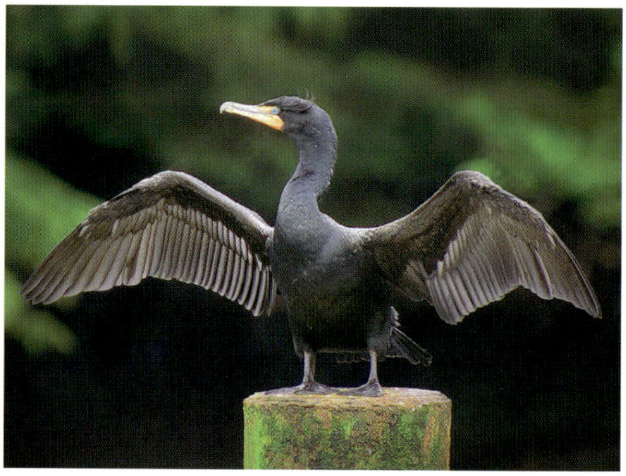

Figure 5.15 Cormorant drying wings.
Shlomo Shalev on Unsplash.

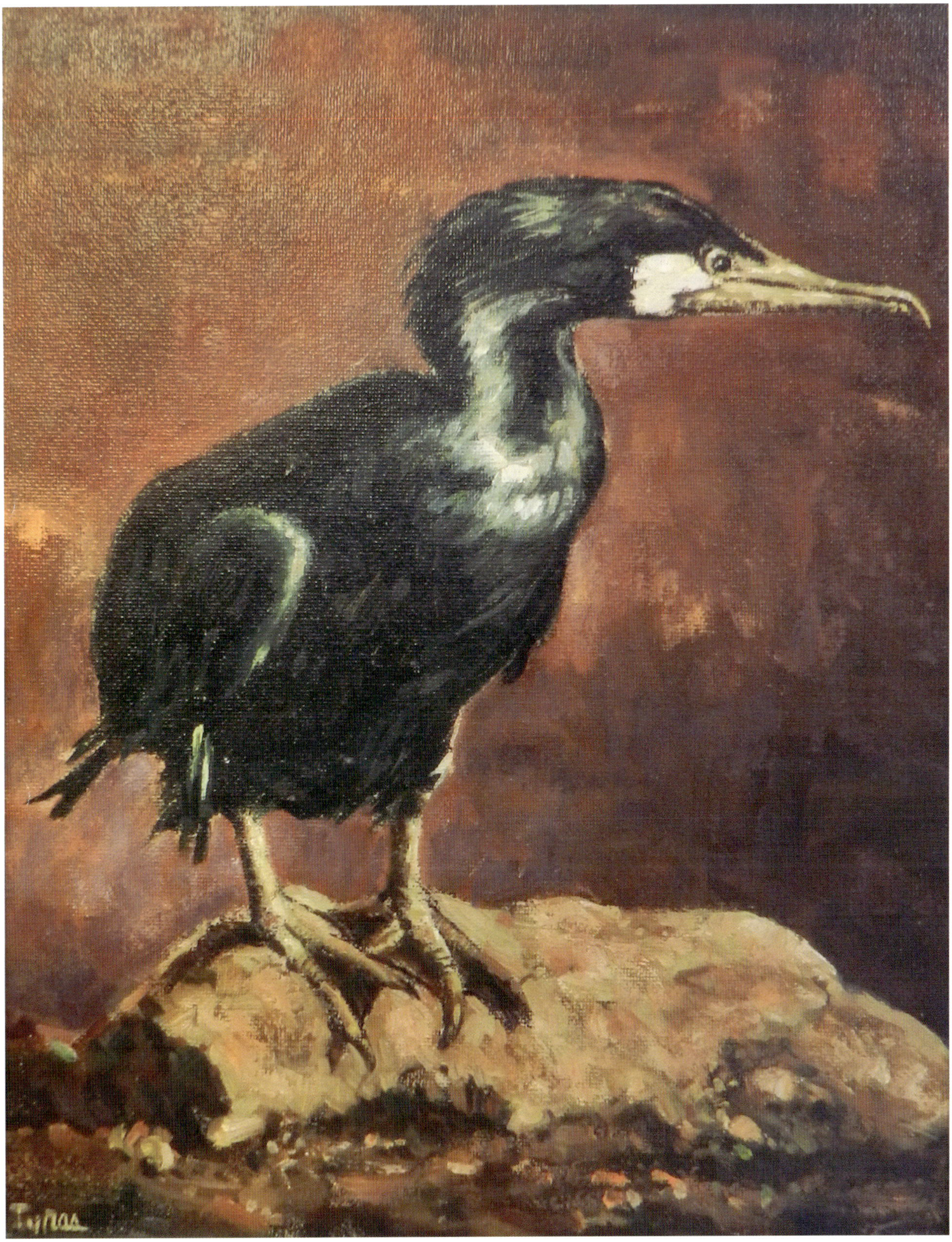

Figure 5.16 *Cormorant*, Ralph Burke Tyree. 1973, oil on board, 20" x 16", Tyree family.

Thanksgiving Dinner, Anyone?

The **wild turkey** (*Meleagris gallopavo*) of the *Phasianidae* family is native to North America (figure 5.17). While usually found roosting in wooded forests or foraging in open pastures for nuts and insects, it is adaptable to many habitats. This likely ancestor to the domestic turkey is a similarly large bird with proportionately small wings. Unlike its obese domesticated relative, it is a surprisingly agile flyer. Although it has strong sight, its night vision is poor. Because of this, wild turkeys move in their flocks to the upper tree branches at twilight for overnight comfort and protection. While the turkey makes many different calls, it is best known for the male's gobble sound, which can be heard from a mile away. Occasionally, turkeys join with squirrels or deer, hunting, playing, and watching for predators together.

The iconic wild turkey is often associated with the United States due to the tradition of eating turkey when celebrating American Thanksgiving. An intriguing and little-known fact, Founding Father Benjamin Franklin expressed his preference for the turkey over the bald eagle for the United States' national bird in a private letter to his sister. He portrays the bald eagle as an unglamorous dead fish scavenger. He even calls it "a bird of bad moral character" before comparing it to the turkey, which he describes as "much more respectable" and a true native to the country. This is a sensible viewpoint as the wild turkey has long inhabited the region and held an important place in many Native American cultures. For some tribes, the wild turkey was an important food source, and some used its feathers for headdresses or cloaks. Other tribes took inspiration from its movements for traditional dances, and some viewed it as sacred.

When Europeans arrived in North America, the wild turkey roamed widely across the continent. However, by early in the twentieth century, hunting and habitat loss had caused a sharp decline in the species' numbers. Reintroduction programs began in the 1940s and were so successful that wild turkeys now number around seven million and may be more populous than ever before. Today, they are found throughout Europe, North America, and even Hawaii. It is safe to say wild turkeys are not scarce or endangered.

Tyree painted a lone wild turkey foraging in a grassy environment using oils on board (figure 5.18). Based on the brown color of its breast feathers and smaller body size, it appears to be a female. Captured mid-step, she observes her surroundings, using her sharp vision to scan for predators. The vibrant blue and red of her head immediately draw the eye. Tyree further incorporates blue throughout her plumage, helping her stand out against the yellow-green grasses that surround her.

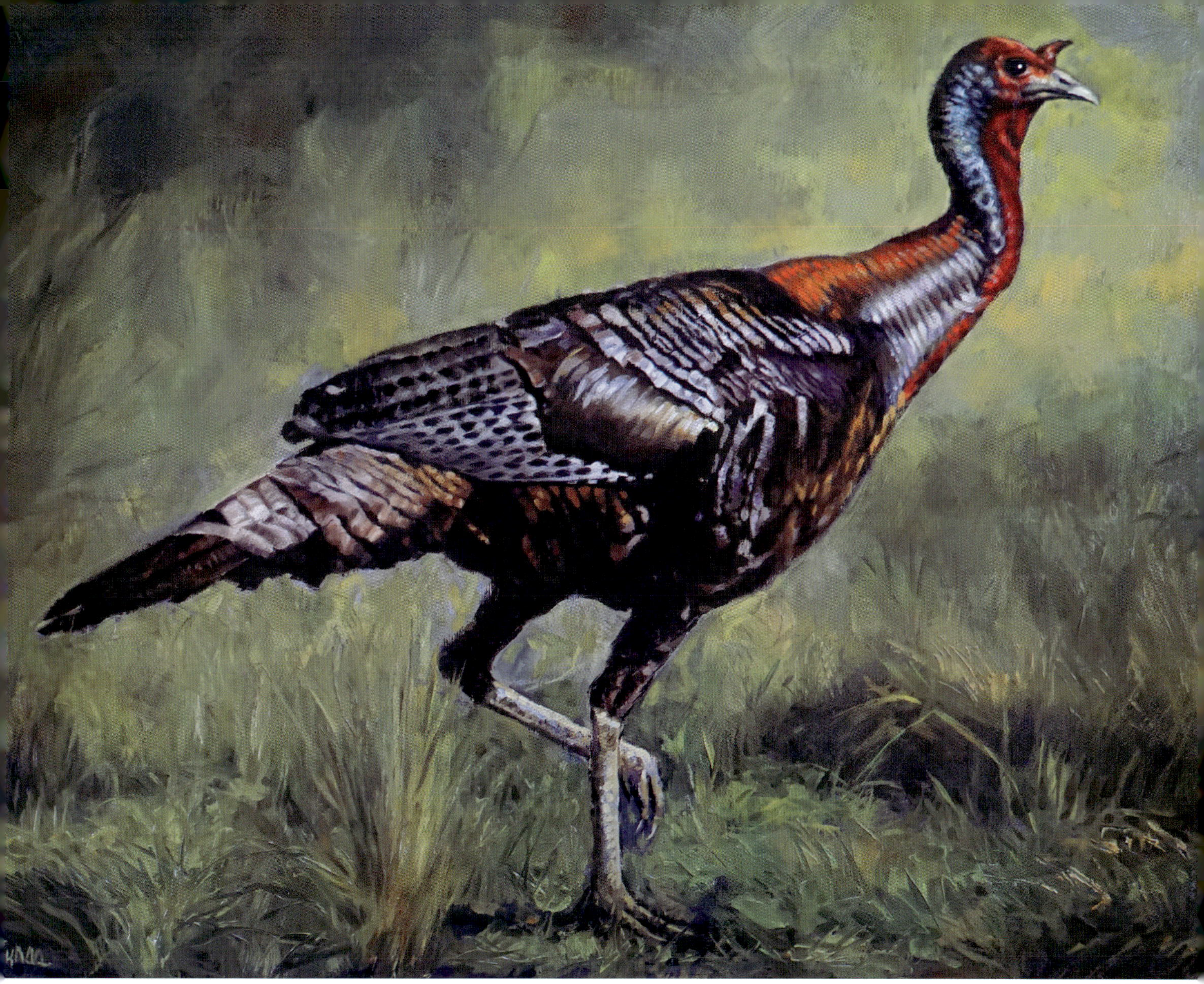

Figure 5.18 *Wild Turkey*, Ralph Burke Tyree. 1973, oil on board, 20″ x 24″, author's collection.

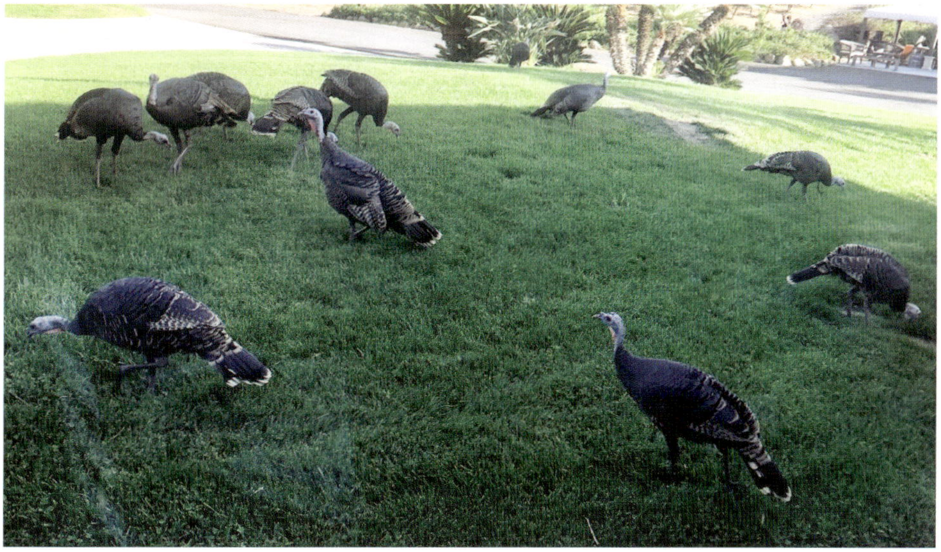

Figure 5.17 Wild turkeys on our front lawn. 2022, author's photo.

Great Blue Heron

The **great blue heron** (*Ardea erodias*) of the *Ardeidae* family is a large North American bird at about 50 inches tall with a roughly 70-inch wingspan. When flying, this slate-colored heron quacks hoarsely while curling its neck back, resembling a pterodactyl (figure 5.19). While it primarily eats small fish, it is not picky and also enjoys crabs, frogs, insects, reptiles, rodents, and even ducklings. The heron slowly wades through shallow water, visually locating prey before quickly spearing it with its long, sharp bill and swallowing it whole. It is a dreaded threat to koi owners, as this bird is known to wipe out koi ponds seemingly for fun. When nesting, the blue heron forms colonies of between five and 500 nests in trees around waterways, well away from mammal predators.

While the great blue heron is common to shorelines and wetlands across North and Central America, the Caribbean, and the Galápagos Islands, even southern coastal areas of Europe have documented rare sightings of individuals. With this widespread distribution, the IUCN does not consider it at risk. An even larger all-white variety, the great white heron, is native to South Florida. However, there is debate over if this is a color morph, a subspecies of the great blue heron, or a separate species altogether.

Tyree captures the great blue heron wading stealthily in shallow waters in a small, oil-on-board painting using shades of grey (figure 5.20). As the bird takes a step, it searches for its fish dinner, ready to strike. He paints the piece with stunning detail, using the black and white color palette to focus on contrast and value. Since the fifteenth century, artists have worked in black and white to study their subjects and compositions, focusing on light and shadow before moving to the full-color final work. Tyree continues this practice while making it his own.

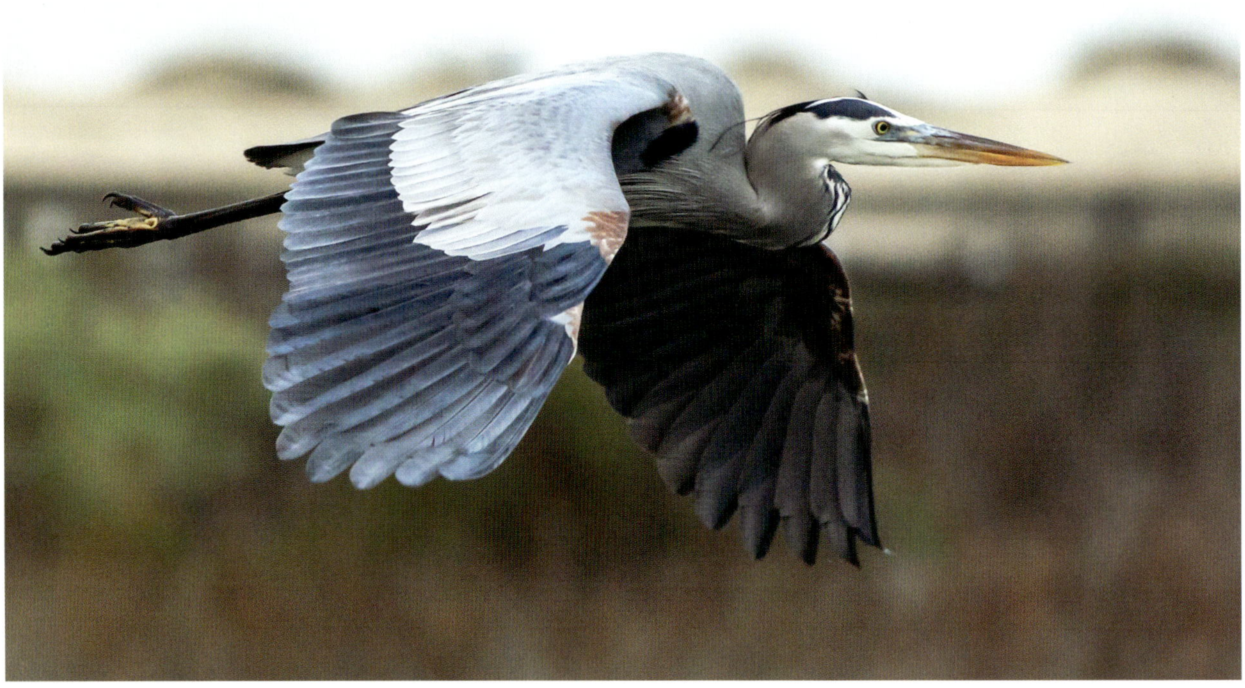

Figure 5.19 Great blue heron flying. William Foley on Unsplash.

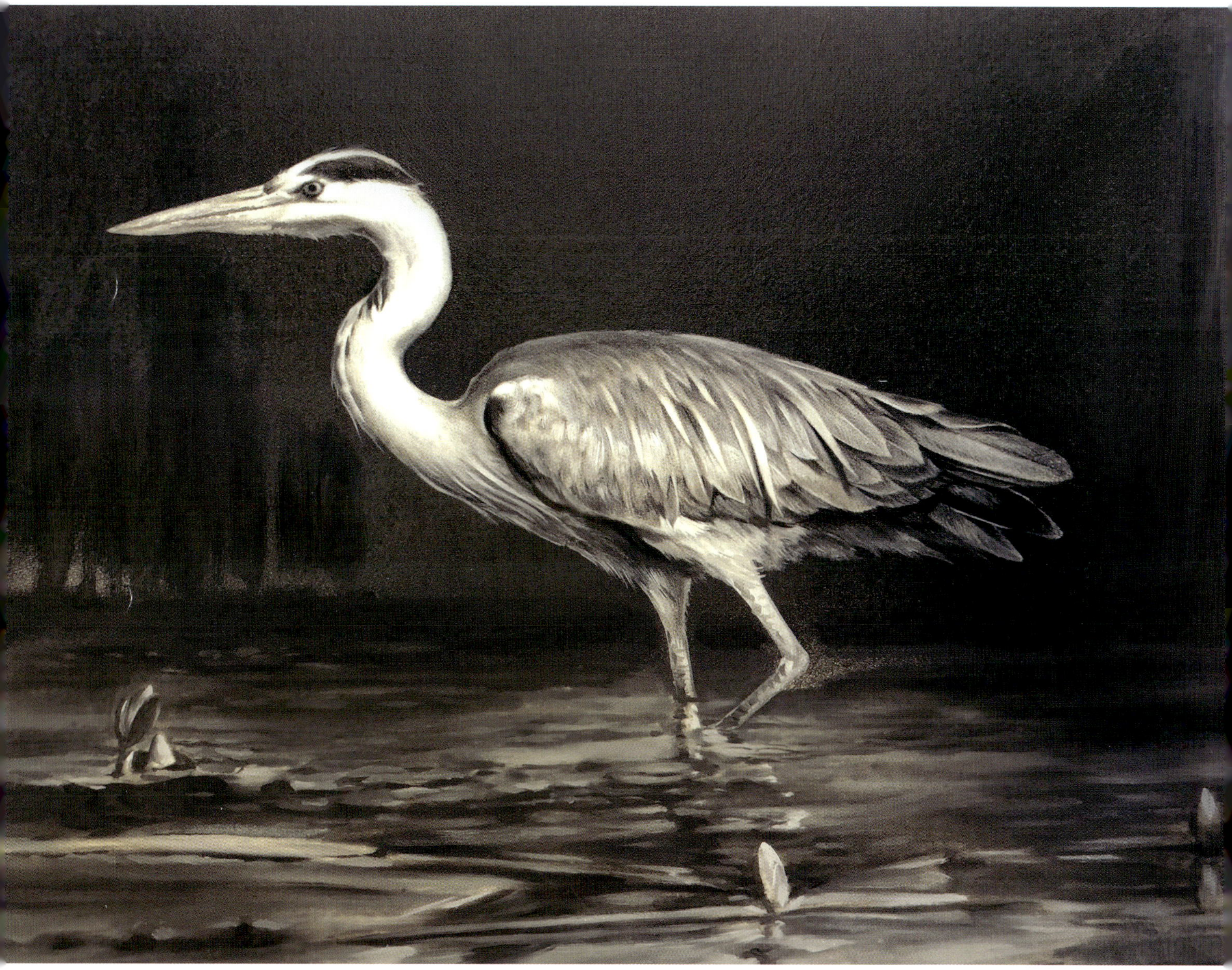

Figure 5.20 *Great Blue Heron*, Ralph Burke Tyree. Oil on board, 12" x 16", Tyree family.

Sandhill Crane

The **sandhill crane** (*Antigone Canadensis*) is a large and ancient crane of the *Gruidae* family native to North America and Siberia (figure 5.21). One of the oldest bird species, sandhill fossils date back at least two million years, with some scientists arguing they are closer to six million years old. Today, this bird is found throughout freshwater wetlands and is the most numerous and wide-ranging species of crane. Adult sandhills are primarily gray with red foreheads and white cheeks. Thanks to its large, often over six feet wingspan, the sandhill crane is skilled at soaring, similar to a hawk or eagle. This migratory bird forms large flocks, which fly in a V formation for up to 400 miles a day during spring and fall (figure 5.22). The bird's name refers to Nebraska's Sandhills along the Platte River At the edge of the Great American Plains, where over 500,000 sandhills congregate during spring migration for the largest gathering of these birds. The sandhill crane is perhaps most famous for its dancing ritual of bows, leaps, and calls. Mated pairs are known for synchronizing their distinctive loud, guttural cries in what's known as unison calling. This unique sound is one of at least 18 calls the crane makes and can be heard from miles away.

As the world's most populous crane species, the sandhill crane is not threatened, but it has not always been so common and widespread. After the widespread destruction of wetlands and heavy hunting of this crane during the first half of the twentieth century, the sandhill's numbers dropped very low, particularly the greater sandhill crane subspecies. Restoration of wetlands in the decades since the 1960s has allowed the sandhill's numbers to rebound and increase. Despite this, hunting and loss of habitat still threaten it, and two subspecies, the Mississippi sandhill and the Cuban sandhill, are endangered.

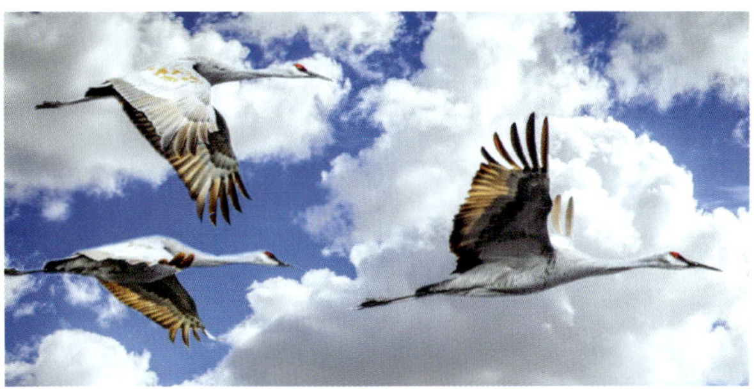

Figure 5.21 Sandhill cranes. Chris Briggs on Unsplash.

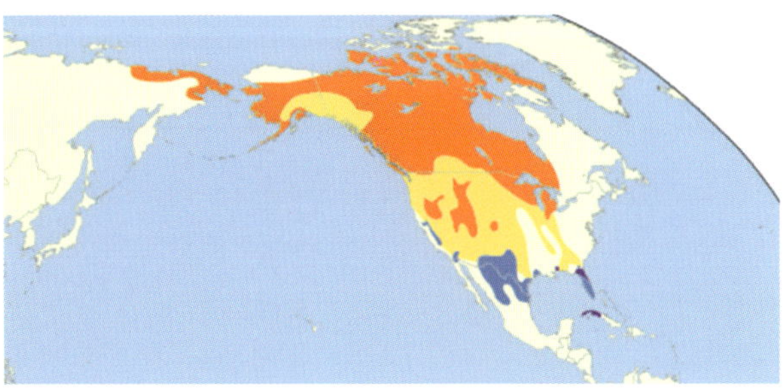

Figure 5.22 Geographical distribution of sandhill cranes.
Orange=breeding **Yellow**=migration
Purple=year-round **Blue**=nonbreeding

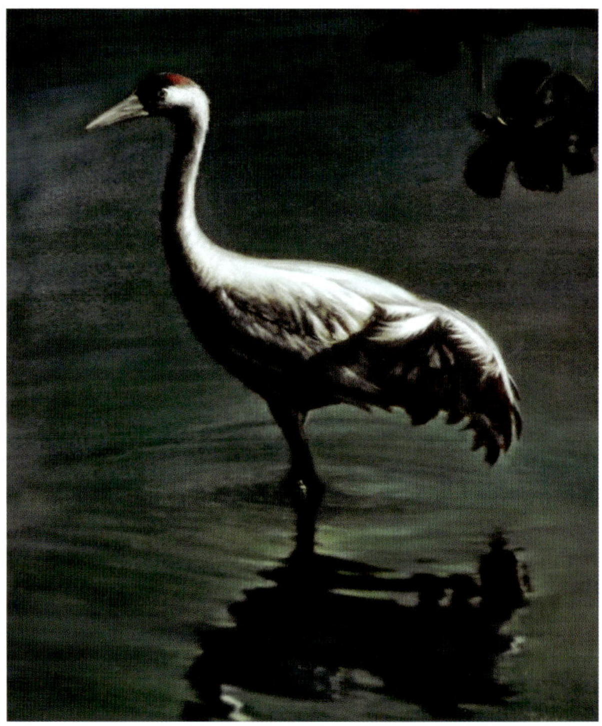

Figure 5.24 *Sandhill Crane*, Ralph Burke Tyree. 1978, oil on velvet, 24" x 18", Tyree family.

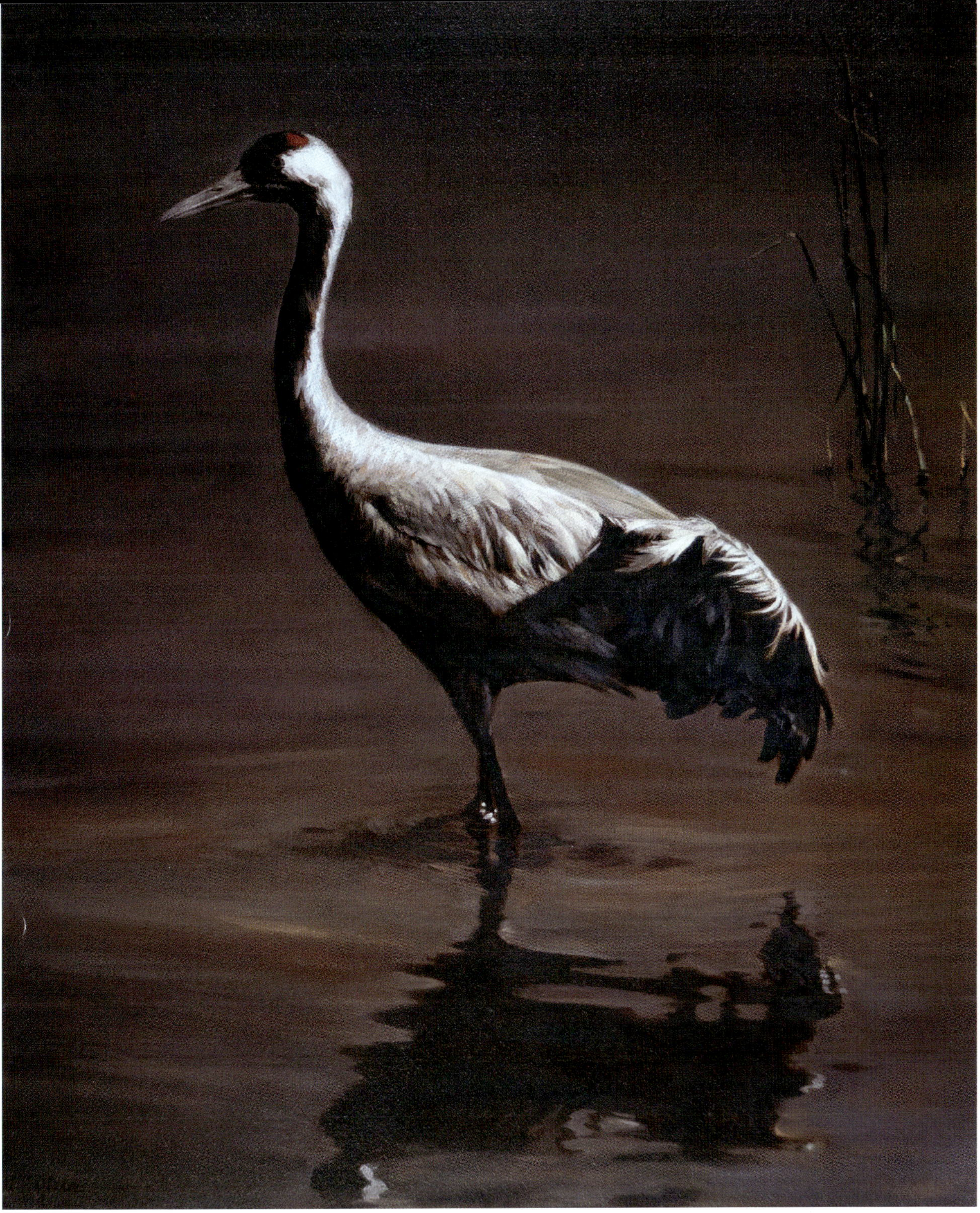

Figure 5.23 *Sandhill Crane*, Ralph Burke Tyree. 1976, oil on board, 24" x 18", author's collection.

A singular sandhill crane peacefully wades in both of Tyree's twin oil paintings of this bird. First painting the scene on board, he utilizes browns and reds to color the water (figure 5.23). Several years later, he painted it on velvet, this time depicting the water in greens and greys (figure 5.24). Both works show the bird brightly lit from above and behind, with dark shadows cast across the front and bottom of its body, highlighting its long, elegant neck and red crown. The still water ripples around the crane's legs as it slowly walks through the wetlands. The two pieces exhibit Tyree's skill in creating realistic textures and using light and shadow.

RAPTORS (VULTURES, HAWKS, EAGLES, AND OWLS)

Birds of prey, also called raptors, are an order characterized by keen eyesight used to spot game and strong talons with which to grab it. The word raptor is derived from the Latin root *rapere*, which means to take an object by force, which also gives us the word velociraptor, a genus of dinosaurs and an ancient ancestor to modern birds. Eagles, hawks, vultures, and Tyree's favorite, the owl, are among the best-known raptors. As a reticent artist with sharp observational skills, he related to the wise owl. This silent nighttime flyer with brilliant hearing and eyesight was his most cherished feathery friend. He had long dreamed of opening the Tyree Owl Sanctuary—a dream never fulfilled. The largest raptors are the California Condor and the Andean Condor, the giants of the sky with wingspans of nearly ten feet.

Like woodpeckers, cockatoos, and parrots (found in Chapter 10), raptors have zygodactyl feet, meaning the inner two toes point forward, and the outer two toes point backward (figure 5.25). This clever foot formation, which can look like an X or a K, helps birds grip more efficiently, and different species use it for various purposes. Raptors use it to grasp struggling prey with a powerful hold. In

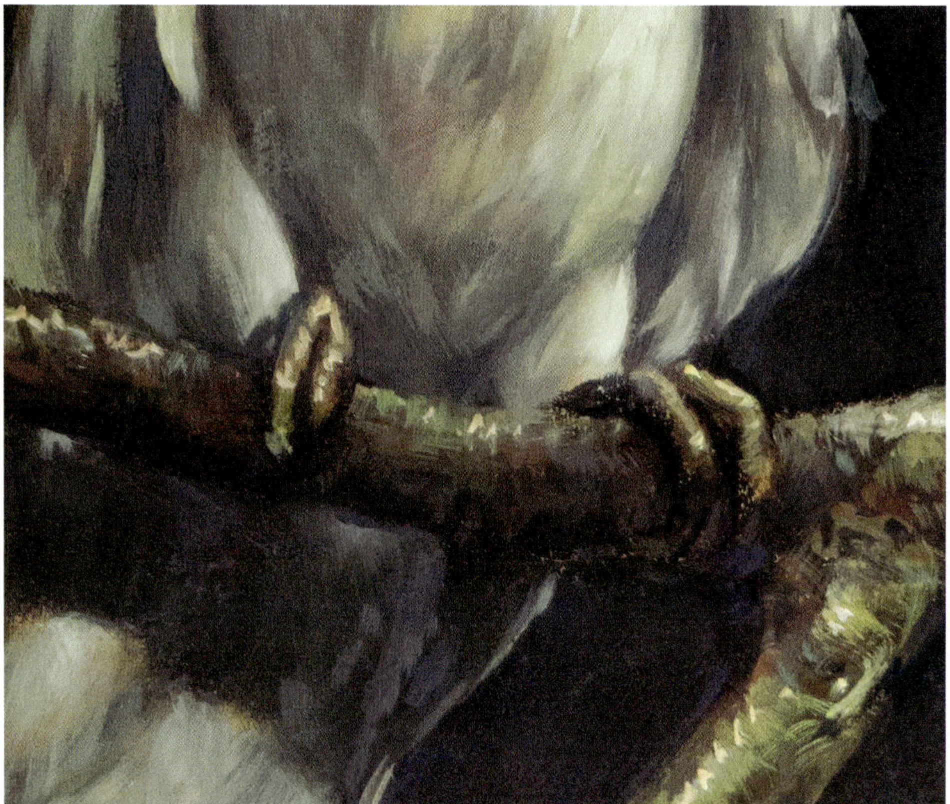

Figure 5.25 Zygodactyl feet in Tyree's *Major Mitchell's or Pink Cockatoo* (figure 10.16 detail).

contrast, it facilitates woodpeckers' ability to cling to vertical tree trunks. And thanks to zygodactyl feet, cuckoos, parrots, and cockatoos have fantastic dexterity when handling food items and navigating the branches of tree canopies.

In the 1970s, many raptors were on the brink of extinction due to the widespread use of the previously mentioned insecticide, DDT, over the preceding decades (figure 5.26). DDT interferes with the calcium metabolism of raptors, causing issues like sterility and brittle eggshells. In 1962, Rachel Carson's book *Silent Spring* questioned the logic of using potentially dangerous chemicals in the atmosphere (figure 5.27). The book's publication brought attention to this insecticide and its environmental impact. The subsequent public outcry was significant and eventually, in 1972, led to a complete ban on DDT use across the US. Around this time, Tyree painted many of these majestic birds of prey. Thirty years after his death, this ban has allowed populations of impacted raptors to rebound impressively.

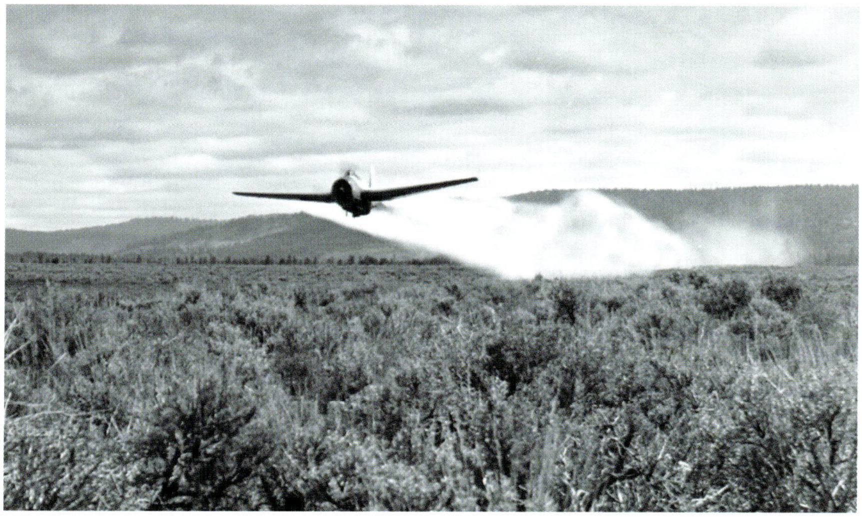

Figure 5.26 A plane spraying DDT. Courtesy of US Bureau of Land Management.

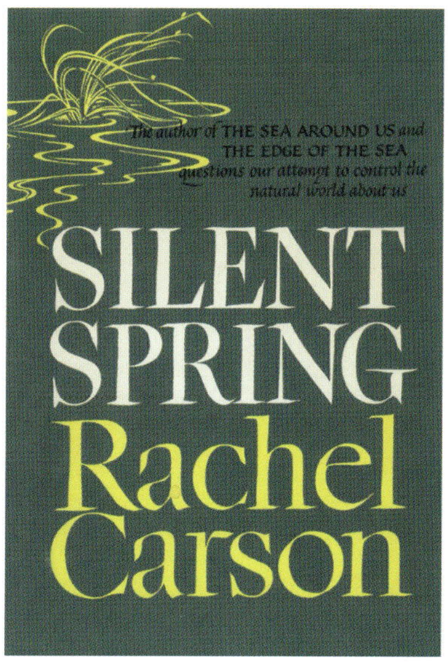

Figure 5.27 Cover of Rachel Carson's book *Silent Spring*, 1962.

California Condor—A Great Save by Conservationists

The **California condor** (*Gymnogyps californianus*) of the *Cathartidae* family is one of North America's most enormous birds (figure 5.28). It prefers rocky, forested habitats throughout the Western US. Mostly black, with white patches and a recognizable bald head, it uses its massive wings to soar, sometimes going miles between flaps and reaching speeds over 50 miles per hour. This scavenger raptor regularly travels 160 miles daily looking for carrion and has a long lifespan of up to 60 years. The California condor is significant to many Native American tribes throughout its native region, holding the role of creator to some and destroyer to others. One tribe tells of the condor causing eclipses and the lunar cycle. Some tribes participated in the ritualistic sacrifice of the condor, using its feathers for clothes or headdresses. Condor bones have even been found at burial sites.

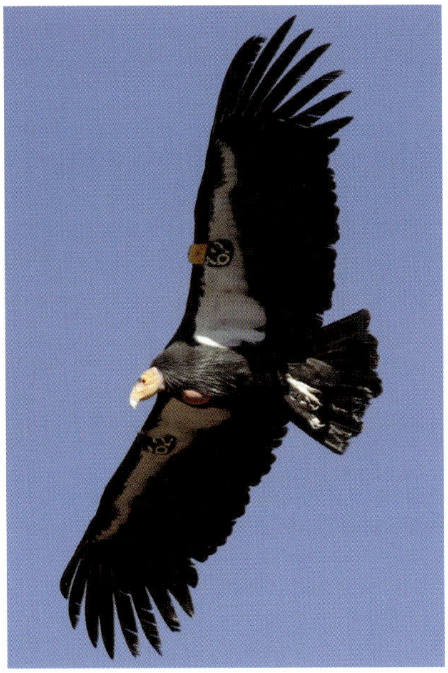

Figure 5.28 Adult condor in flight with tracking tags on both wings. Phil Armitage, 2007.

The California condor is a remarkable conservation success story. After drastic declines in condor numbers during the twentieth century due to a mix of hunting, lead and DDT poisoning, egg collecting, the increasing presence of power lines, and escalating habitat destruction, in 1987, it became extinct in the wild when conservationists captured all 22 remaining condors and took them to zoos in an effort to preserve the species.

An elaborate captive breeding program was employed to raise the species' numbers, even utilizing condor puppets to prevent human bonding, as seen in figure 5.29.

Four years later, these zoos began to release some condors into the wild. Since then, numbers have steadily grown (figure 5.30), and as of 2019, there were 518 California condors recorded in the wild and captivity. Remarkably, in the summer of 2022, four birds were released into Northern California's Redwood National and State Parks. With this, the condor returned to its native redwood forests for the first time in over a century. Despite increases in the population, the species still has a long way to go, and the IUCN continues to categorize it as critically endangered.

Tyree painted a full-body portrait of his home state's iconic California condor using oil paints on board (figure 5.31). The raptor is shown in a three-quarter profile, resting on a snag, a dead tree left upright to decompose. The viewer looks up at the bird, indicating how physically high the bird is perched and giving it a majestic or regal air. Looking out over a nighttime vista, the condor's red head and chest stand out against the vibrant blues of the starry sky. This redness signifies that the condor is not only a male but is currently searching for a prospective mate. With this comes the implication

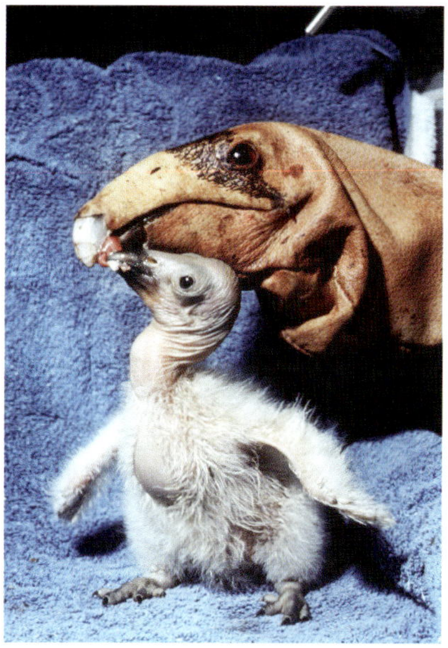

Figure 5.29 Zoo caretakers use a condor puppet to feed a California condor chick. Ron Garrison, 2005, courtesy of San Diego Zoo.

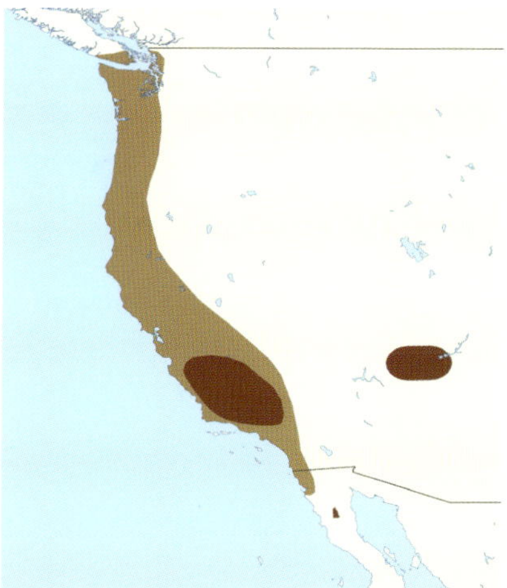

Figure 5.30 California condor range. M.Bitton, 2022.

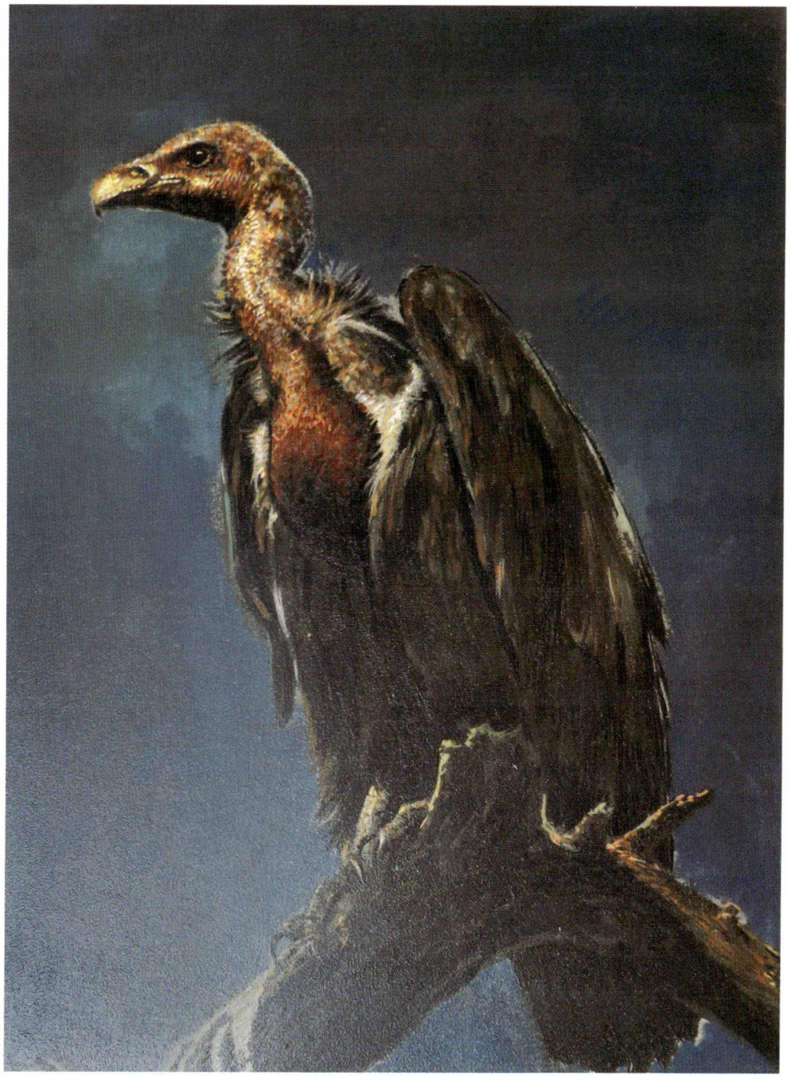

Figure 5.31 *California Condor*, Ralph Burke Tyree. 1975, oil on board, 24" x 16", private collection.

of the continuation of the species. Notably, Tyree completed this painting in 1975, when this condor was on the brink of extinction. Conservation efforts began that same year with the US government's implementation of the California Condor Recovery Plan, following the ban on DDT only three years earlier. At a time when this species' future was uncertain, Tyree brought attention to and preserved an image of this species.

Peregrine Falcon

The **peregrine falcon** (*Falco peregrinus*), also called the duck hawk, reaches 200 miles per hour when it dives, making it one of the world's fastest birds. Consisting of 19 subspecies found almost worldwide, this member of the *Falconidae* family is about the size of a crow. It has grey, brown, black, and white plumage and a hooked yellow and black beak (figure 5.32). Hunting at dawn and dusk, the peregrine falcon plucks its prey, mainly small birds, from the sky as it dives, a move also known as stooping. This falcon performs a mid-air courtship dance that involves complex maneuvers, including the male passing freshly caught prey to the female when choosing a lifelong mate. This incredible flyer has been used in falconry for over 3,000 years because it is easy to train and has a strong hunting instinct. During WWII, both sides employed the peregrine falcon to intercept homing pigeons, and more recently, airports have used them to scare other birds away and improve air safety.

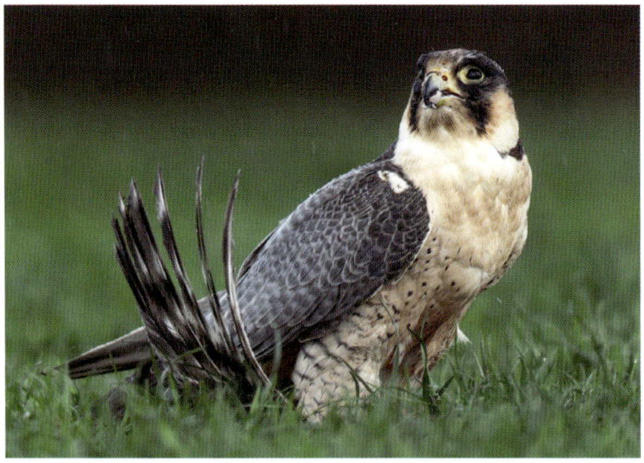

Figure 5.32 Peregrine falcon.
Vincent van Zalinge on Unsplash.

Another victim of DDT, peregrine falcon populations crashed during the middle of the twentieth century. Since the ban on this insecticide, this speed champ's numbers have slowly recovered, and today, it is not classified as endangered. Like the California condor, recovery efforts after the DDT ban involved breeding in captivity using unique puppets to stop the bird from bonding with humans. These efforts were extraordinarily successful, and in 1999, the US removed it from the Endangered Species list. Today, the peregrine falcon is the world's most widespread raptor (figure 5.33). It has even become a common sight in urban habitats, where it takes advantage of the wind streams created between tall buildings and nests in high vantage points.

Like the California condor, Tyree painted the peregrine falcon soon after the ban on DDT while the bird's population was at an all-time low (figure 5.34). He shows the falcon from the breast up, facing the viewer, with its head turned in profile. This framing is essentially the bird version of a head and shoulder portrait. To further enhance this quality, Tyree shows the bird against a deep blue background that could just as easily be a dark sky as a fabric backdrop. This humanizes the bird while also placing all the focus on the precise details and lifelike texture.

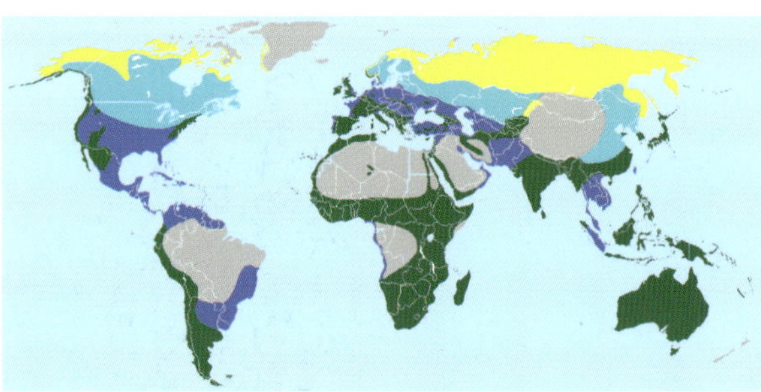

Figure 5.33 Peregrine falcon distribution. MPF, 2007.
Yellow=breeding summer **Green**=breeding resident
Blue=winter visitor **Teal**=migration

Figure 5.34 *Peregrine Falcon*, Ralph Burke Tyree. 1979, oil on velvet, 16" x 12", Tyree family.

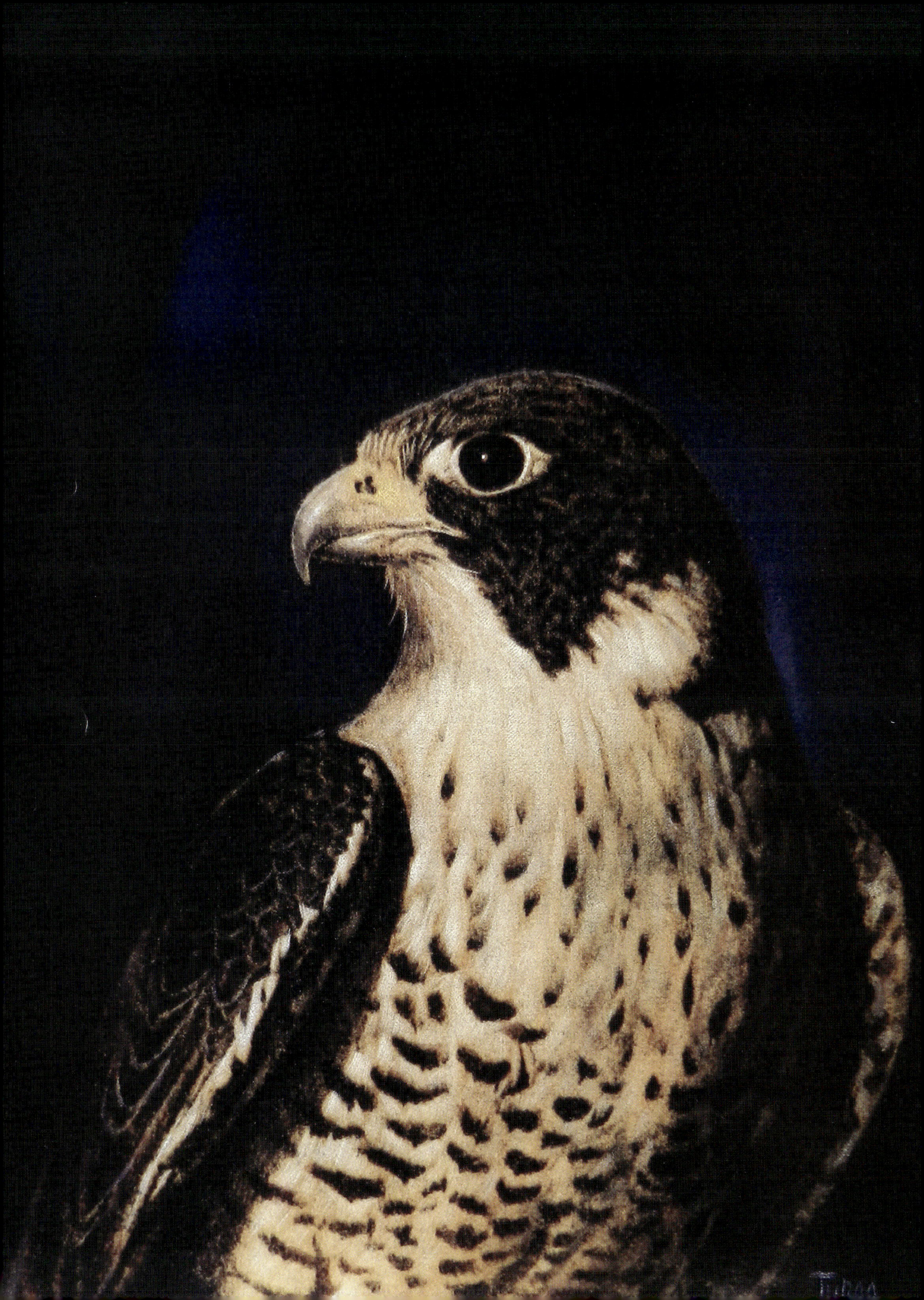

Rough-Legged Hawk

The **rough-legged hawk** (*Buteo lagopus*), also known as the rough-legged buzzard or falcon, is a bird of prey from the *Accipitridae* family (figure 5.35). This medium-sized raptor is brown with white and black speckling and long tail feathers. It lives in subarctic regions of North America, Europe, and Russia during the breeding season before migrating south to marshes and prairies for the winter, where it hunts for small mammals and birds (figure 5.36). Hunting in open spaces, the rough-legged hawk will often hover over its prey before diving for the kill. Nests are typically located on cliffs or trees to avoid the hawk's many predators, like arctic foxes, brown bears, and wolverines. With a stable population and wide range, the IUCN does not classify it as threatened.

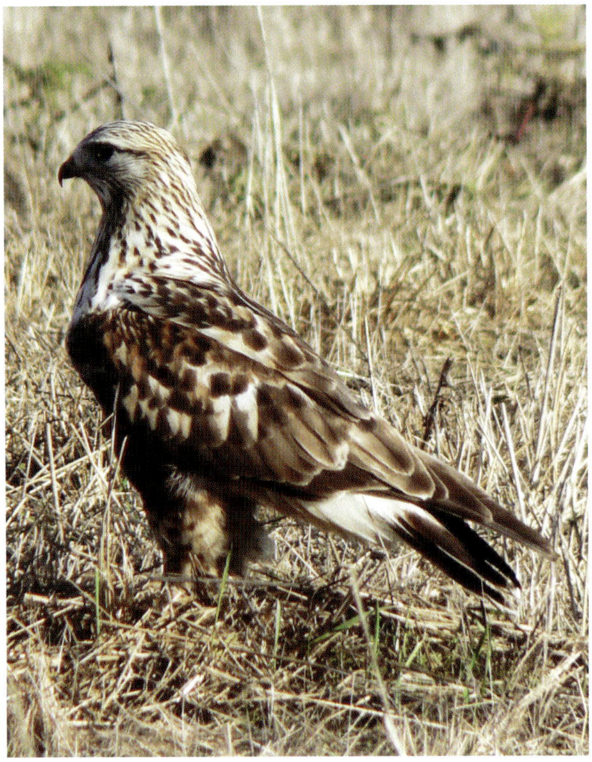

Figure 5.35 Rough-legged hawk. Walter Siegmund, 2007.

Completed the year before Tyree painted a peregrine falcon, his oil on velvet painting of a rough-legged hawk similarly depicts a head and shoulder portrait of the bird (figure 5.37). He painted the hawk on velvet with oil paints, showing it in profile with incredible detail given to its hooked beak, glowing amber eye, and soft plumage. Rarely viewed so closely, the plain blue background and zoomed-in perspective provide the viewer with a close view and personal connection to this magnificent bird.

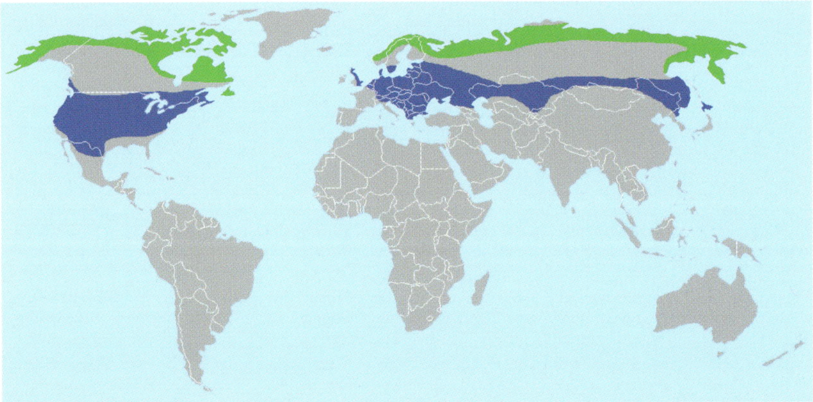

Figure 5.36 Rough-legged hawk distribution. The Engineer, 2005.
Green=nesting **Blue**=wintering

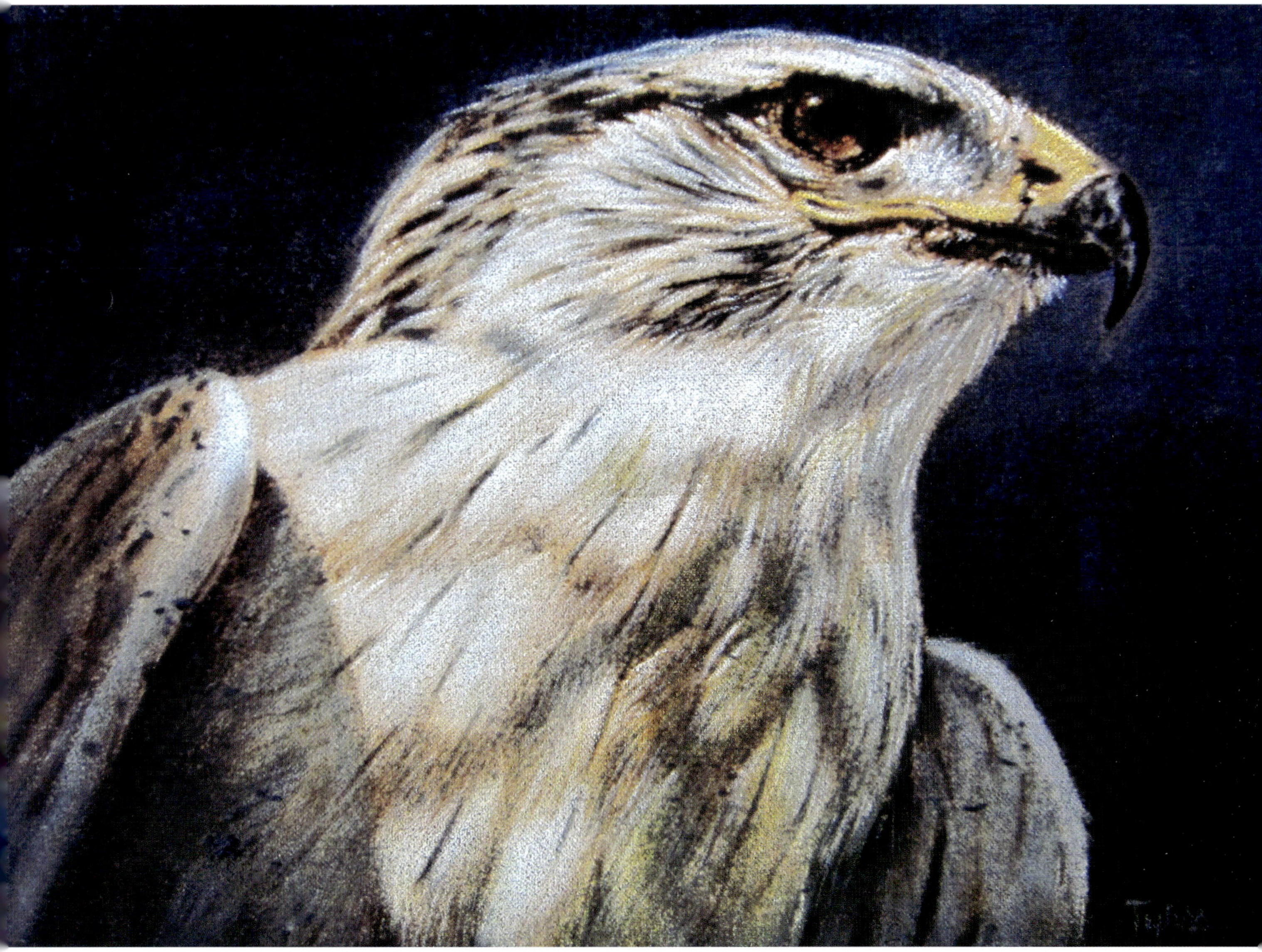

Figure 5.37 *Rough-Legged Hawk*, Ralph Burke Tyree. 1978, oil on velvet, 12" x 16", Tyree family.

Golden Eagle

The **golden eagle** (*Aquila chrysaetos*) from the *Accipitridae* family is the most widespread eagle on the planet. It is found across the Northern Hemisphere because of its adaptability across various environments (figure 5.39). Still, it is most numerous in rocky mountain habitats. The six subspecies of this large raptor are all characterized by their recognizable golden and dark brown plumage, hooked beaks with dark tips, and impressive size (figure 5.38). It is one of the larger raptors at up to 40 inches long, with a wingspan that can approach eight feet. Mostly silent, the golden eagle has a weak call it usually uses during nesting, for which it prefers high locations like cliffsides. This agile bird is often considered the best flyer of all eagles. Gliding and diving while hunting may even beat the peregrine falcon's impressive speeds, also reaching up to 200 miles per hour. It predominately eats rodents and rabbits but has also been known to go after larger prey like seals, coyotes, geese, and even other raptors. These eagles mate for life, performing an elaborate courtship in which both sexes repeatedly drop a rock or stick while in flight before diving down to catch it again.

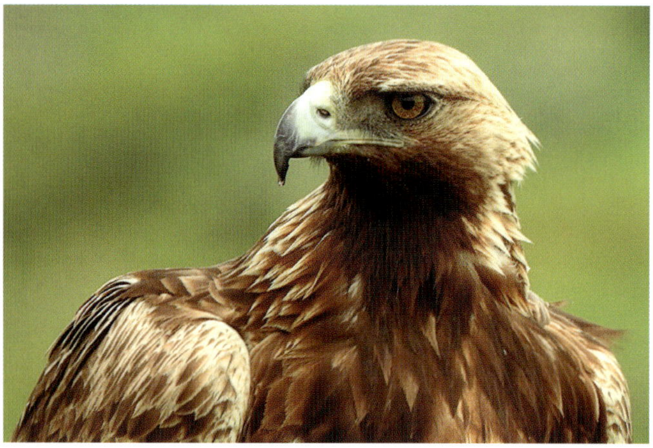

Figure 5.38 Golden eagle. Mark van Jaarsveld on Unsplash.

The golden eagle has long been used symbolically by humans. It is central to the story of Huītzilōpōchtli, the Aztec god of the sun, war, and fire, who told the Aztec people to build the city that became Tenochtitlan when they saw the sun in the form of an eagle perched on a heart-shaped

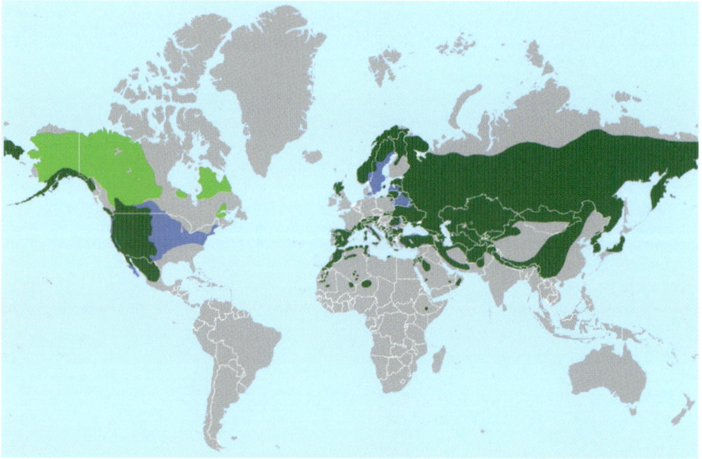

Figure 5.39 Golden eagle range. SanoAK: Alexander Kürthy, 2018.
Light green=nesting summer
Dark green=nesting year-round
Blue=non-nesting

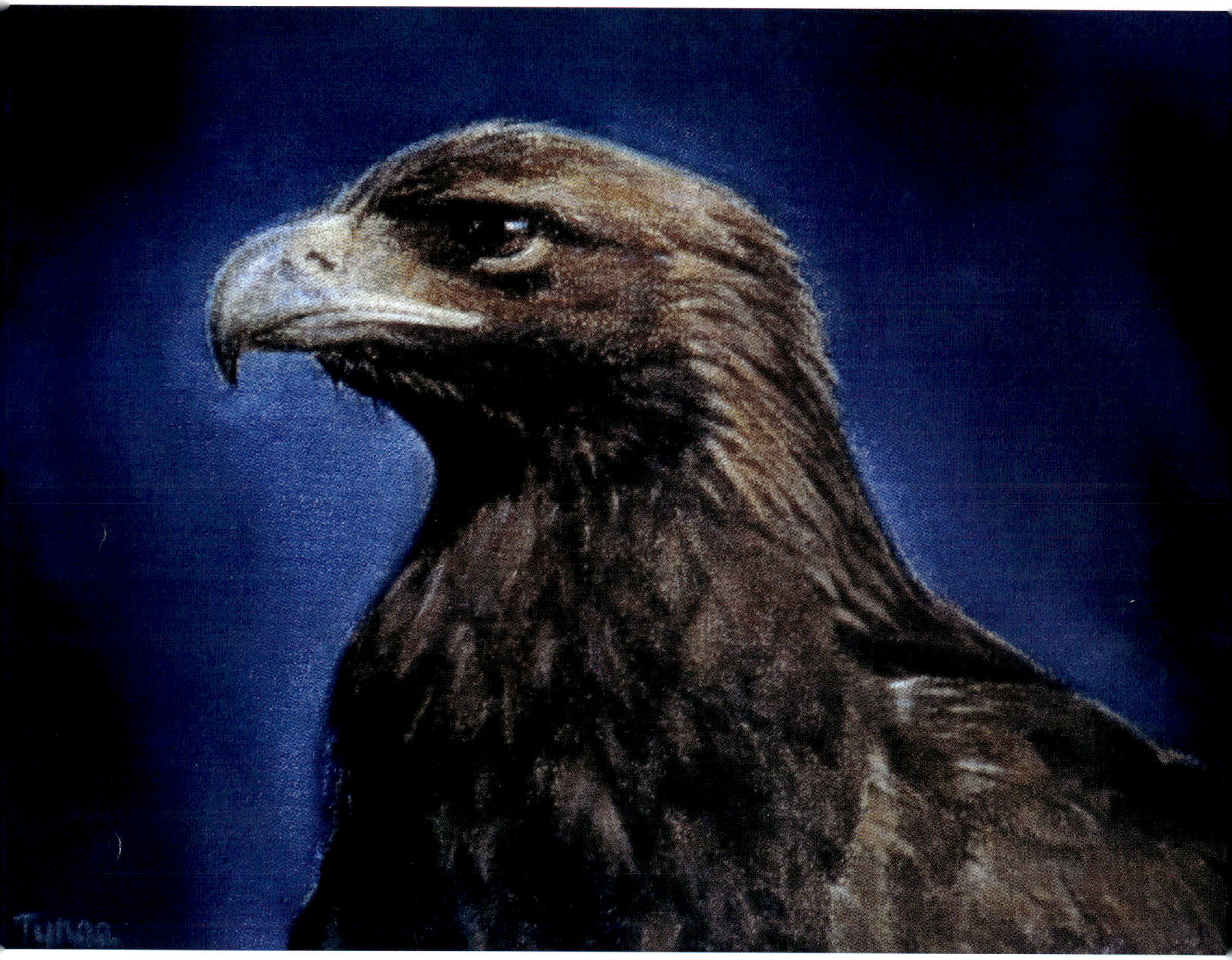

Figure 5.40 *Golden Eagle*, Ralph Burke Tyree. 1978, oil on velvet, 9" x 12", private collection.

cactus with red fruit. This scene is so significant it is depicted on the current Mexican flag. After the Industrial Revolution, the golden eagle became highly regarded in falconry for its hunting prowess and speed, especially when targeting larger game like foxes and wolves. In 2017, the French army impressively trained this bird to catch drones.

While the golden eagle is relatively populous and not endangered, it has experienced a notable decrease in some regions. Still, it is one of the most numerous eagles found worldwide and is not at risk.

The same year Tyree painted the rough-legged hawk, he also depicted a golden eagle with oils on velvet (figure 5.40). Like his paintings of the hawk and the peregrine falcon, he shows a head and shoulder portrait of the regal bird, humanizing the bird and ensuring it is the center of attention. The bird's feathers appear fluffy and soft, illustrating Tyree's skill and attention to detail. Again, he places the bird against a deep blue background reminiscent of a night sky and a portrait photo backdrop, but this time the eagle is positioned in a three-quarter view.

Bald Eagle—An American Symbol Saved

The **bald eagle** (*Haliaeetus leucocephalus*) is a huge North American raptor from the *Accipitridae* family that famously acts as the national bird for the US. Easily identified by its great size, prototypical white head, and hooked yellow beak, the rest of its body is mostly brown (figure 5.41). It is usually seen near open water, searching for its primary food source, large fish, which it snatches out of the water with its massive talons. Still, it is an opportunistic predator that will eat almost any animal it sets its sights on. The bald eagle mates for life and maintains the same nest year after year, rebuilding if needed. These nests are impressive and weigh about a ton, the biggest recorded for any animal. This eagle is indigenous only to North America and, in 1782, was declared the national bird and symbol of the United States of America because of its elegance, power, and long life. Today, six states feature it on their flags.

In the 1970s, the bald eagle was on the brink of extinction, in large part due to DDT. Since DDT was banned, the bald eagle population has rebounded with great success, and it is no longer a threatened species. Today, there are more bald eagles than at any other time over the last 400 years. This iconic bird is widespread throughout North America, including around the lakes near my home in dry Southern California (figure 5.42).

Tyree painted the bald eagle twice in 1974, two years after DDT was banned. Like the California condor and the peregrine falcon, he brought attention to the bird when it was most endangered. In figure 5.43, Tyree captures the toughness and fighting determination of the American bald eagle as it glares at the approaching discord. The eagle is shown in profile in impressive detail, standing in a grassy environment. Despite the distinct outdoor setting, the background is a vague, blended mix of patriotic red and blue. Similar to his paintings of the peregrine falcon and the golden eagle, it is reminiscent of a portrait backdrop. In this case, it even has the appearance of draped fabric, reminiscent not only of a fabric backdrop but also of an American flag because of the colors.

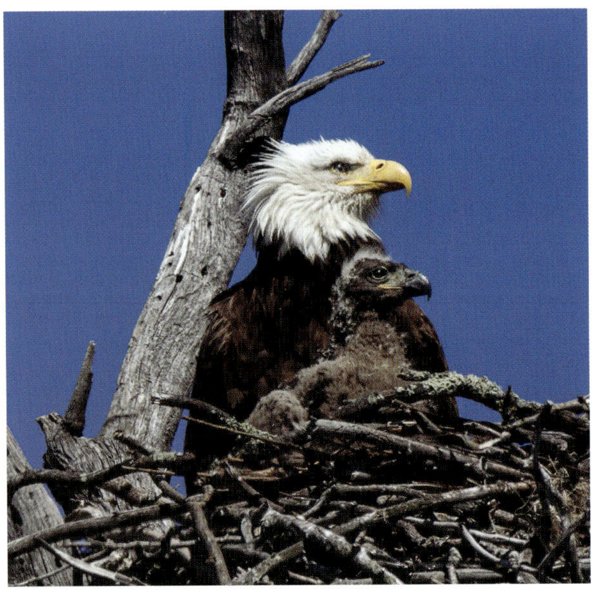

Figure 5.41 Mother bald eagle and eaglet. Courtesy of Brian Caldwell, California, 2022, www.lakehodgestours.com.

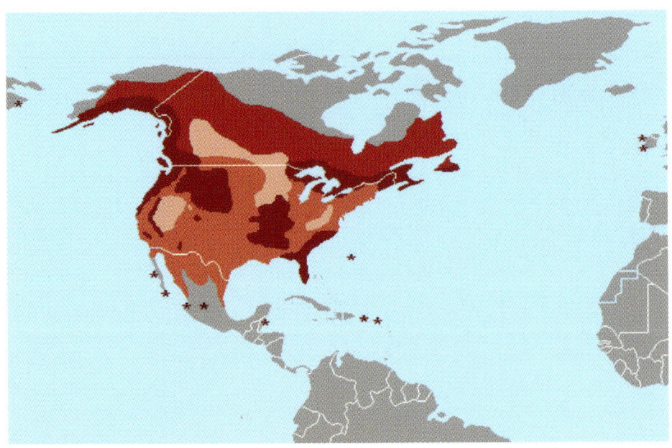

Figure 5.42 Bald eagle distribution. Mad Max.
Dark red=breeding resident
Medium red=breeding summer visitor
Orange=winter visitor
Light orange=migration only

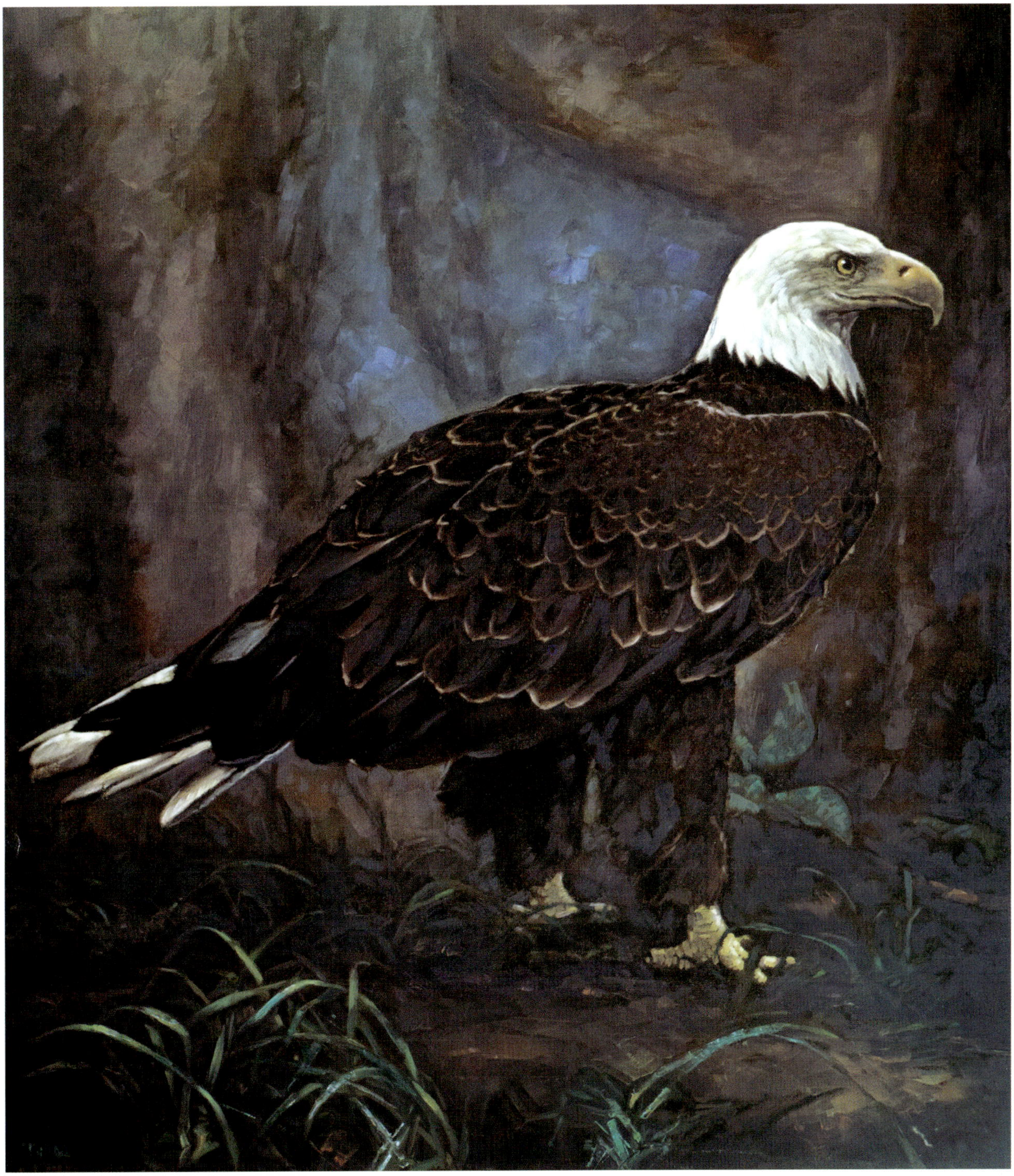

Figure 5.43 *Bald Eagle*, Ralph Burke Tyree. 1974, oil on board, 20" x 24", author's collection.

The second of Tyree's bald eagle paintings features a young bird (figure 0.2). With feathers in various shades of brown, the juvenile eagle does not yet have its iconic white markings. It perches high in a bare tree, likely scouring the landscape below for prey. Painting it on velvet, Tyree forgoes his usual hyper-realism, instead opting for hard angles and large brushstrokes. This almost cubist style gives the bird added toughness and edginess. Like his painting of a male condor, his choice to paint a young bald eagle brings explicit focus to the species' future at a time when it was in dire straits.

Barn Owls Don't Give a Hoot—Shhh, Don't Tell

"The little owls call to each other with tremulous, quavering voices throughout the livelong night, as they sit in the creaking trees."

—THEODORE ROOSEVELT

The **barn owl** (*Tyto alba*) of the *Tytonidae* family is found on every continent other than Antarctica and is the most widespread owl species. Recognizable by its heart-shaped, usually white face, the rest of its plumage is typically a mix of grey, brown, and white (figure 5.44). The many subspecies vary considerably in size, but on average, the owl is about a foot long with a wingspan of nearly three feet. It prefers open grasslands or woodlands, where rodents and other small animals are plentiful. Like most owls, it is a remarkably silent flier. It hunts at night using its exceptional and precise hearing and low-light sight. Unlike other owls, it does not hoot, instead calling out with "shsh" sounds, raspy hisses, and the occasional shriek. Many cultures have considered the barn owl a terrible omen, likely because of this sound and its ghost-like appearance. This owl usually mates for life, with the male hunting as the female sits on the clutch. Once the babies have feathered out, both parents act as hunters for the family.

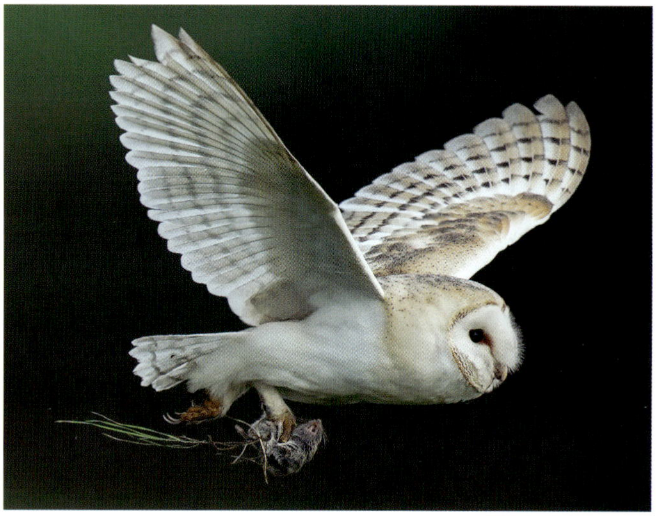

Figure 5.44 Barn owl, a silent flyer with dinner. Bob Brewer on Unsplash.

Like many raptors, the barn owl also suffered losses due to the use of DDT. These losses were mostly isolated to European and North American populations, which have since largely recovered. The IUCN does not consider it at risk, although certain regions have experienced local endangerment where the owl has become rare. This is usually due to the loss of foraging grounds to agriculture, the ingestion of rodenticides through prey, and difficulties created by harsh winters.

The barn owl was Tyree's favorite among his best-loved owl friends. He painted it often, and three of these works, all oils on black velvet, are included. Beyond this, the three compositions are entirely unique. Figure 5.46 is a head and shoulder portrait, much like those previously discussed in this chapter. The owl stares straight at the viewer, enhancing this portrait-like quality and creating a sense of intimacy between the owl and the viewer. The other two paintings both show barn owls in their natural habitat. In figure 5.1, the owl lands on a branch with its wings still outstretched, the beautiful details of its markings on full display. It faces the viewer, looking out into the night with its huge dark eyes. In figure 5.45, a smaller barn owl stands on a log facing the viewer in a moment of rest. The commonality between three works lays in their focus on this owl's unique, heart-shaped face and Tyree's expert depiction of soft, feathery textures.

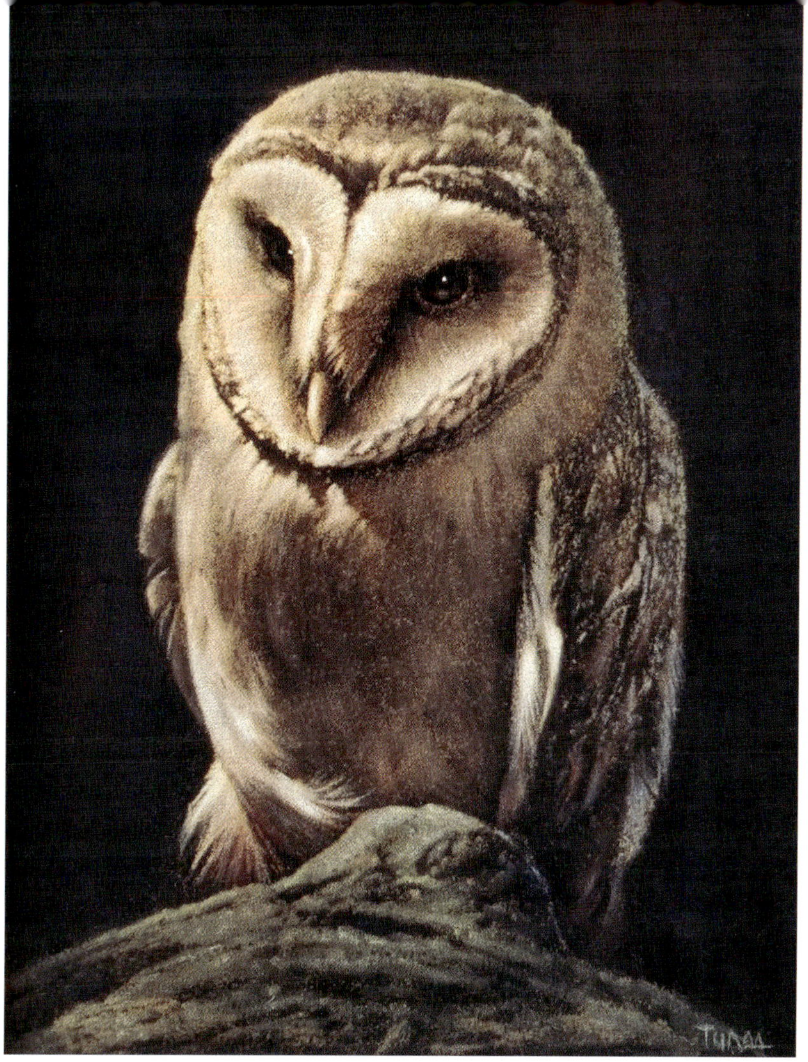

Figure 5.45
Barn Owl, Ralph Burke Tyree. 1978, oil on velvet, 16" x 12", courtesy of Anna Chavez.

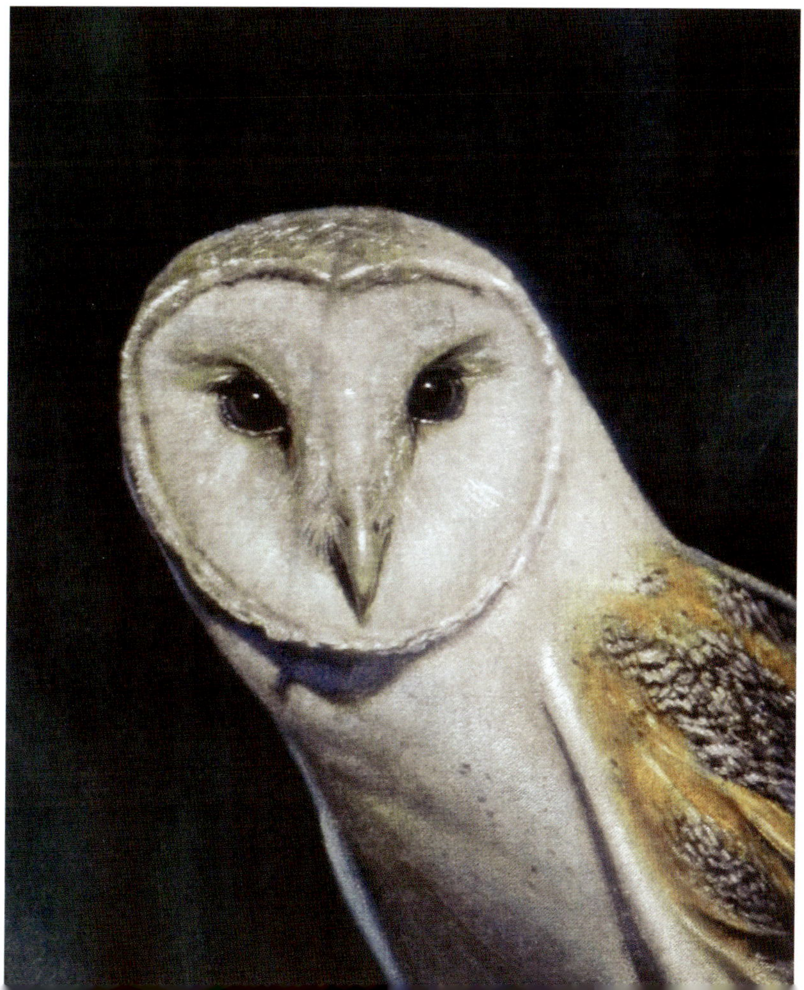

Figure 5.46
Barn Owl, Ralph Burke Tyree. 1976, oil on velvet, 20" x 16", Tyree family.

Screech Owl

Consisting of 23 species across the Americas, **screech owls** (*Megascops*) are a genus in the *Strigidae* family of true owls (figure 5.47). These owls get their name from their piercing calls that often resemble screams. Approximately ten inches tall with a wingspan of about 20 inches, these small owls are colored like tree bark, with distinctive pointed ear tufts and bright yellow-green eyes. Screech owls usually nest in tree hollows to raise their babies. Like most owls, they hunt from perches in semi-open habitats, relying on sharp hearing to locate prey and keen eyes for the final approach. Their primary food source is small animals, most often bats, mice, and birds.

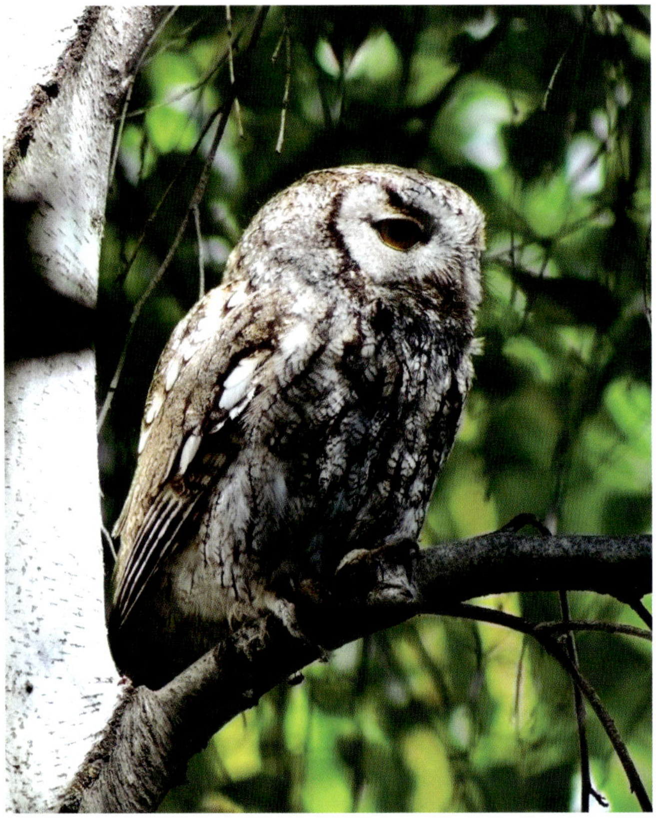

Figure 5.47 Screech Owl. Ethan Grey on Unsplash.

Most species of screech owls have stable populations and are, thus, not threatened. Despite this, the IUCN classifies two species at some level of risk. The one threatened species is the Santa Maria screech owl, listed as vulnerable on the IUCN Red List. This screech owl only lives in a small area of Columbia, and much of this region has been replaced by agriculture and livestock, leading to a marked decline in its numbers. Similarly, the cloud-forest screech owl is only found in Peru and Bolivia's cloud forest and has a small range. Because of this, the IUCN considers it near threatened, as any threat to its habitat will be detrimental to the species.

In Tyree's oil painting of a winking screech owl, the small owl sits perched on a mossy log, facing the viewer in a three-quarter profile (figure 5.48). Painted on velvet, the rough texture of the log contrasts with the bird's soft feathers. Winking with one golden eye, it gazes at the audience with the other, aloof and indifferent. The owl stands against a dark blue night sky encircled in glowing cobalt light, reminiscent of the halos common to Christian art throughout the Middle Ages and early Renaissance prior to the development and popularization of naturalism during the Renaissance, around the fifteenth century. This was a common technique in which a radiant crown or disk, often gold, yellow, or white, was painted surrounding a person to symbolize divine light and indicate holiness. In Tyree's work, an owl, a symbol of wisdom, dispassionately winks while glowing with light, giving it a mysterious, all-knowing aura.

Figure 5.48 *Winking Screech Owl*, Ralph Burke Tyree. 1978, oil on velvet, 16" x 12", author's collection.

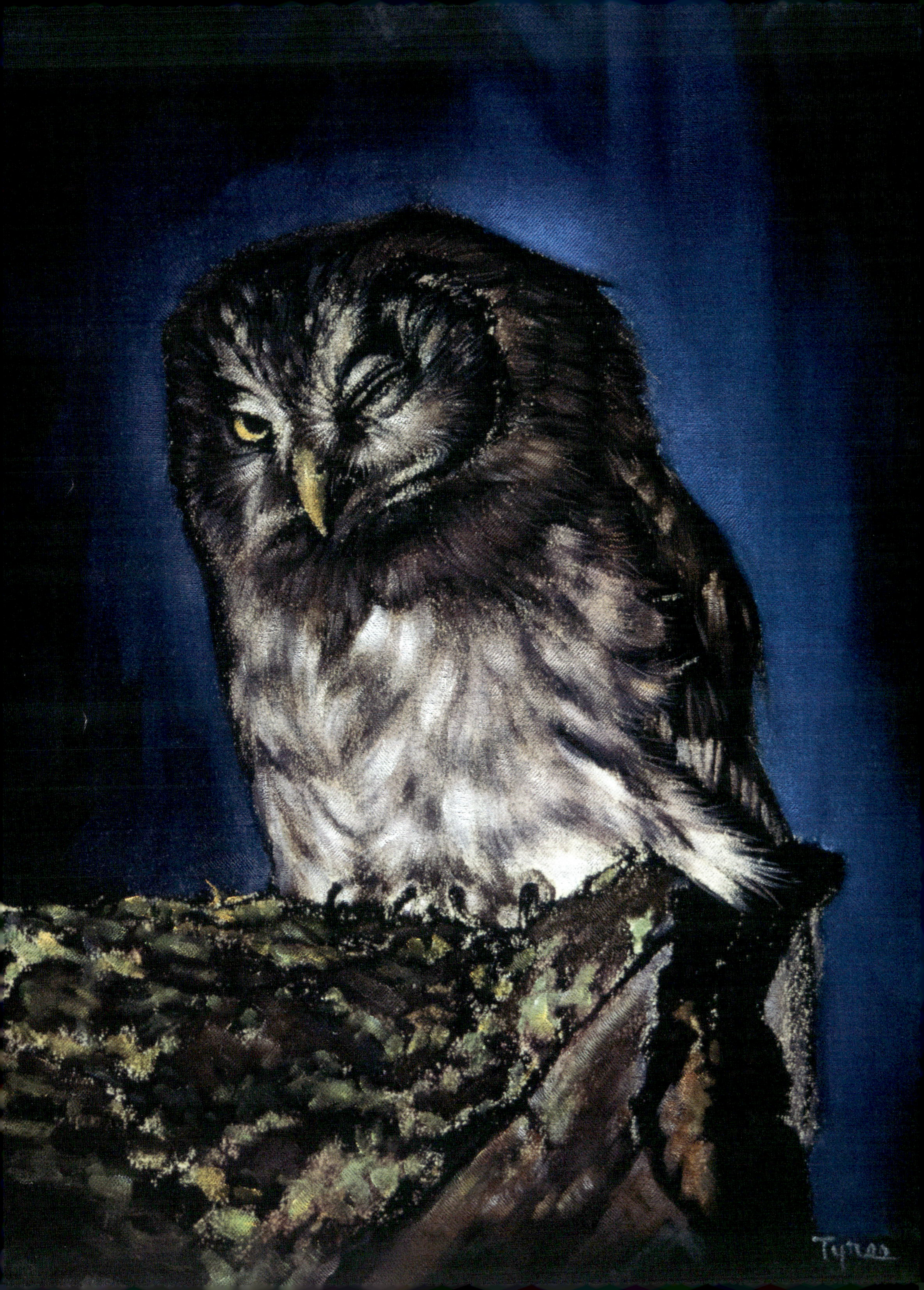
Tyree

Boreal Owl

The **boreal owl** (*Aegolius funereus*) is a small nocturnal owl of the *Strigidae* family native to the dense northern forests across North America, Europe, and Asia (figure 5.50). In Europe, it is called Tengmalm's owl for Swedish naturalist Peter Gustaf Tengmalm. Dappled with brown and white (figure 5.49), it is similar in size to the screech owl. The shape of its facial disk gives it a perpetually surprised appearance. Like other owls, it waits on a perch to hear prey pass by before silently swooping in to snatch it up. The boreal owl is particularly fond of voles and other small mammals but will also eat birds, amphibians, and insects. This owl usually nests in a hole in a tree, often created by a woodpecker. Humans rarely spot it because of its inaccessible habitat, nocturnal nature, and general shyness. While not threatened by extinction, logging and agriculture threaten this owl's habitats.

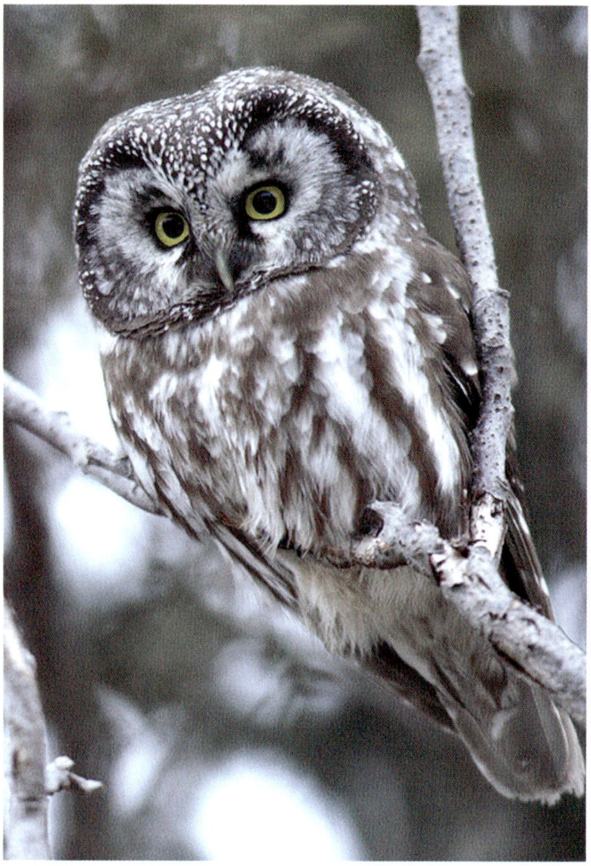

Figure 5.49 Boreal owl. Mdf, Ontario, 2005.

Tyree painted a boreal owl sitting in its nest in the hollow of a tree, alert as it observes its environment (figure 5.51). Seen from the breast up, the owl has its head slightly turned as if it has just heard a sound. With huge yellow eyes, it scans the forest for the source of the sound. The owl is not on the hunt but instead is likely looking for a predator, like a larger owl or a hawk. The tree fills the rest of the frame with the trunk painted predominantly in yellows and greens, picking up the yellow of the owl's eyes and, thus, emphasizing them. Like his painting of the winking screech owl, Tyree uses his velvet surface to create beautifully contrasting textures between the tree and the bird.

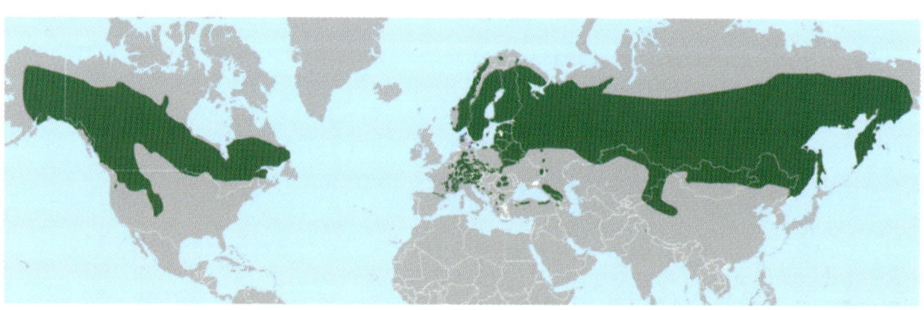

Figure 5.50 Boreal owl distribution. SanoAK: Alexander Kürthy, 2019.

Figure 5.51 *Boreal Owl*, Ralph Burke Tyree. 1978, oil on velvet, 16″ x 12″, author's collection.

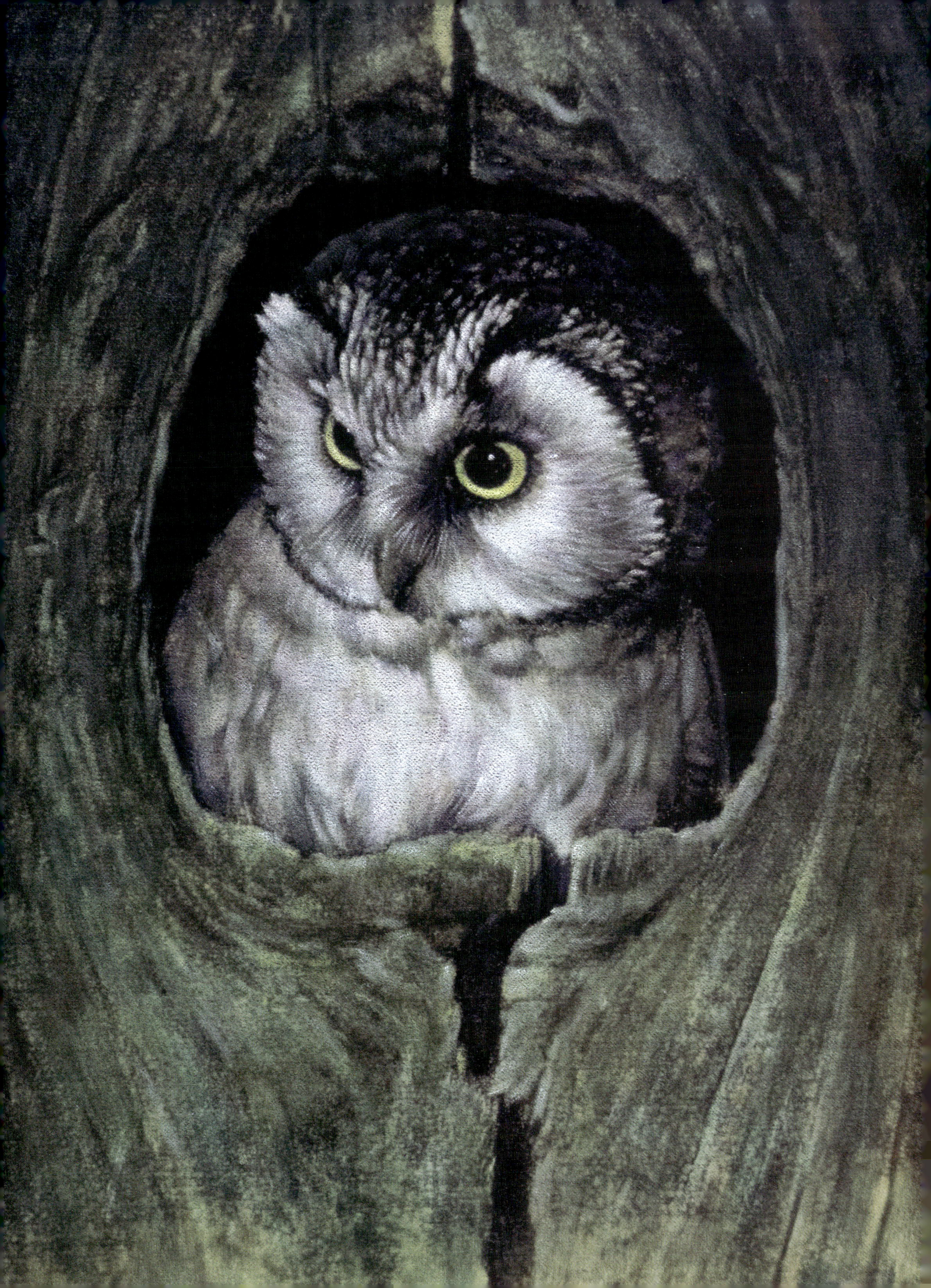

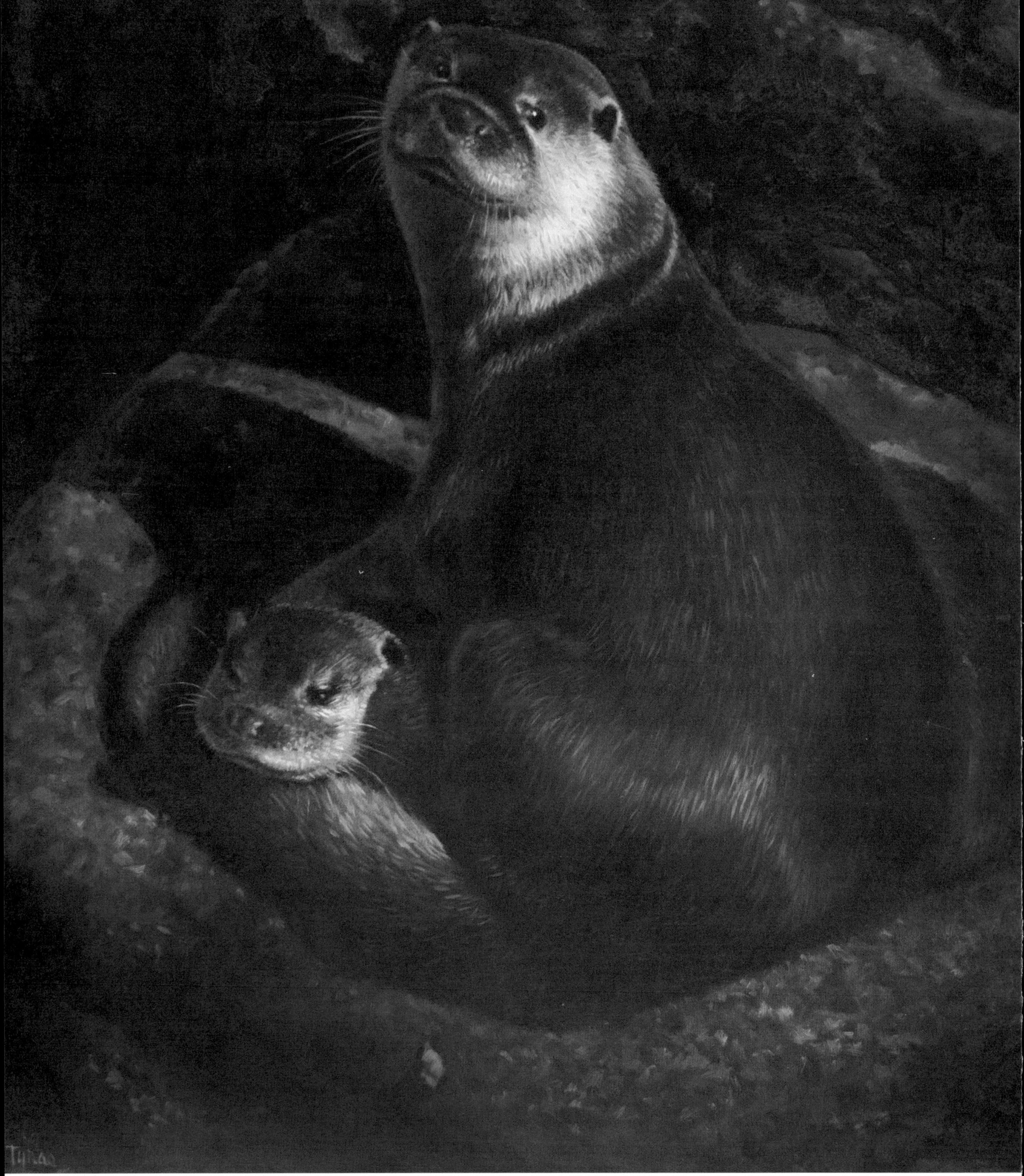

Figure 6.1 *Mother Otter and Baby*, Ralph Burke Tyree. Oil on board, 16″ x 12″, private collection.

CHAPTER 6

Rodents, Rabbits, & Carnivores

"If a man aspires towards a righteous life,
his first act of abstinence is from injury to animals."

—LEO TOLSTOY

Although squirrels, chipmunks, and rabbits are generally not thought of as endangered species, Ralph Burke Tyree nonetheless painted these ubiquitous creatures. He brilliantly captured their natural textures using lightly diluted oil paints to penetrate the velvet fiber's depths without losing the material's softness and pliability. He occasionally painted these animals on board, even sometimes painting an animal on both velvet and board, providing a striking comparison. Painting on board allowed him to perfect his compositions and play with the light and shadow in a work. Besides rodents, Tyree painted carnivores of all sizes, from the *Mustelidae* family of weasels, martins, and badgers, to giant beasts like the cougar and grizzly, both at risk due to habitat loss.

STARTING WITH RODENTS

Squirrel

Squirrels (*Sciuridae*) are a family within the *Rodentia* order that includes almost 300 species of squirrels as well as chipmunks and prairie dogs (figure 6.2). This order, characterized by two incisors in the upper and lower jaws, represents about 40% of the world's mammal species, with species native to every continent and large landmass except Antarctica, Australia, New Zealand, and a few oceanic islands, including Hawaii. Squirrels are found in every type of habitat except the most extreme hot and cold locales. They have even thrived in urban human environments. Figure 6.3 shows these usually small and slender familiar rodents with fluffy tails. They prefer to eat nuts, seeds, fruits, fungi, and occasionally small rodents, birds, and insects.

Of the 285 species in the *Sciuridae* family, the IUCN does not classify the vast majority as threatened. However, 34 are considered either vulnerable, endangered, or critically endangered, usually

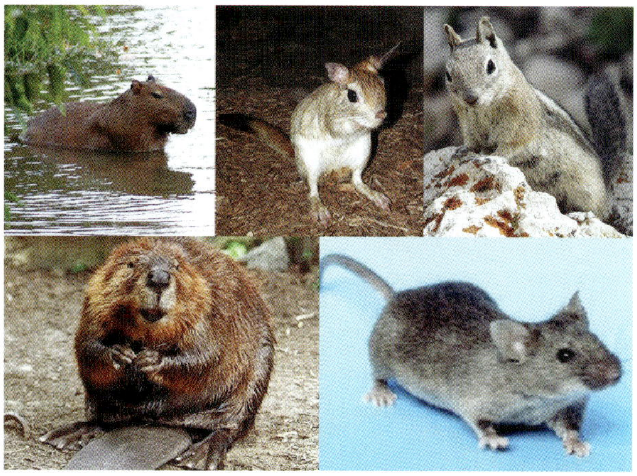

Figure 6.2 Rodents (*clockwise:* capybara, springhare, golden-mantled ground squirrel, house mouse, and North American beaver). Bobisbob, 2013.

because of habitat loss due to the destruction of their native woodland. The two critically endangered species are the Namdapha flying squirrel of India and the Vancouver Island marmot, the latter of which only has about 90 mature individuals remaining.

Tyree painted squirrels on both board and velvet (figures 6.4 and 6.5). In both paintings, the squirrel's bushy tail is on full display. In the piece on board, the squirrel sits on a rock, nibbling at something as the light bounces off its fur. As when Tyree painted the great blue heron, he utilizes a monochrome color palette to depict this furry rodent, focusing on the precise details. In the oil-on-velvet piece, the orange-hued squirrel is pictured mid-movement as it scurries across a log surrounded by green vegetation with red flowers melting into the blackness of the velvet backdrop. The squirrel's fur appears soft and fuzzy, illustrating Tyree's skill in painting on velvet. Which do you prefer—the oil on board or velvet?

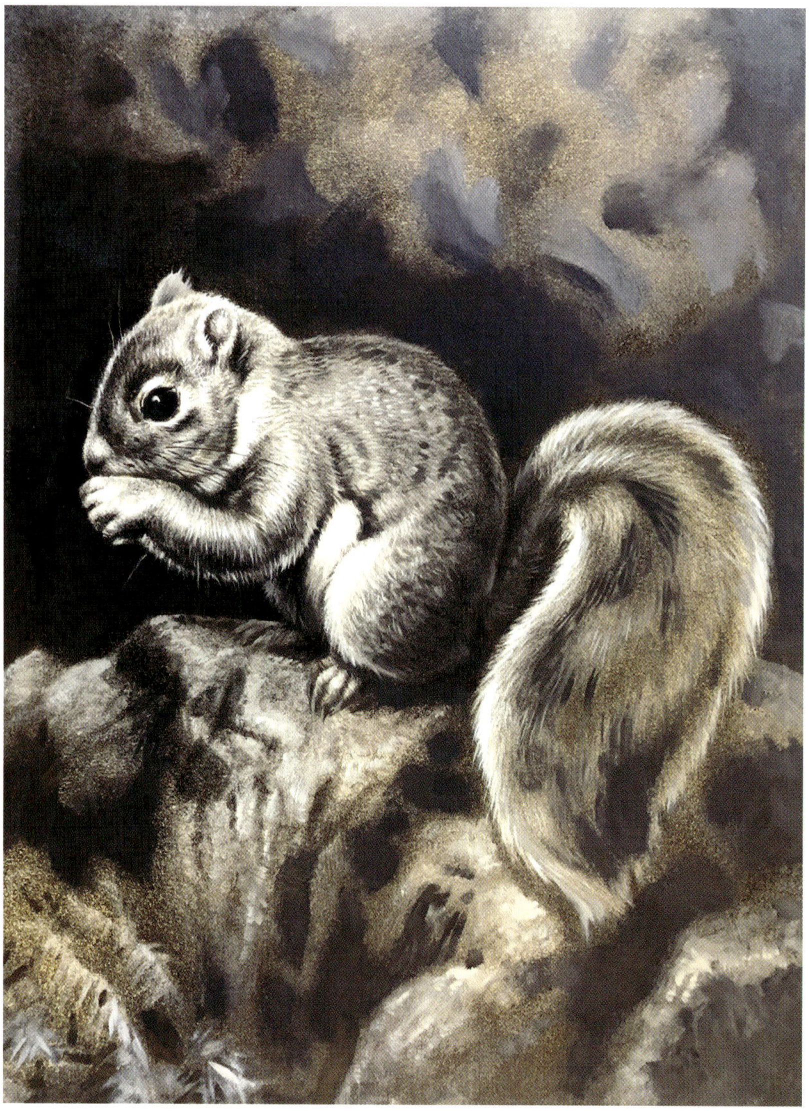

Figure 6.4 *Squirrel*, Ralph Burke Tyree. Oil on board, 14" x 11", private collection.

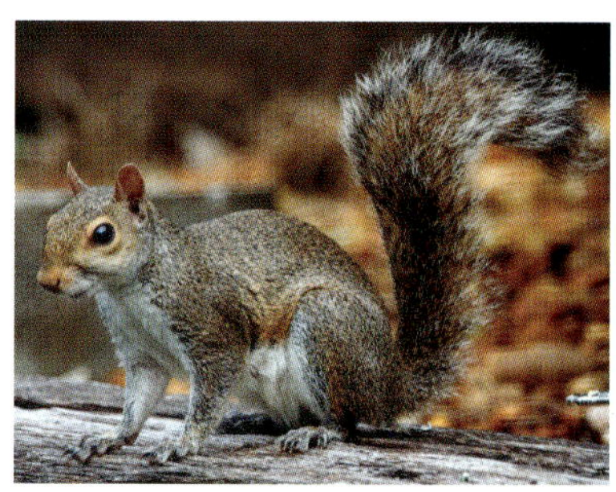

Figure 6.3 Gray squirrel. Joshua J. Cotten on Unsplash.

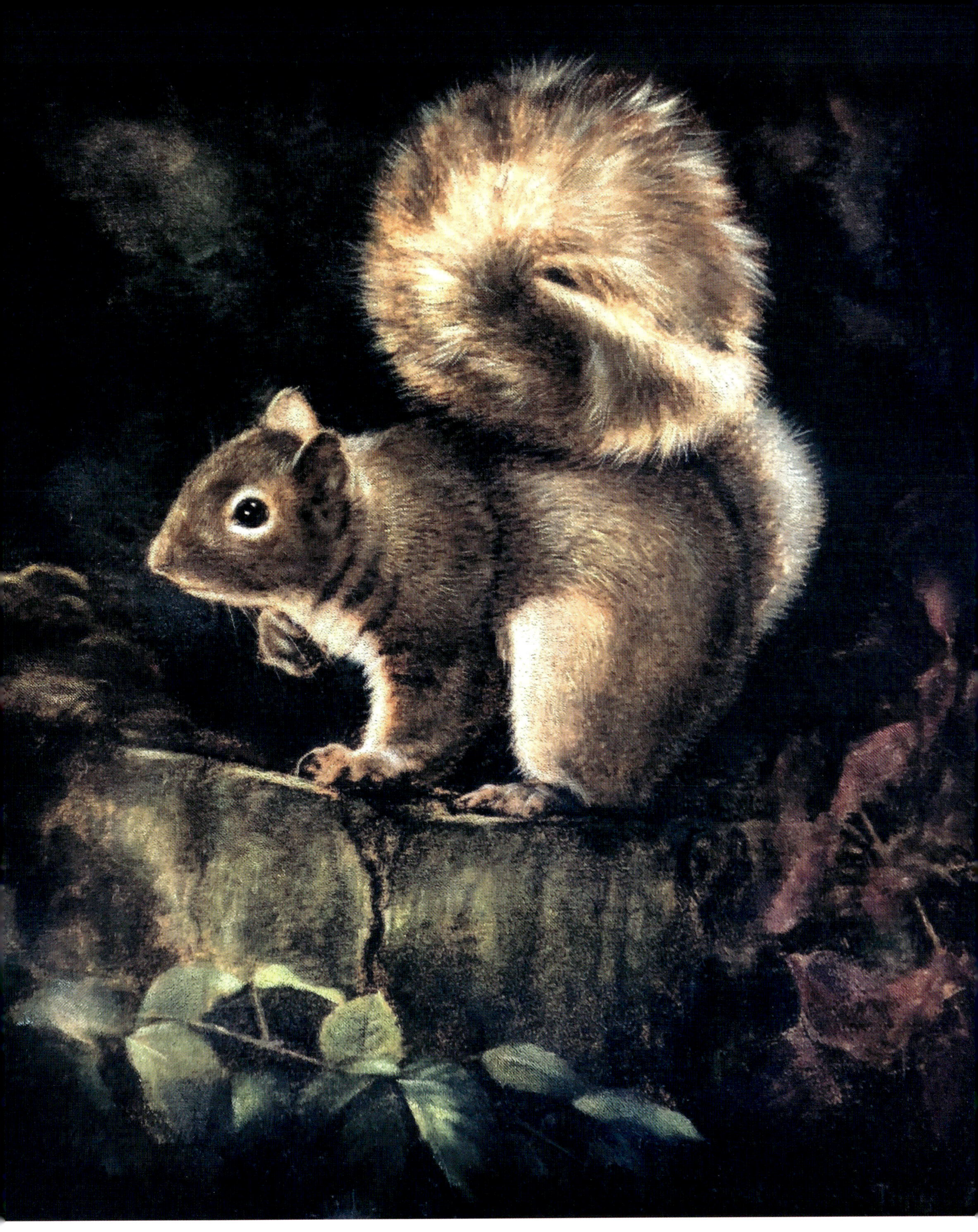

Figure 6.5 *Squirrel*, Ralph Burke Tyree. 1978, oil on velvet, 20" x 16", private collection.

Chipmunk

Chipmunks (*Tamias*), occasionally called striped squirrels or timber tigers, are a genus in the *Sciuridae* family consisting of 25 individual species (figure 6.6). These small, striped rodents are, even at their largest, around ten inches long and under five ounces. Aside from one species native to Asia, chipmunks are almost exclusively native to North America, where they can be found scurrying throughout various habitats. Some chipmunks dig extensive networks of burrows to live in, whereas others build nests in trees or bushes. All species gather seeds, nuts, and whatever else they can forage, eating some and storing extra in their large cheek pouches to take home, creating a stockpile to get through winter hibernation. These critters play a crucial role dispersing seeds and nuts throughout their environments. As one of the cutest rodents around, chipmunks have become a favorite of animators, famously represented as cartoon characters like Chip 'n' Dale and Alvin, Simon, and Theodore of the Chipmunks.

While most species of chipmunks are not at risk, the IUCN categorizes two as threatened. Palmer's chipmunk, found in the Spring Mountains of Nevada, is endangered because of habitat loss associated with Las Vegas' urban sprawl and increased predation by feral dogs and cats. Buller's chipmunk, only found in a small region of Mexico, has also struggled with habitat loss, leading the IUCN to classify it as threatened.

Tyree portrays an acorn-carrying chipmunk with lifelike accuracy against colorful leaves in figure 6.7. The chipmunk sits on a log, ready to stuff the nut into its cheek pouch. The fall leaves indicate that the tiny critter is probably collecting goodies to store for the impending winter. Painted on velvet, this is a prime example of Tyree's use of the medium to create realistic textures as he paints the chipmunk's soft fur with precise detail.

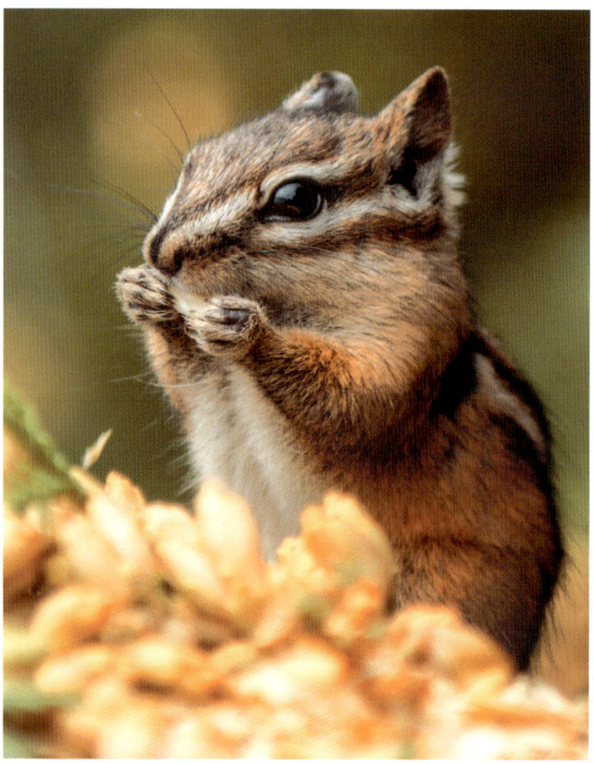

Figure 6.6 Chipmunk eating. Matthieu Pétiard on Unsplash.

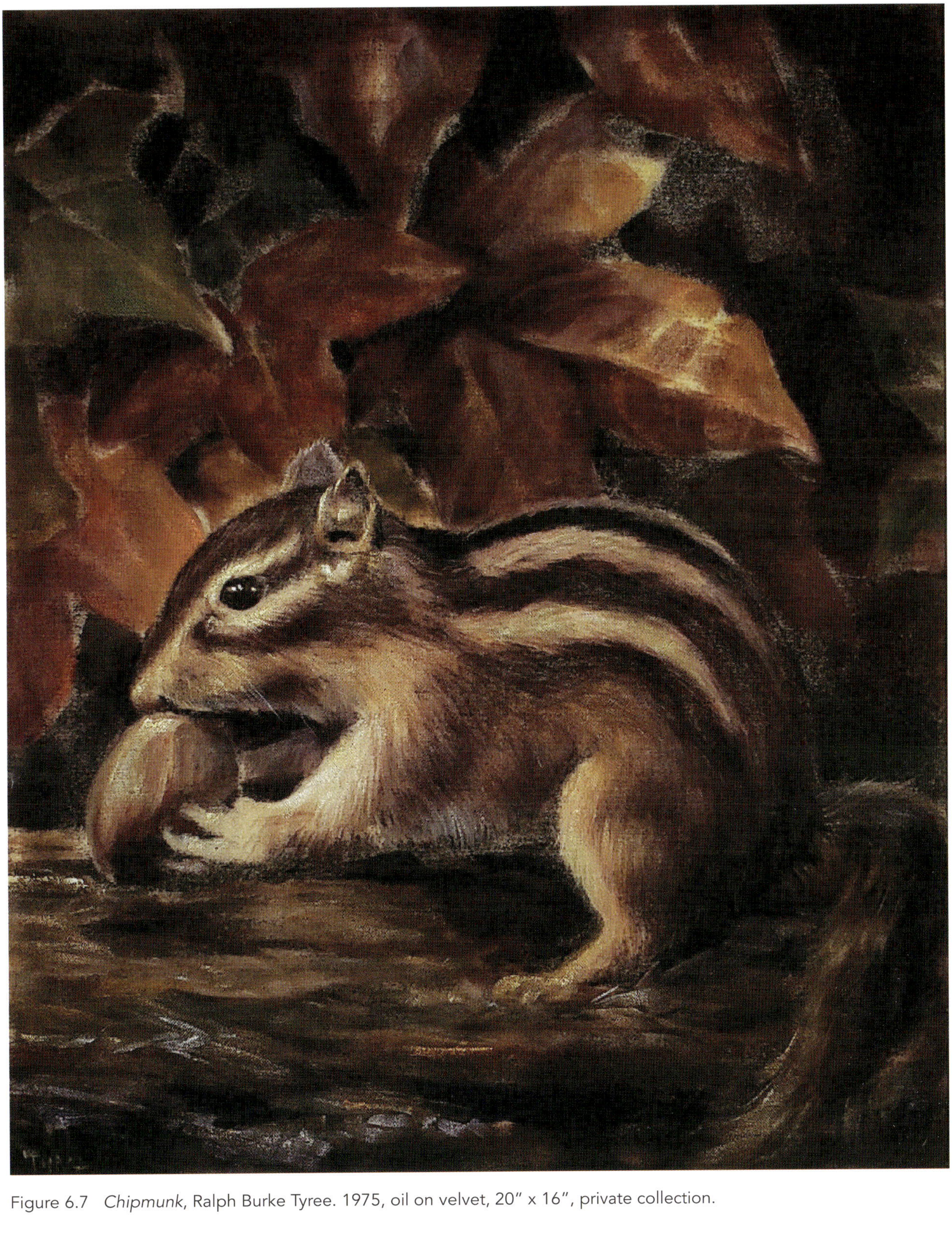

Figure 6.7 *Chipmunk*, Ralph Burke Tyree. 1975, oil on velvet, 20" x 16", private collection.

Prairie Dog

Prairie dogs (*Cynomys*) are another genus from the *Sciuridae* family (figure 6.8). While not actually related to canines, they are named after man's best friend because their call sounds similar to a dog's bark. Technically considered burrowing ground squirrels, the five species of prairie dogs are native throughout the grasslands of North America, eating grasses and seeds. These stout rodents are tan, sometimes with black and white markings. They live in large colonies of up to 26 family groups, creating sophisticated burrows that cover hundreds of acres. This results in little towns with complex systems of communication and social structure.

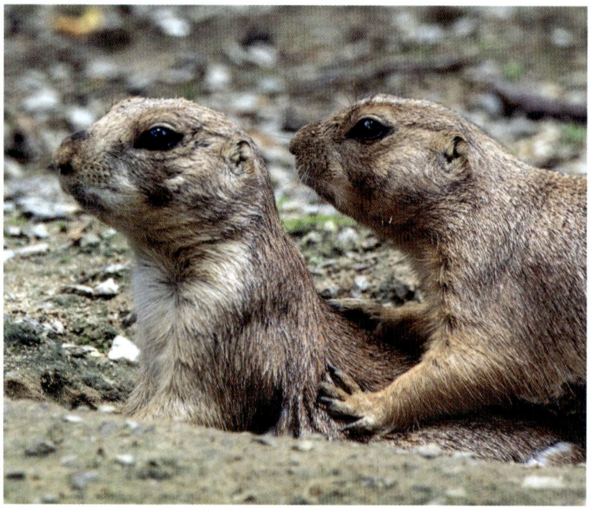

Figure 6.8 Wary prairie dogs. Joshua J. Cotten on Unsplash.

As an essential food source for birds of prey and larger mammals of the plains, prairie dogs are a keystone species. Their mound-building also serves an important role, encouraging topsoil renewal and grass development. Despite this, they are often considered pests by farmers who seek to eliminate colonies by killing these rodents. In addition to their removal by farmers, one of the greatest threats to prairie dogs is urban development, which impacts their habitats and can result in dramatic population loss. Bubonic plague was another major threat to prairie dogs due to their high susceptibility until 2017, when the US government distributed an oral vaccine, with 70% of wild prairie dogs successfully ingesting it.

Painting two different prairie dogs in 1975, Tyree depicts both standing up on their back legs, looking around. In figure 6.9, the prairie dog stands in rolling, grassy hills, shown in a full-body profile, with its front paws relaxed and its tail draped behind it. Again, Tyree works with oils in shades of black and white on board, specifically focusing on light and shadow. Lit from the front, the prairie dog's belly is bathed in bright light, and its back is hidden in dark shadow. In the same year, Tyree painted another prairie dog, this time on velvet in full color (figure 6.10). Also standing upright, this prairie dog pauses mid-snack to look around, perhaps listening for a family member's call or looking out for potential predators. It holds an orange flower with a long, partially eaten stem and seems to have made a meal out of these stems, as similar flowers lay scattered around its feet, all stemless. The environment it inhabits is abstracted, with blue tree bark and orange flowers framing the little prairie dog like fireworks. While both paintings are extraordinarily realistic, the texture Tyree achieves when painting on velvet is unmatched.

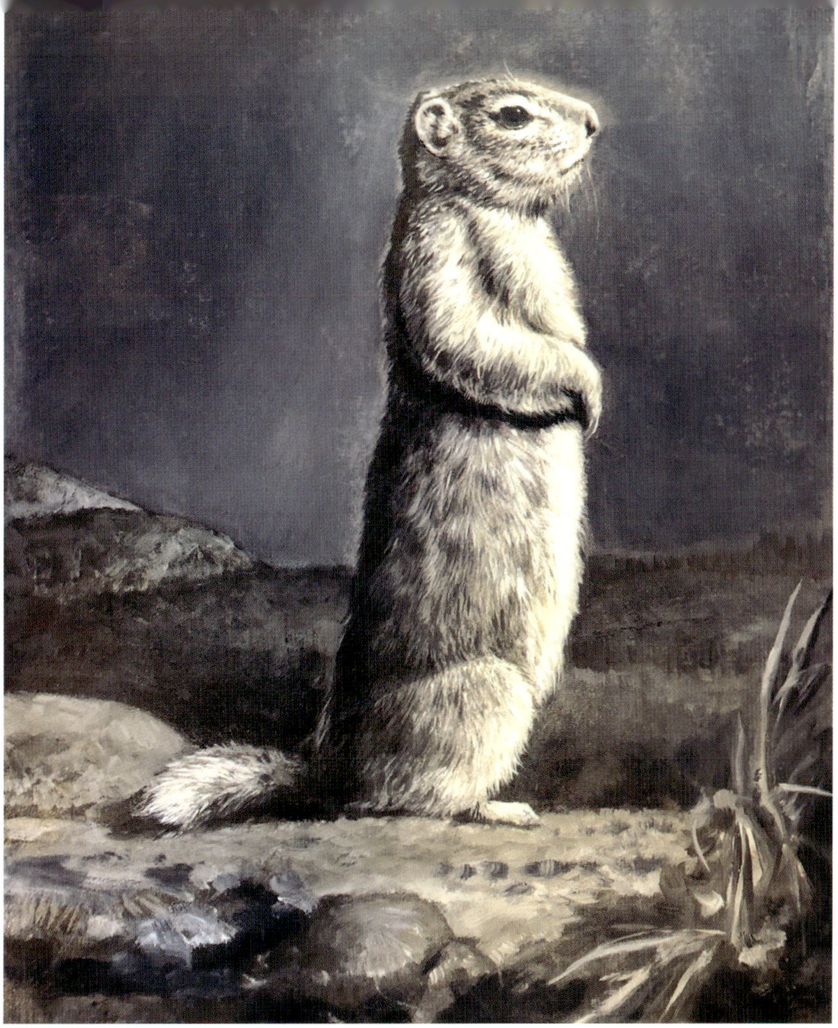

Figure 6.9
Prairie Dog,
Ralph Burke
Tyree. 1975, oil
on board, 14"
x 11", private
collection.

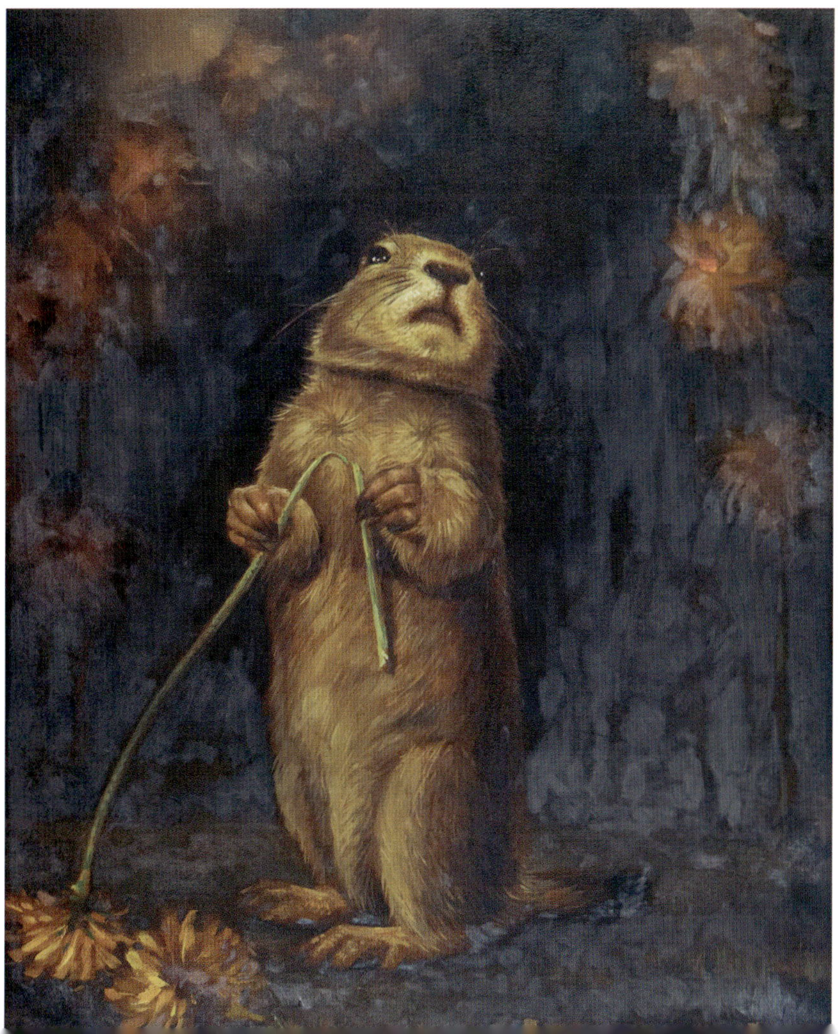

Figure 6.10
Prairie Dog,
Ralph Burke
Tyree. 1975, oil
on velvet, 20"
x 16", private
collection.

White-Tailed Jackrabbit

The **white-tailed jackrabbit** (*Lepus townsendipri*), also known as the prairie hare, is the most common jackrabbit from the *Leporidae* family of rabbits and hares (figure 6.11). Although once included in the order *Rodentia*, rabbits, hares, and pikas now make up a separate order, *Lagomorpha*, which consists of 110 species. As the largest jackrabbit species, the white-tailed jackrabbit can weigh nearly ten pounds and measure over two feet long. It is most identifiable for its distinctive white stub tail and long, black-tipped ears. Its superb eyesight, large ears for hearing, and sensitive whiskers give this rabbit a rapid response time and an edge when fleeing to survive. With impressively powerful rear legs, this jackrabbit can run up to 40 miles per hour and leap up to ten feet when threatened. The white-tailed jackrabbit is an important food source for many predators, including badgers, foxes, and eagles. As shown in figure 6.12, it is native to Canada and the US, where this nocturnal rabbit emerges at dusk to feed on grasses and plants.

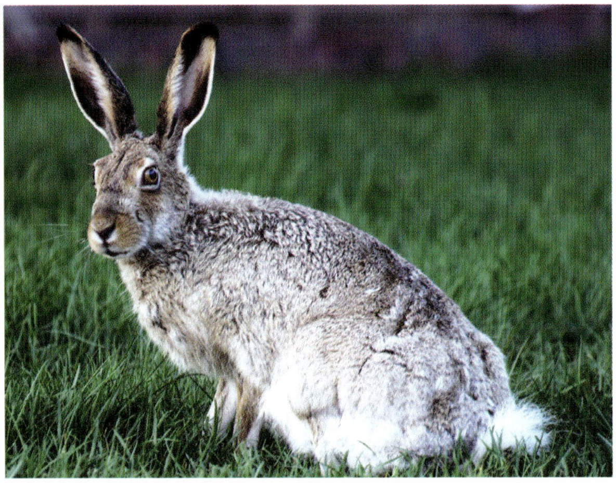

Figure 6.11 White-tailed jackrabbit. Connormah, Canada, 2014.

Habitat changes have caused the white-tailed jackrabbit's numbers to decline. In 2008, widespread reporting suggested that this once-abundant hare had all but disappeared from Yellowstone and Grand Teton National Parks over the course of the latter half of the twentieth century. Although a mystery to conservationists, this was likely due to a mix of habitat degradation, competition with black-tailed jackrabbits, and the shifting climate. Still, the IUCN does not consider this jackrabbit at risk.

Tyree painted a white-tailed jackrabbit on velvet, showing it mid-jump, about to land on its front paws (figure 6.13). With this composition, he captures the movement of the jackrabbit and highlights its agility. It inhabits a beautifully abstracted environment of red soil and bright green grass below a darkening blue, grey, and purple sky. The dark sky seems to suggest that the sun has recently set. With the setting sun, the jackrabbit has awoken and is out foraging or possibly running from a predator.

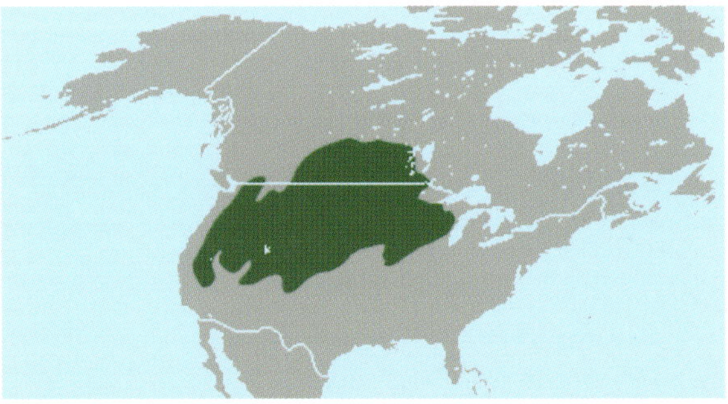

Figure 6.12 White-tailed jackrabbit distribution. Chermundy, 2010.

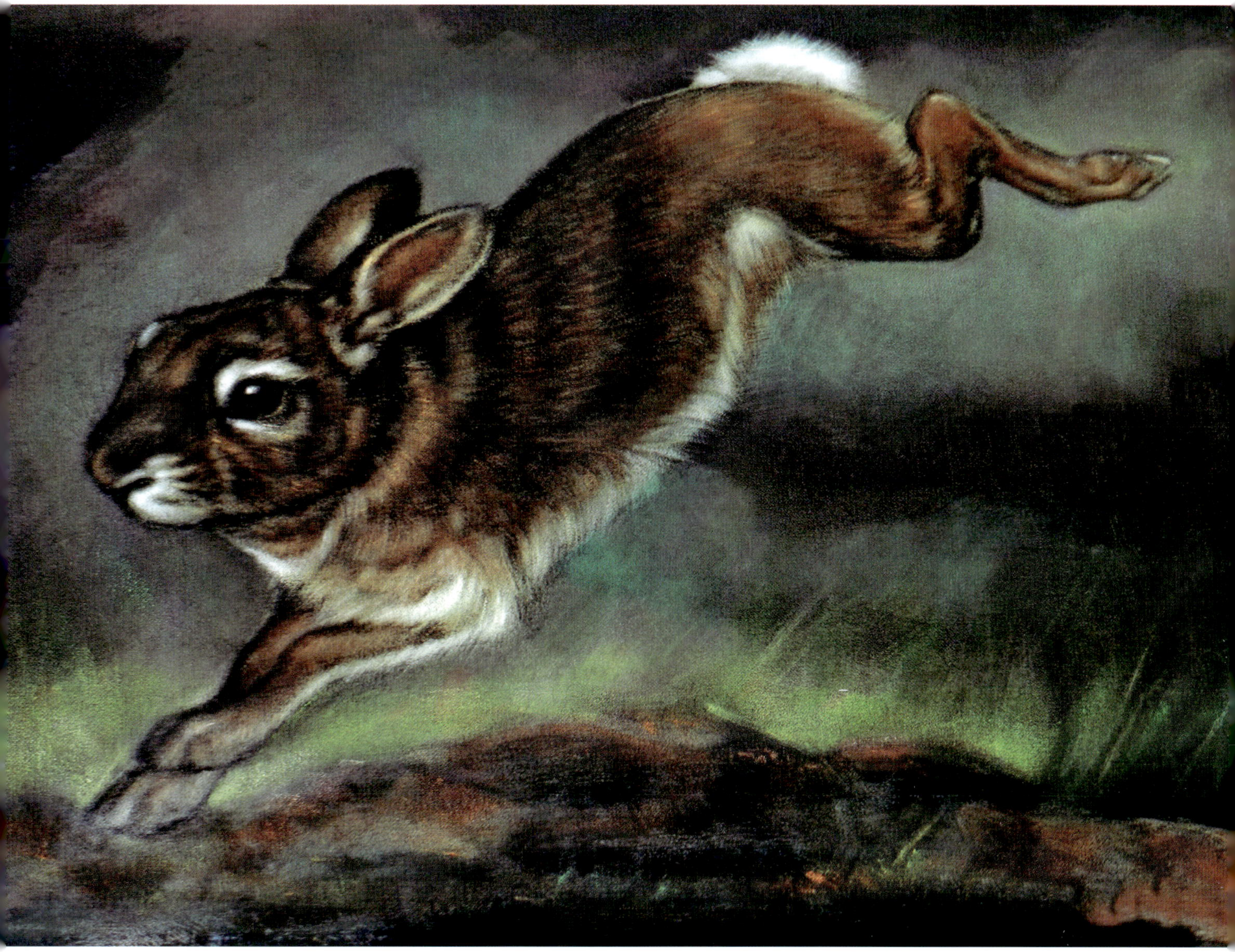

Figure 6.13 *White-Tailed Jackrabbit*, Ralph Burke Tyree. 1978, oil on velvet, 16″ x 20″, private collection.

TO CATCH A WEASEL

Weasels, ferrets, minks, and otters are all members of the *Mustelidae* family of carnivorous mammals. This family also includes badgers and wolverines—the tough guys of the group. As the largest family in the order *Carnivora*, it consists of about 70 species across nine subfamilies that have significant variances in size and behavior. The land-dwelling common weasel is the smallest family member at under a foot from head to tail. In comparison, the largest species of the family, the giant river otter of the Amazon, is even more impressive at nearly six feet long. Despite these differences, all mustelids share certain characteristics, including elongated bodies, short legs, small round ears, and thick fur. Most of these mammals are also solitary, nocturnal, and ferocious, with strong jaws that can challenge larger animals. Notably, wolverines are said to be, pound-for-pound, the most formidable animal in America. The ferret is the only member to have been domesticated as a pet, but several, such as the mink and sable, have been raised for their fur.

Weasel

The **weasel** (*Mustela nivalis*), also known as the least weasel, the common weasel, and the little weasel, varies in length over its geographic range but is usually under a foot (figure 6.14). Native to Europe, Asia, North America, and North Africa, it also varies in color throughout this range, however, it is typically a shade of brown with a white underbelly. It makes many different sounds depending on the situation, including hisses, barks, wails, and squeals. This fierce little hunter has a long, flexible body and short legs, which enable it to pursue prey, like small rodents and rabbits, into burrows. With a forceful bite even stronger than that of apex predators like the lion and tiger, this small weasel can take down prey up to ten times heavier than itself. The weasel is often considered a nuisance to farmers because it can kill rabbits and poultry. Thanks to the weasel's wide range and relatively large population, the IUCN does not classify it as threatened.

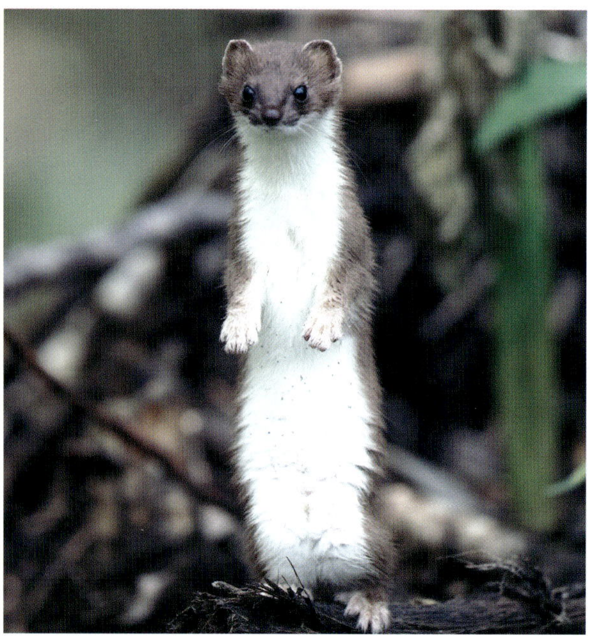

Figure 6.14 Weasel. Brent Jones on Unsplash.

Tyree's oil painting on velvet of a weasel illustrates this little carnivore's climbing ability and curious personality (figure 6.15). Sitting in a tree hollow, the nimble climber uses its front legs to hoist itself up and peek at its surroundings. Tyree beautifully juxtaposes the textures of the weasel's soft, sleek fur, the tree's rough bark, and the background's atmospheric forest.

Figure 6.15 *Weasel*, Ralph Burke Tyree. 1974, oil on velvet, 20" x 16", author's collection.

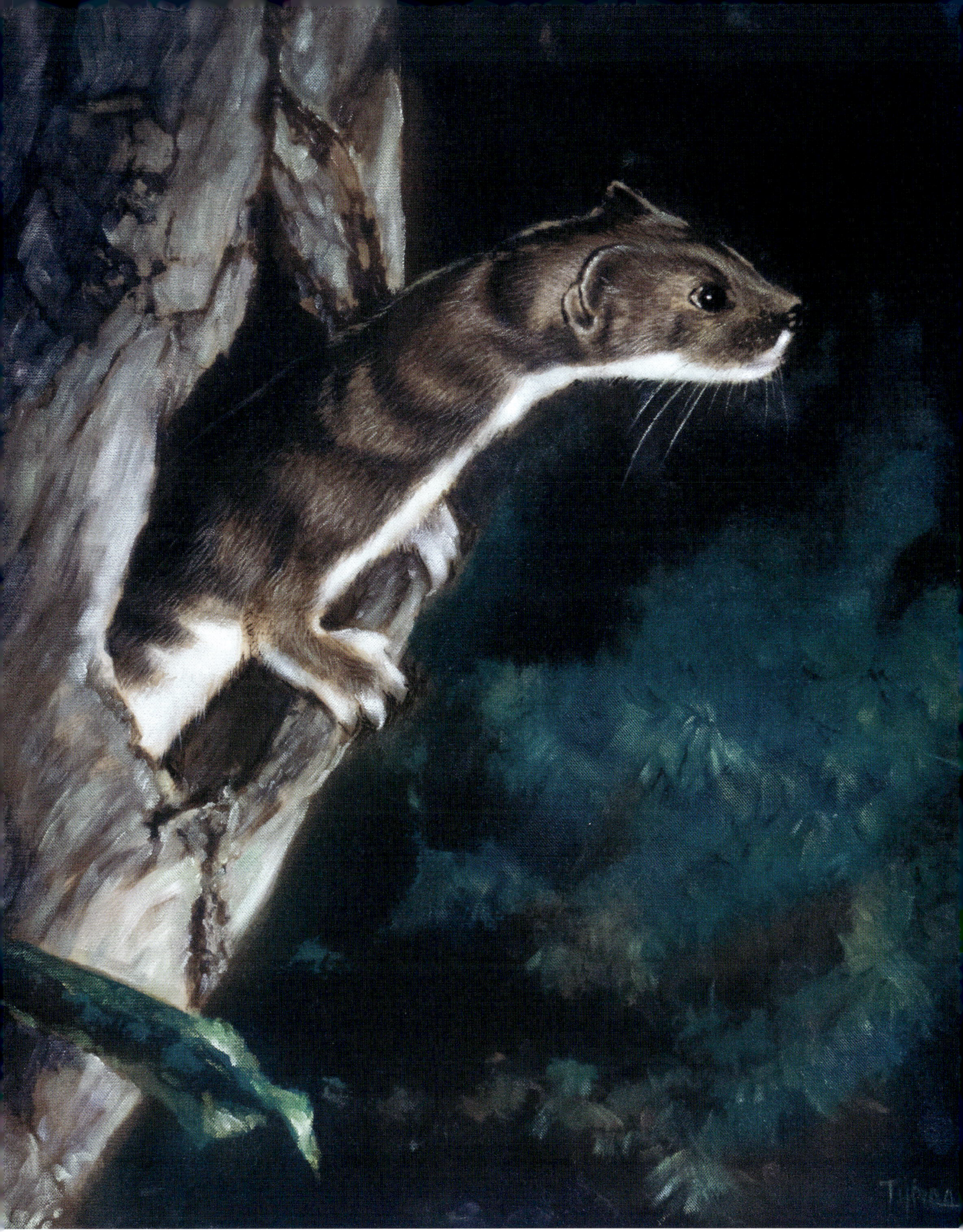

American Marten

Another species of the *Mustelidae* family, the **American marten** (*Martes americana*), or American pine marten, is found throughout the forests of Canada, Alaska, and the northern US (figure 6.16). Like the weasel, it has a long, slender body, but its fur ranges from yellowish-brown to almost black with a light-colored bib and ears. This tree-dwelling marten is also slightly larger with longer legs than the weasel, about the size of a house cat. While it is an opportunistic omnivore, it primarily eats small mammals, specifically voles, for nutrition. It consumes more fruits and vegetation during the summer, playing an essential role as a seed disperser.

Due to the American marten's fine fur, it was highly sought out during the height of the North American fur trade. Trapping peaked in 1820, and by the turn of the nineteenth century, the marten's population had been nearly depleted. Luckily, as the fur trade has declined, its numbers have rebounded, although hunting is still its primary threat. Habitat loss due to deforestation is another major threat to this marten. Although it is not considered at risk today, its population is declining.

Like the weasel, Tyree painted the American marten on velvet, sitting in a tree hollow, gazing out at its environment (figure 6.17). Instead of inquisitively extending itself out of the hole like the weasel, the marten observes calmly from the safety of its hiding spot. Only its cute cat-like head is visible as it stares straight at the viewer with soulful eyes. Tyree creates the illusion of a formal portrait by placing the marten against the darkness of the tree's hollow, which he centers in the composition, with the tree trunk filling the remainder of the frame. These elements mimic a framed canvas and serve to humanize the once-endangered marten.

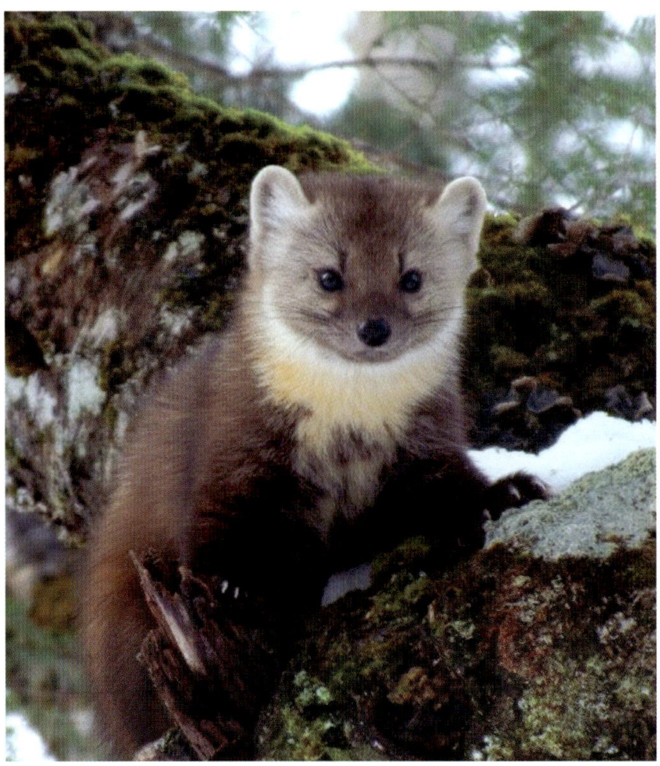

Figure 6.16 Pine marten. Bailey Parsons, 2013.

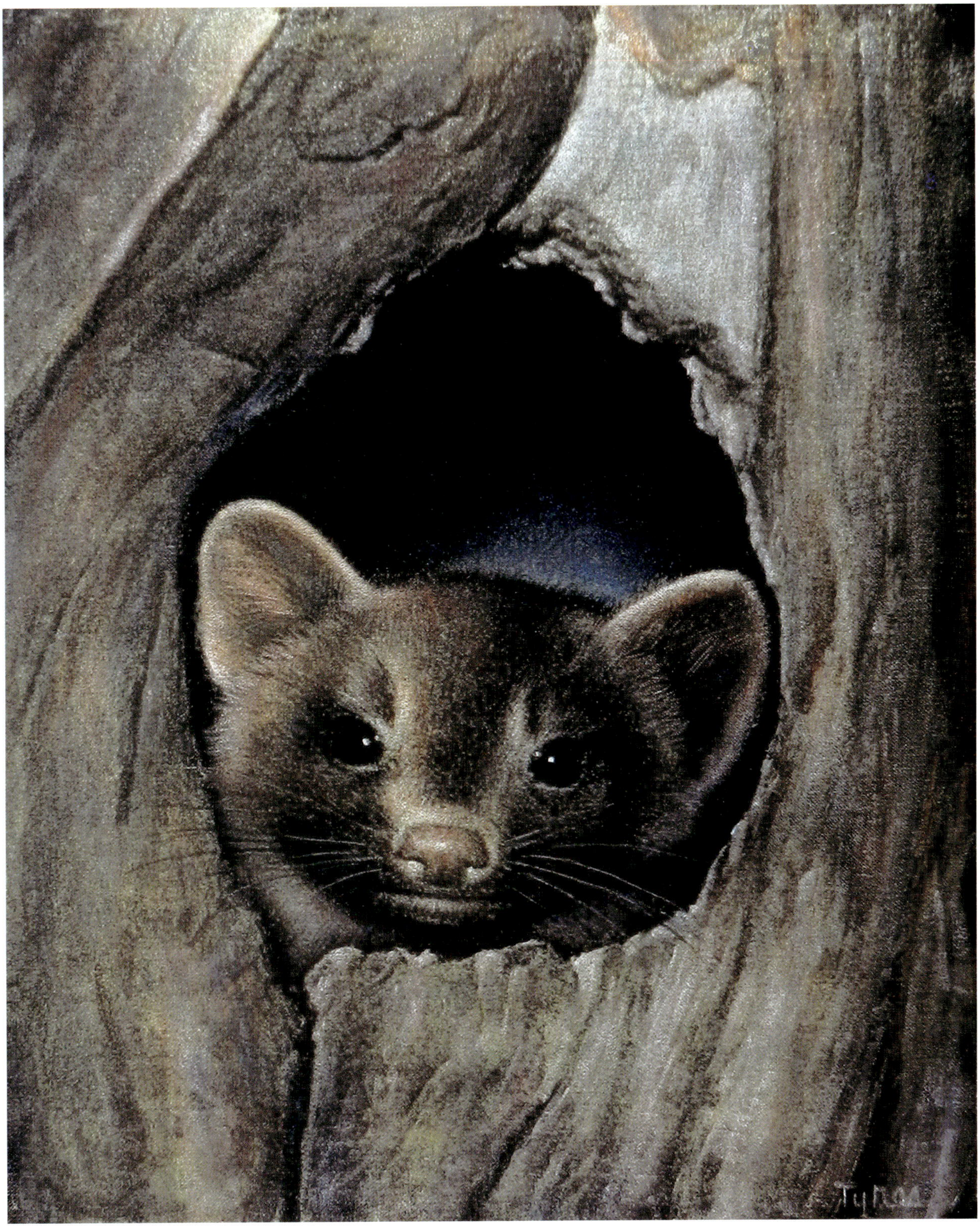

Figure 6.17 *American Martin*, Ralph Burke Tyree. 1978, oil on velvet, 16" x 12", Tyree family.

Sea Otters, Utterly Adorable

The **sea otter** (*Enhydra lutris*) of the *Mustelidae* family is native to rocky areas of the northern coasts of the Pacific Ocean. This marine mammal has the densest fur of any animal, which protects it from cold ocean waters. Weighing up to 100 pounds, it is both the heaviest member of its family and the smallest marine mammal. The sea otter dwells in thick kelp forests near shorelines where its favorite foods, sea urchins, mollusks, and crustaceans, are plentiful. It is a keystone species, keeping sea urchin and mussel populations under control and helping to maintain kelp forests.

One of few mammals that use tools, it holds a rock in a special pouch on its chest, which it uses to open shells and access prey.

This strong swimmer is notably buoyant and is often spotted floating on its back, eating, grooming, nursing, and resting (figure 6.18). Together, otters form groups called rafts by holding paws and wrapping themselves in kelp, linking together as many as 100 individuals.

Like the American marten, the sea otter's fabulous fur made it particularly desirable during the nineteenth-century fur trade. Russian fur traders, hunting for Chinese customers, were the predominant dealers of the otter pelts. Soon, sailors throughout the North Pacific saw the value of these pelts and, because of the high prices they could fetch, started calling them "soft gold." In 1911, with less than 2,000 sea otters estimated to remain in the wild, Russia, Japan, Great Britain, and the US signed the North Pacific Fur Seal Treaty, banning open-water seal hunting. Since this ban, otter numbers have risen to over 200,000, about two-thirds of historical population highs. While widely considered a success story, the IUCN still classifies the sea otter as endangered due to significant threats from oil pollution, predation, and conflicts with fisheries.

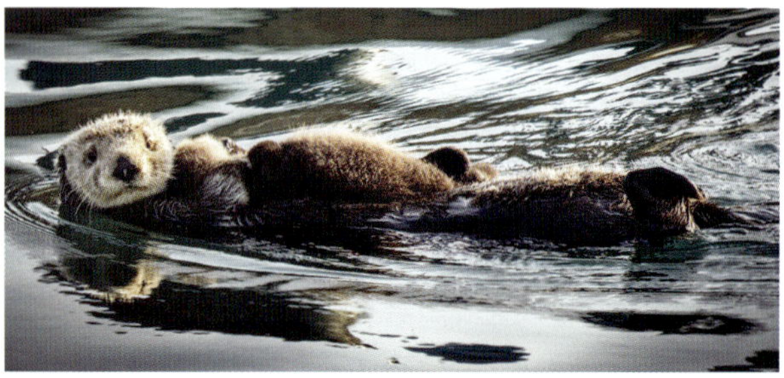

Figure 6.18 Sea otter mother nursing her pups. \Anchor Lee on Unsplash.

Tyree completed two paintings of sea otter pairs, one on velvet and one on board. In figure 6.19, the otter pair bobs upright in the ocean, almost as if one is sitting on the other's shoulders. They fill the frame, with the rest of the composition filled with dark grey-blue sea. Light bounces off the back otter's glistening wet fur. In figure 6.1, Tyree again shows an otter pair, but this time, a mother and baby. They rest together on a rock, with the mother curled up and the pup tucked under her tail. Tyree fills the rest of the frame with an obscure rocky environment—again, the otters occupy most of the composition in a vague environment as their wet fur gleams. Both works focus on the otter's thick, dense fur, which has long made it a target of hunters. To further emphasize the fur in the latter work, the mother is also angled so that her coat takes up much of the composition.

Figure 6.19 *Two Otters*, Ralph Burke Tyree. 1974, oil on velvet, 20" x 16", private collection.

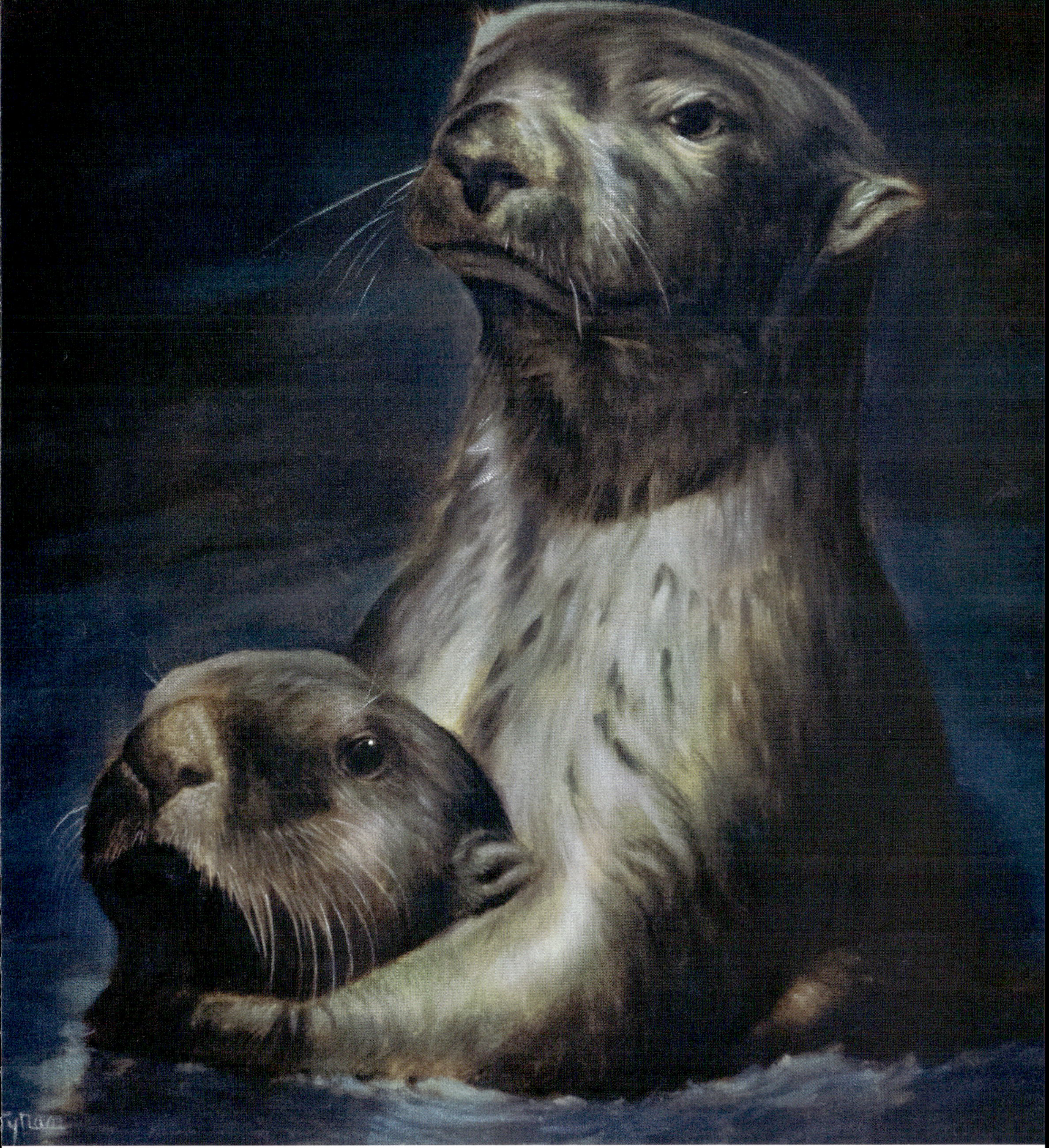

Don't Badger Me

The **American badger** (*Taxidea taxus*), another member of the *Mustelidae* family, is a squat and sturdy burrowing carnivore found throughout the open grasslands of North America. It has a blended black, white, and brown coat and a triangular face with recognizable white and black markings (figure 6.20). Its broad body, short, strong legs, and huge foreclaws are made for digging. This aggressive predator often tunnels after prey, primarily small rodents, mammals, snakes, and birds, and can run at about 20 miles per hour. Despite its fearsome nature, it is prey to the golden eagle, the cougar, and the coyote, its occasional ally. Usually solitary, the badger is sometimes known to form short-term alliances with coyotes, hunting alongside them. It is a nocturnal creature that generally remains in its burrow until dark. However, it has been observed throughout the day in areas away from human activity.

While the IUCN does not consider the American badger at risk, its numbers are decreasing, and in some places, it is a species of particular concern. Trappers have long targeted this badger for its pelts, which were used to make shaving and paint brushes due to the fur's ability to retain water. Most of these practices have stopped, and today, its most significant threats are habitat loss and loss of food as prairie dog populations decrease.

Electing to paint the American badger in shades of grey, Tyree shows off its distinctive facial markings (figure 6.21). Another smaller monochrome oil painting on board, the size and palette again allow the artist to focus on composition and value. The badger's recognizable black-and-white face is unmistakable, especially in this limited color palette. As it climbs out of its burrow surrounded by scattered grasses, it gazes at the viewer, interested but not afraid.

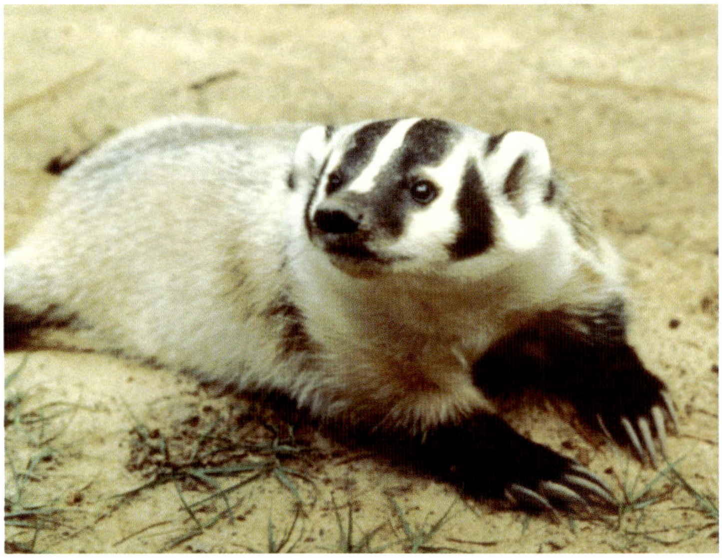

Figure 6.20 American badger. Jamie Fuentes, courtesy of the National Wildlife Research Center.

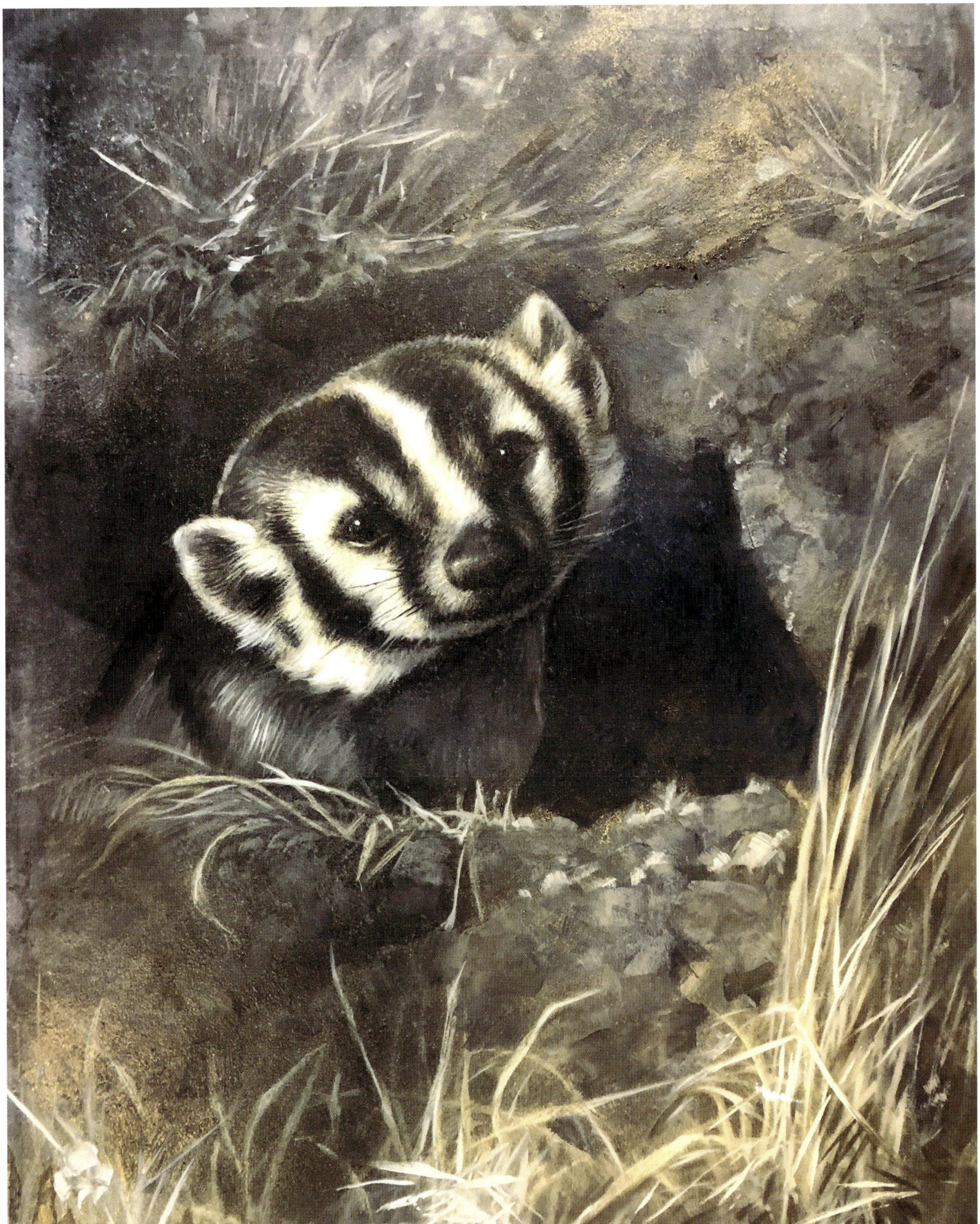

Figure 6.21 *Badger*, Ralph Burke Tyree. 1974, oil on board, 16" x 12", private collection.

Ringtail

The **ringtail cat** (*Bassariscus astutus*), or bassarisk, from the *Procyonidae* family of raccoons, is a small mammal with a long black and white striped tail, as suggested by its name (figure 6.22). Early western settlers and miners domesticated this critter to keep their homes vermin free, calling it the miner's cat. Just under the size of a house cat, it typically weighs less than four pounds, making it one of the smallest members of its family. The ringtail is an excellent climber, using its tail, flexible ankles, and tiny claws to maneuver complex pathways in the trees. This striking tail also distracts and confuses predators, allowing the ringtail time to escape. This solitary critter is found throughout the US and Mexico in shrublands and rocky desert habitats as long as trees are prevalent (figure 6.23). During the winter, it primarily eats small animals like rodents, birds, and reptiles, but it shifts to mostly fruits, berries, and insects when warmer weather comes.

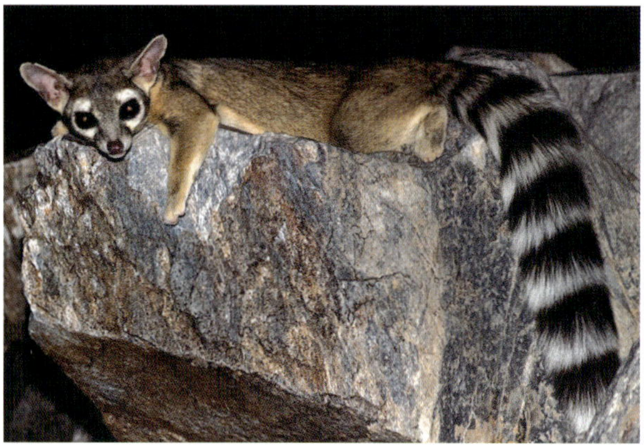

Figure 6.22 Ringtail. Robert Body, 2008.

The IUCN does not consider the ringtail cat vulnerable, but it faces threats of habitat loss. The destruction and fragmentation of its habitats can harm its populations because it requires trees to climb. Additionally, fur traders can legally trap the ringtail for its fur. Luckily, its pelts are not particularly desirable or valuable.

With an adorable face and eye-catching tail, it's no surprise Tyree chose to paint the ringtail cat. Figure 6.24 shows his oil on velvet painting of a ringtail. The cat balances at the base of a tree limb in the darkness of night, illustrating its nocturnal nature. Its characteristic striped tail appears soft and fluffy as it hangs down below the branch, stealing the show. With an alert expression, the ringtail may be on the lookout for prey. Alternatively, it may have spotted a predator and is swinging its tail to divert the threat's attention.

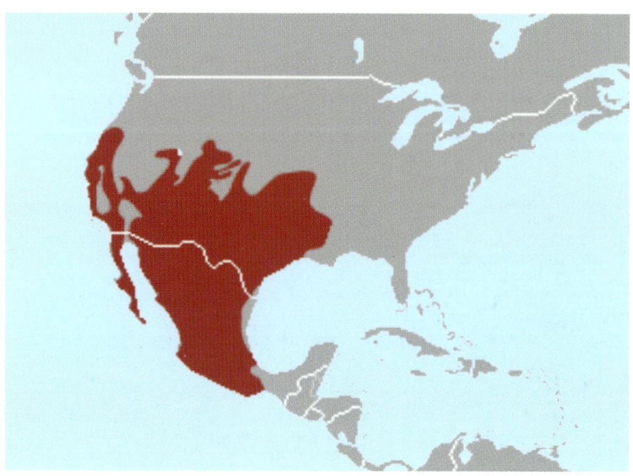

Figure 6.23 Ringtail distribution. Chermundy, 2010.

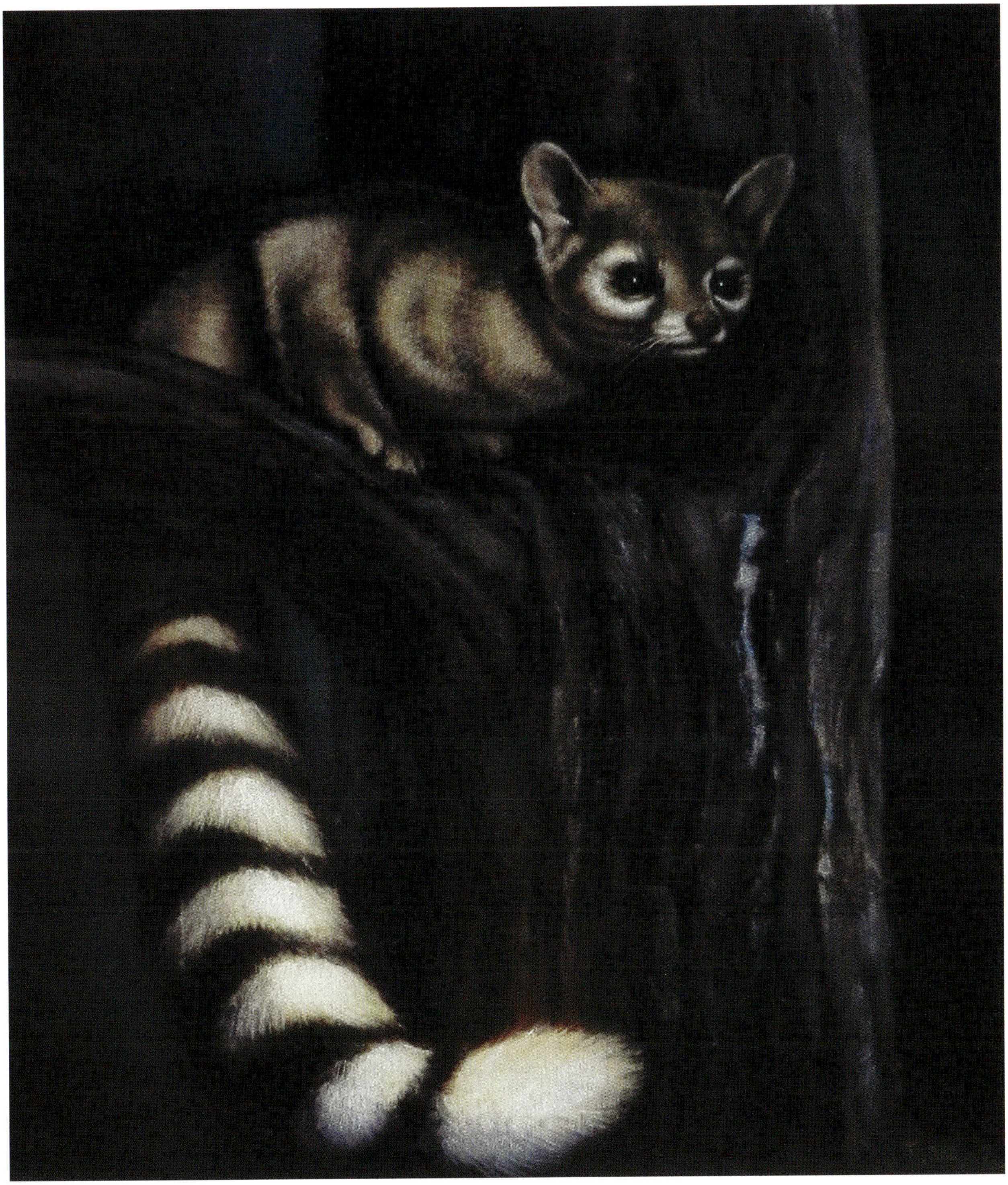

Figure 6.24 *Ringtail*, Ralph Burke Tyree. 1975, oil on velvet, 16" x 12", private collection.

BIG CATS

Bobcat

The lynx genus in the *Felidae* family of cats includes four medium-sized wild cats—the bobcat and Canada lynx of North America and the Eurasian lynx and Iberian lynx from Europe. The **bobcat** (*Lynx rufus*), sometimes called the red lynx, is the smallest of the four and gets its name because of its stubby or bobbed tail. About twice the size of a standard domestic cat, it has a grey-brown coat banded with dark marks and black-tipped ears (figure 6.25). It roams the woodlands of North America (figure 6.26), where this patterning allows it to camouflage. Most active at twilight, the bobcat preys on whatever is local to its region, including rabbits, rodents, reptiles, and birds. When prey is scarce, this fierce hunter is known to take down much larger game like small deer. This widespread wild cat features prominently in many stories of the native peoples of the Americas.

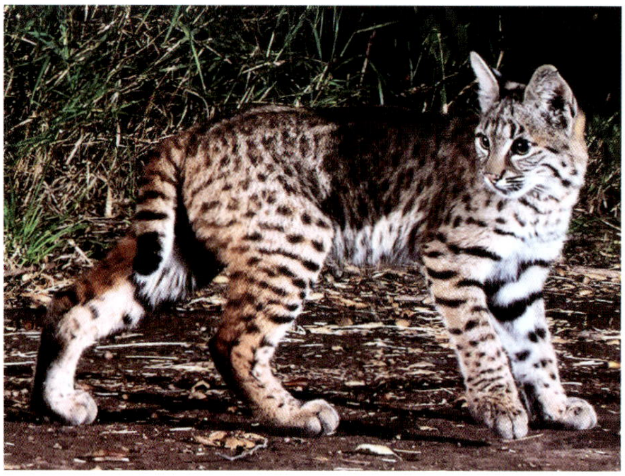

Figure 6.25 Bobcat. Jim Bremner, 2021, courtesy of the photographer, www.jfbphotos.com.

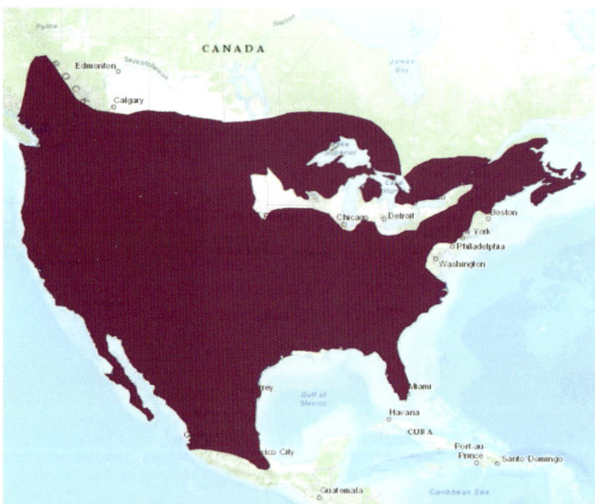

Figure 6.26 Bobcat distribution. BhagyaMani, 2019.

With few natural predators, the bobcat is threatened mainly by human activities. Habitat loss and fragmentation are notable concerns, both causing decreased genetic diversity. As humans have encroached on the bobcat's habitat, it has adapted and now frequents areas where development and nature meet. Hunters have long hunted this cat for its fur and sport, and this accounts for half of all bobcat deaths in some regions. Still, although certain states have classified it as endangered, the IUCN does not consider it at risk.

Figure 6.27 displays Tyree's painting of a young bobcat cub climbing over a boulder against an inky black sky. While the bobcat would be camouflaged in its natural habitat, this composition allows its stunning striped coat to be the center of attention. Painted in profile, the cat's stub tail makes it easy to identify. Tyree completed this painting in 1976, as hunting was on the rise due to the high value of the bobcat's fur. He chose to show a young bobcat, with all attention on its coat, bringing attention to this issue and the species' future.

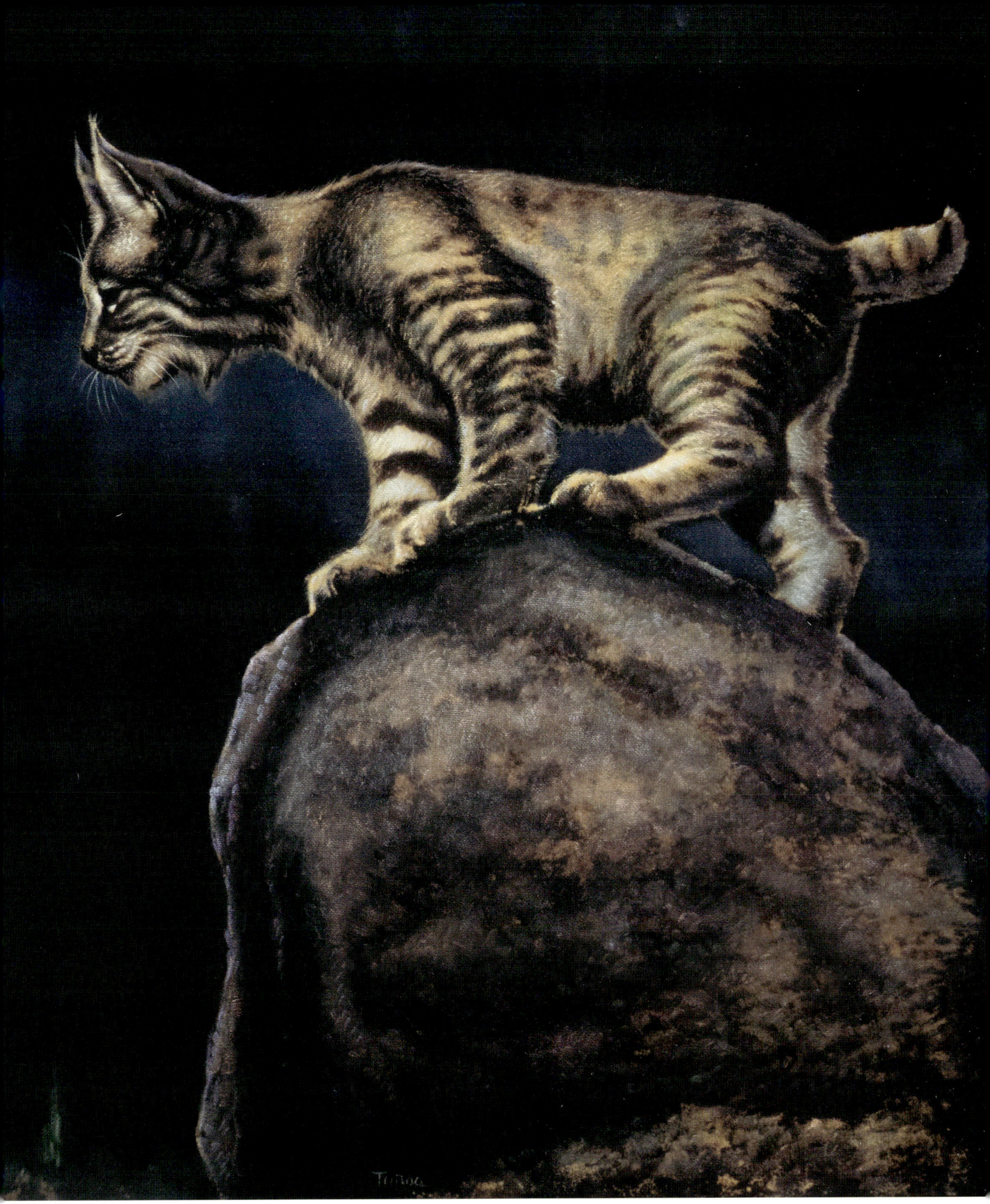

Figure 6.27 *Young Bobcat*, Ralph Burke Tyree. 1976, oil on velvet, 24" x 20", author's collection.

Canada Lynx

Another species of lynx in the *Felidae* family, the **Canada lynx** (*Lynx canadensis*), resembles a large bobcat (figure 6.28). However, it is distinguishable by its longer ear tufts, broader paws, shorter, black-tipped tail, long legs, and greyer, more uniform coat. This skilled climber and swimmer ranges the dense, snowy boreal forests across Canada, Alaska, and the northern US (figure 6.30). The cat's thick fur coat protects it from the cold, while snowshoe-like paws support its weight on the snow. It creates shelters under fallen trees and rocky ledges. A nocturnal hunter, the Canada lynx will go after ducks, rodents, and young ungulates. Still, it relies so much on the snowshoe hare (figure 6.29), its favorite prey, that the two have nearly identical ranges.

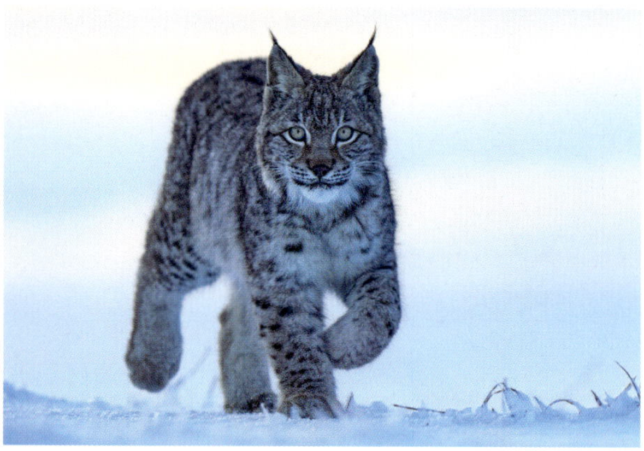

Figure 6.28 Canada Lynx. Zdeněk Macháček on Unsplash.

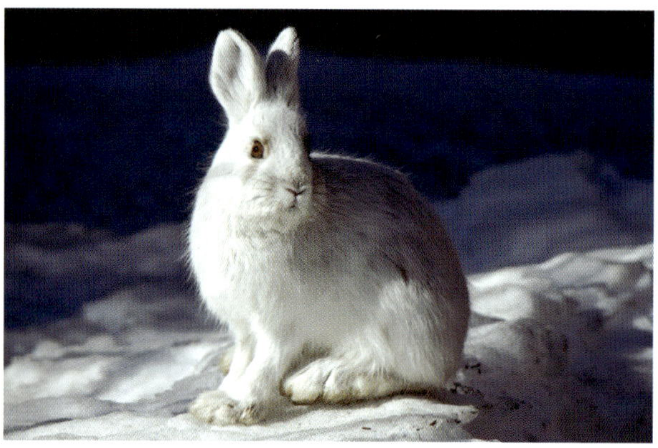

Figure 6.29 Snowshoe hare. Denali National Park and Preserve, 2011.

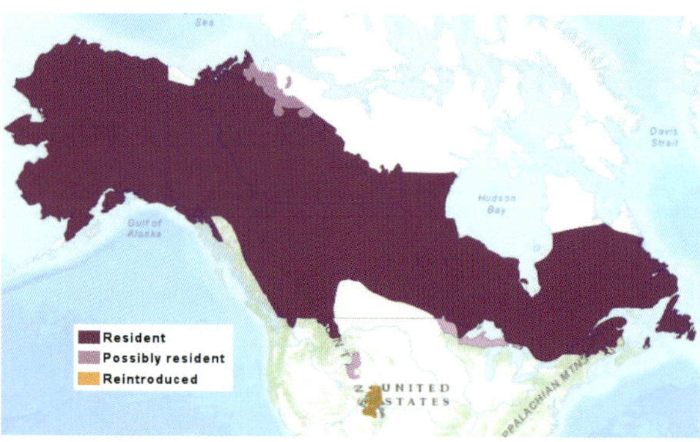

Figure 6.30 Canada lynx distribution. Bhagya Mani, 2016.

The IUCN does not consider the Canada lynx at risk due to its high population numbers, although the cat faces threats from humans, competition, and habitat loss. A long-time target of the fur trade due to its thick coat, it is still trapped for its pelt throughout Canada and Alaska. This practice is prohibited in the southern half of its range, but unfortunately, this is because of the extensive habitat loss the species has experienced in this region. In eastern Canada, competition with the coyote is the cat's primary threat, as the two compete for resources.

Tyree's head and shoulder portrait of the Canada lynx highlights the cat's incredible facial markings (figure 6.31). The zoomed-in framing brings every detail into focus, down to each whisker. The artist depicts the nocturnal cat against a nearly solid, dark background, with its shining green eyes locked on the observer. The Canada lynx's beautiful, thick fur appears soft to the touch. The lighter coat and longer ear tufts that distinguish it from the bobcat are easily discernable. With a proud, knowing look, this lynx has a regal, even mystical air. Again, Tyree completed this work when demand for the Canada lynx's fur was high, threatening this magnificent cat's existence.

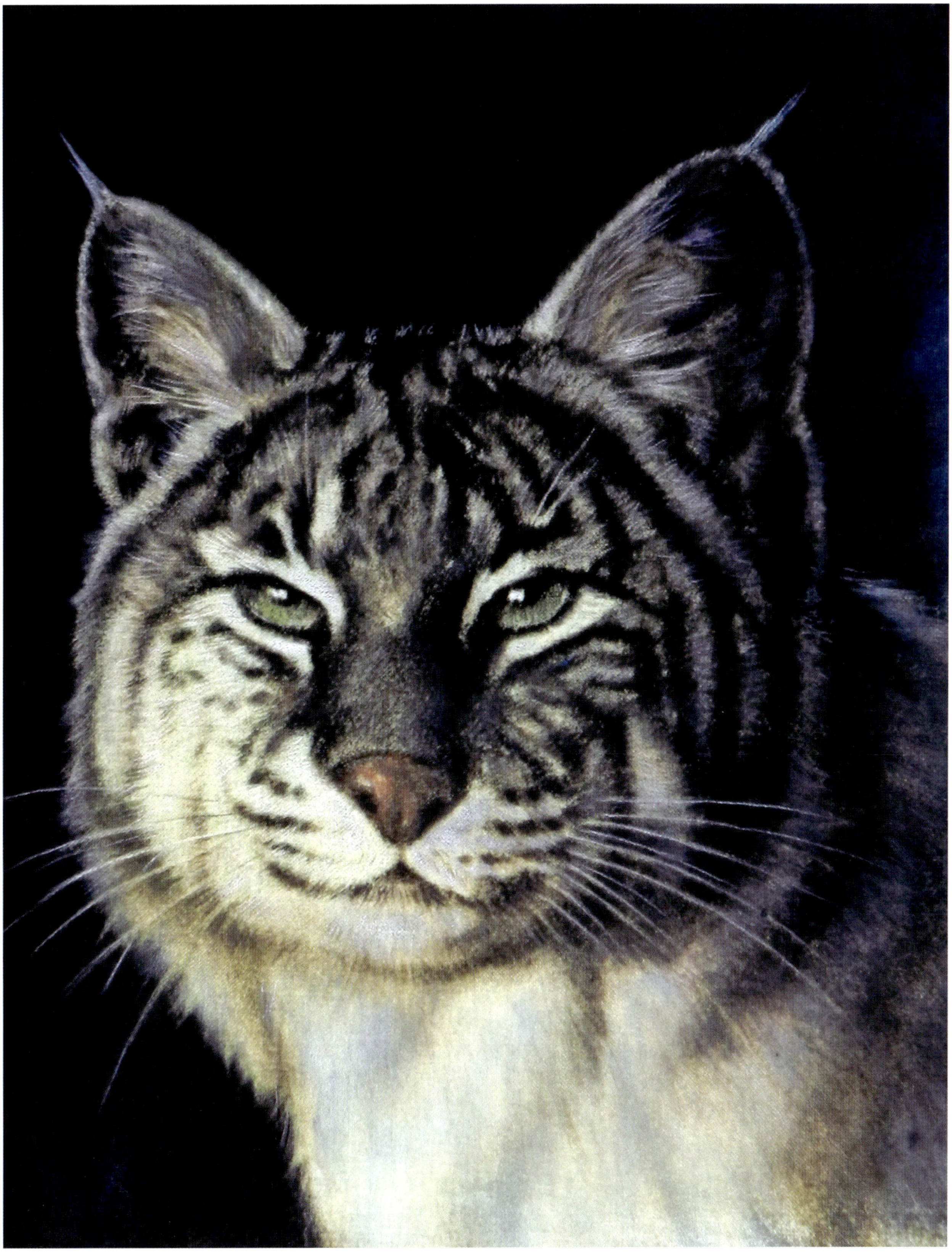

Figure 6.31 *Canada Lynx*, Ralph Burke Tyree. 1978, oil on velvet, 16" x 12", private collection.

The Cougar

The **cougar** (*Puma concolor*) goes by many names across its wide range that covers the Americas from the Yukon to the southern Andes (figure 6.33). These names include puma, mountain lion, catamount, and panther. With a coat that varies widely in color, this large member of the *Felidae* family is slender and agile, weighing up to 150 pounds (figure 6.32). Solitary, except when caring for cubs, it searches for prey at all times of day, covering an extensive range of around 60 miles. This keystone species and opportunistic hunter prefers large mammals such as deer, elk, goats, and capybaras but often eats smaller animals like rodents, rabbits, and birds.

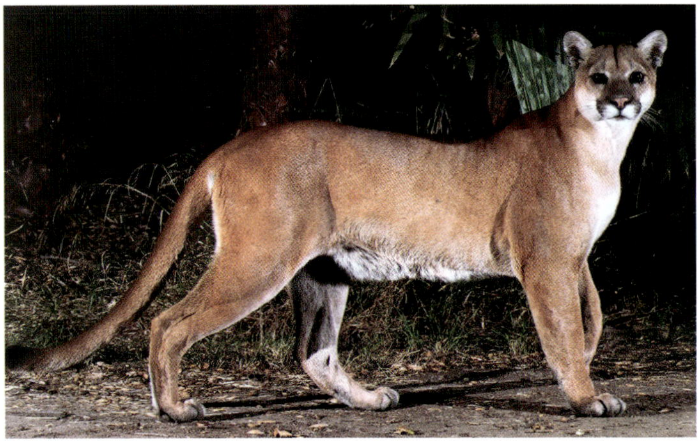

Figure 6.32 Cougar in my neighborhood. Jim Bremner, 2020, courtesy of the photographer, www.jfbphotos.com.

Occasionally, it is known to go after livestock or domestic pets. Like the bobcat, another common cat across the Americas, the cougar features heavily in the stories of the native peoples.

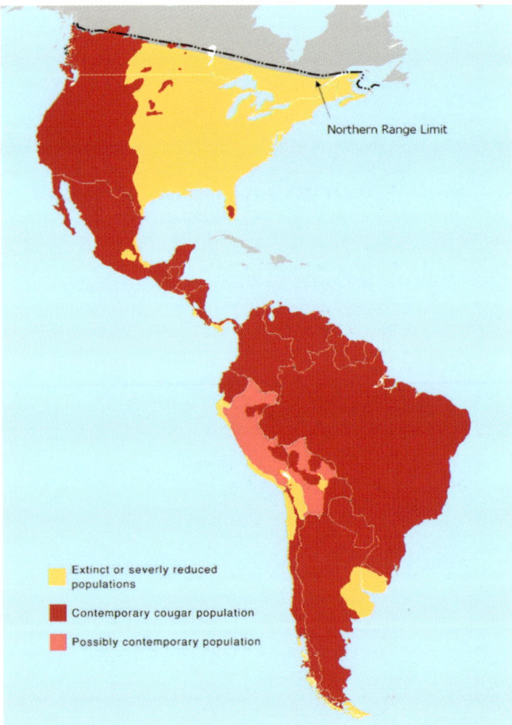

Figure 6.33 Cougar distribution. Kokosdieb, 2011.

Cougar populations drastically dropped with the arrival of the Europeans, especially in the eastern half of the US, as hunting and habitat loss escalated over the following centuries. Today, the cougar is widely protected, and its hunting is restricted or regulated. Additionally, conflict with humans continues to increase as humans encroach more and more on cougar habitats. These conflicts usually lead to the cat's death. Despite this, the cougar remains numerous, and the IUCN does not consider it threatened.

Tyree created two cougar paintings, one on each of his favorite surfaces, board and velvet (figures 6.34 and 6.35). Together, they are mirror images of each other, perfectly illustrating the depth of shadow and realistically soft textures that Tyree could achieve on velvet. Like the Canada lynx, Tyree paints the cats in impressive detail with beautiful glowing eyes. The cougars are shown from the shoulders up in a three-quarter profile, with their mouths slightly open, showing off their teeth. This positioning indicates that the cougar is smelling, likely on the search for prey. What is your preference—board or velvet?

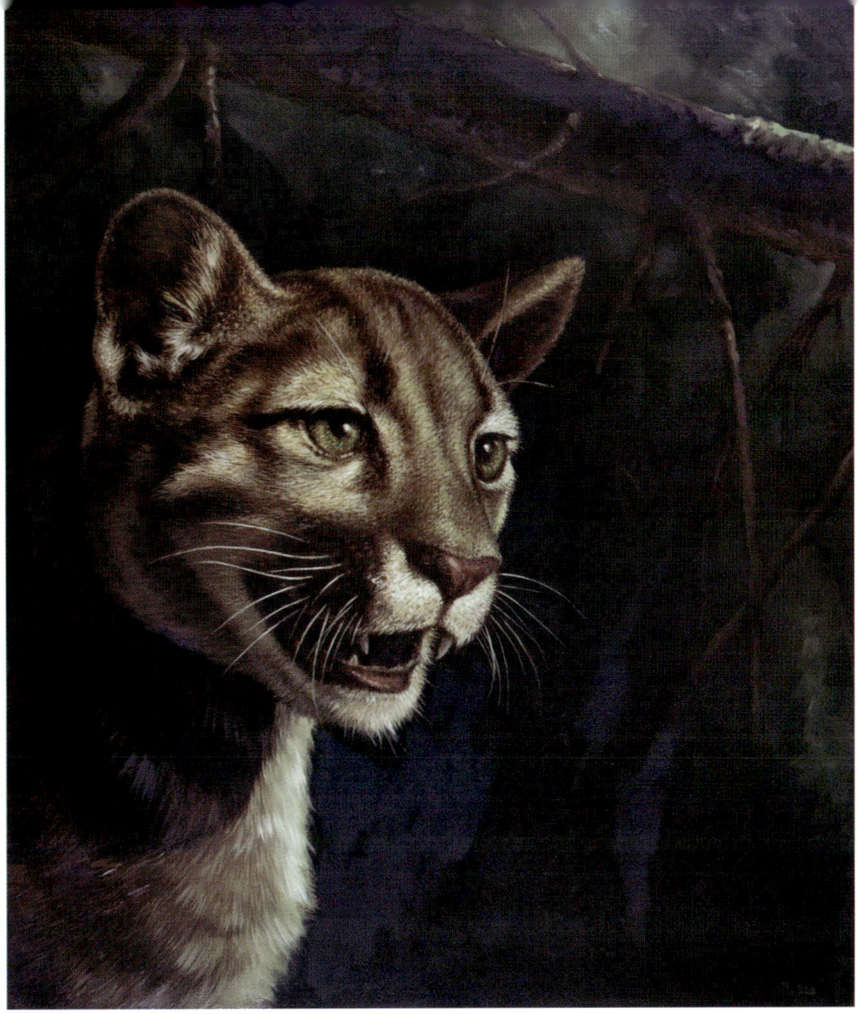

Figure 6.34
Puma (Cougar), Ralph Burke Tyree. 1976, oil on board, 24" x 20", author's collection.

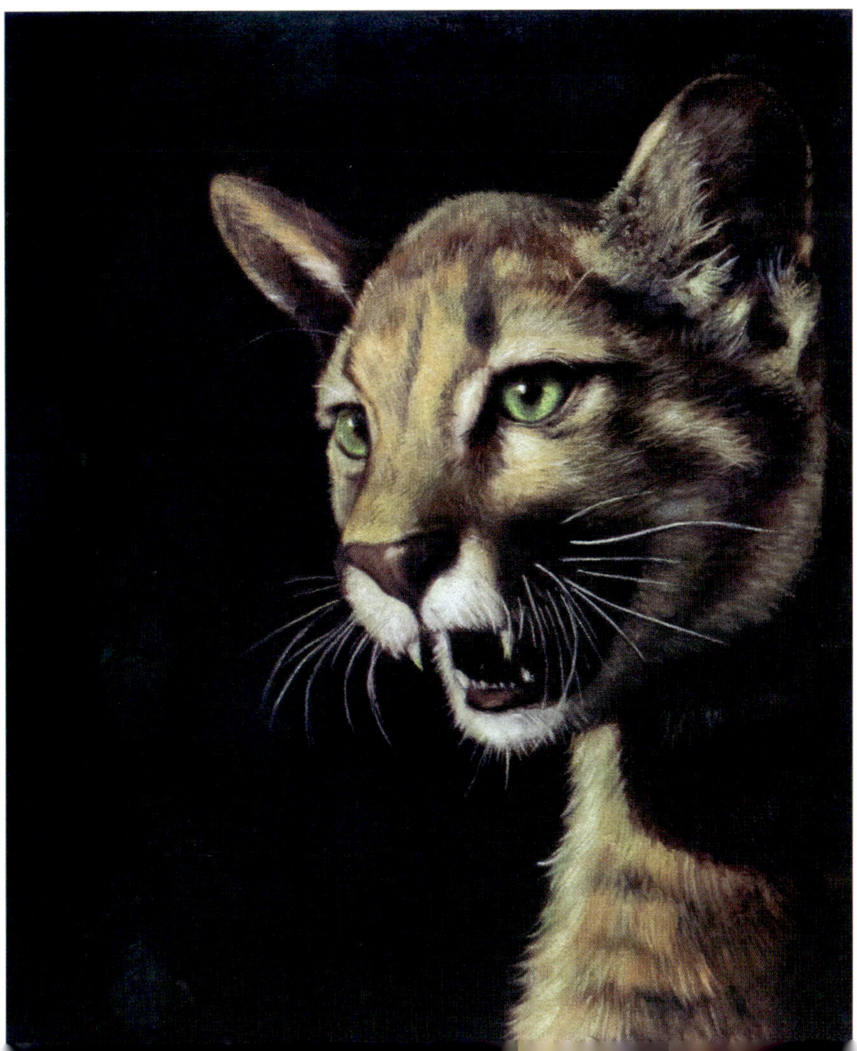

Figure 6.35
Puma (Cougar), Ralph Burke Tyree. 1973, oil on velvet, 20" x 16" author's collection.

Grizzly Bear, Apex Predator of North America

The **grizzly bear** (*Ursus arctos horribilis*) from the *Ursidae* family is a subspecies of the brown bear (figure 6.36). Native to western North America, it is also called the North American brown bear. This typically brown-colored bear is identifiable by adult individuals' muscular shoulder hump. Still, it is closely related to, and sometimes confused with, the Kodiak bear of Alaska's Kodiak islands, the peninsular grizzly of Alaska's southern coasts, and the California grizzly. This massive grizzly bear is over six feet tall, with females weighing around 350 pounds and males around 600 pounds. Although it is a member of the *Carnivora* order, it is actually an omnivore with a diet that includes fruits, salmon, trout, deer, and moose. The bear positions itself in the river rapids, where it catches salmon as they head upstream to spawn. Aggressive when surprised or attacked, the massive and fearsome grizzly makes the black bear seem shy in comparison. Alaskan locals often encounter both bears and use the aphorism, "If black, attack; if brown, get down," to stay safe.

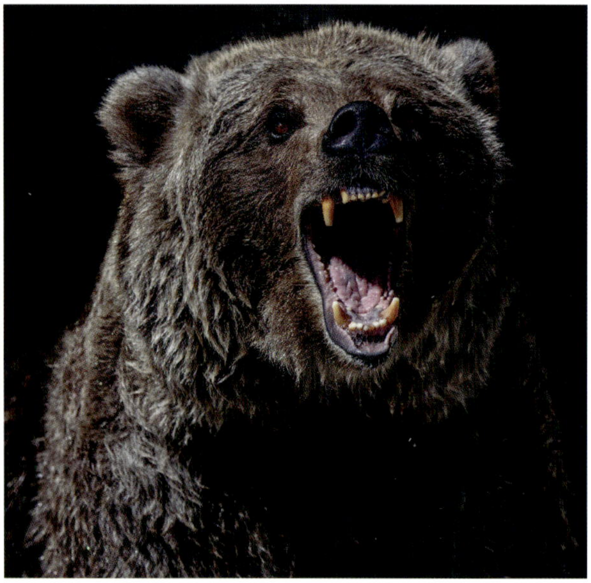

Figure 6.36 *Unhappy Grizzly.* Mana5280 on Unsplash.

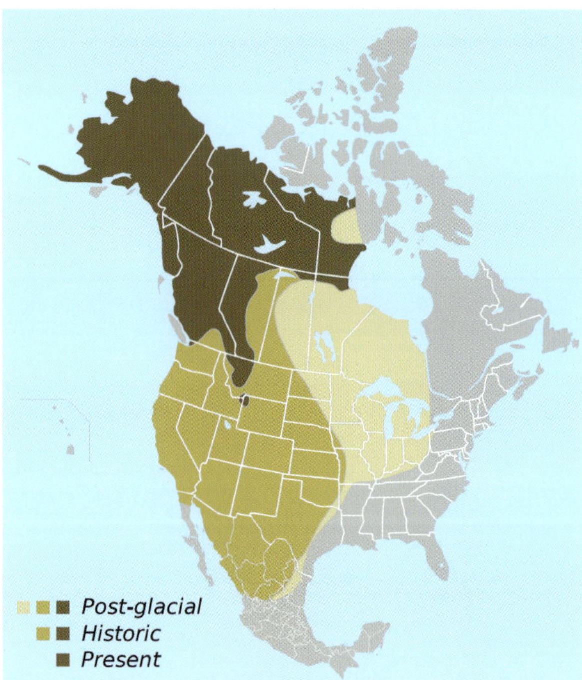

Figure 6.37 The grizzly bear's shrinking distribution since the post-glacial period. Cephas, 2009.

Once roaming much of Canada, the US, and northern Mexico, the grizzly bear's range and population have dramatically declined (figure 6.37). By the early twentieth century, aggressive hunting and human expansion had reduced its population to less than 2% of its historical level. Today, it is found primarily in Alaska, where 30,000 of the 55,000 total grizzly bears reside. In California, where the grizzly was once so abundant that officials chose it for the state flag (figure 6.38), it is now only found in zoos and on the flag. Sadly, this bear is entirely extinct from the forests of the Sierras where it once roamed.

In 1975, the grizzly bear came under the protection of the US government when it was listed as threatened under the Endangered Species Act. Since then, its population has grown notably, making it a conservation success story. Around Yellowstone Park, its numbers have increased so much

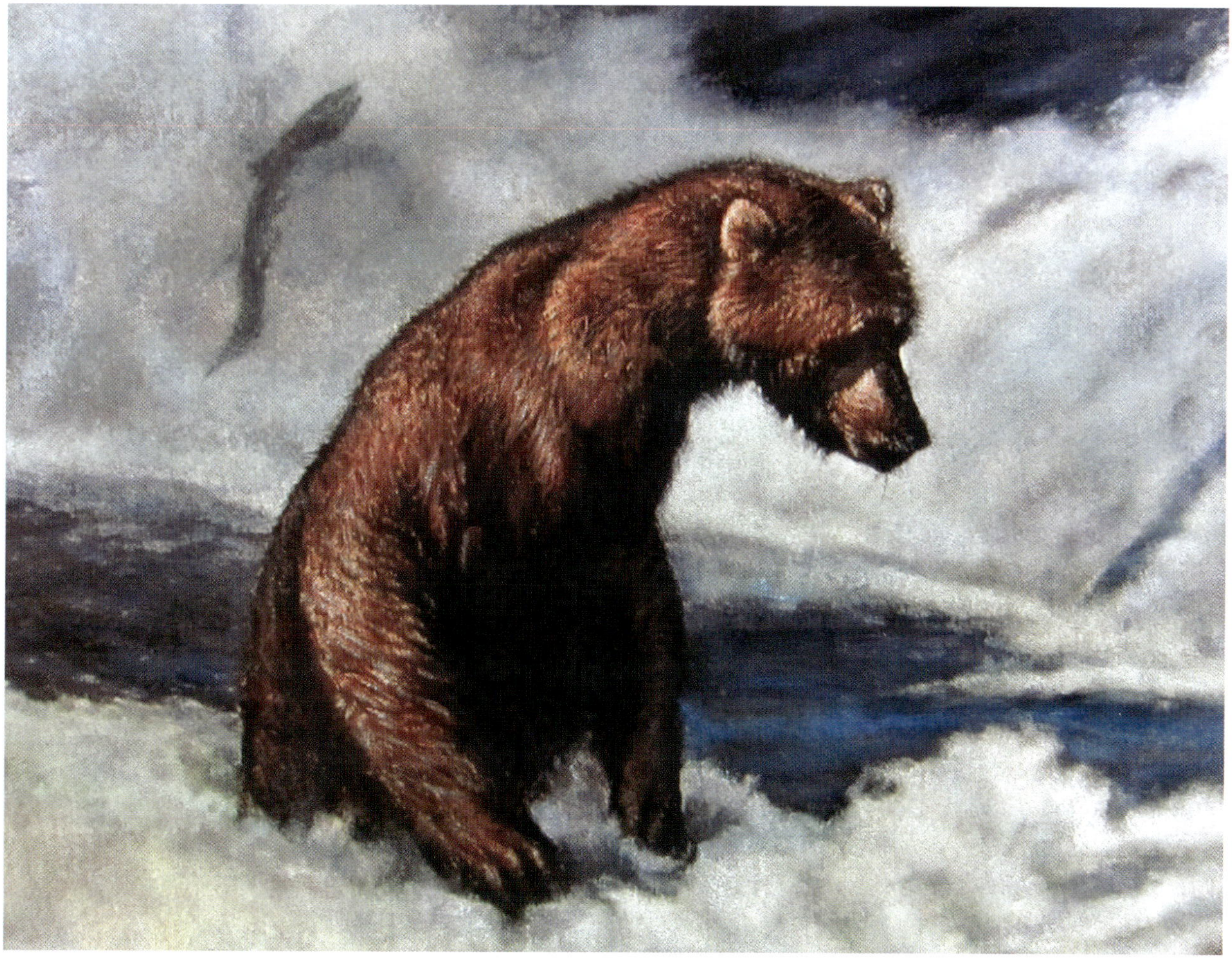

Figure 6.39 *Grizzly Bear Looking for Salmon*, Ralph Burke Tyree. Oil on velvet, 20" x 16", Tyree family.

that in 2017, the US Fish and Wildlife Service delisted this bear, removing it from the oversight of the Endangered Species Act. By 2019, the US government brought it back under federal protection based on arguments that it remains vulnerable. While concerns for the grizzly bear remain, the IUCN does not classify it as threatened.

Figure 6.39 shows Tyree's painting of a grizzly bear standing in a rushing river, hunting for salmon. The bear has staked out a spot at a small rapid and patiently waits for the salmon to leap upstream. Painted on velvet, the bear's soaking wet fur gleams, appearing heavy as the water weighs it down.

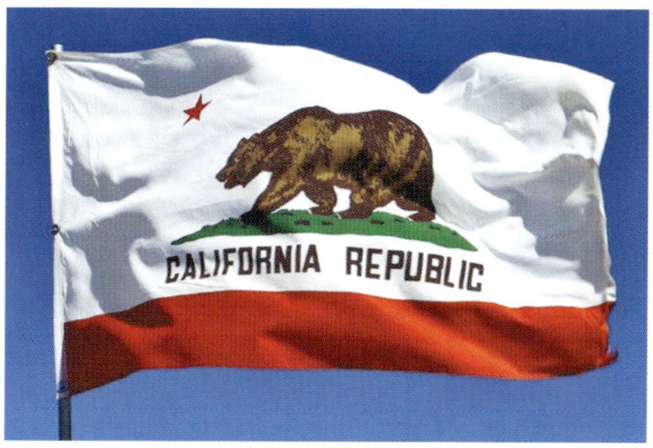

Figure 6.38 California state flag featuring a grizzly bear. Håkan Dahlström, designed in 1953 by Donald Graeme Kelley.

Moose, Food for the Grizzly

The **moose** (*Alces alces*) is the giant of the *Cervidae* family of deer as the largest and heaviest family member (figure 6.40). A male moose can reach almost seven feet tall, weigh over 1,000 pounds, and is topped with massive antlers up to 80 inches wide. Called elk in Europe and Asia, this giant inhabits boreal forests of the Northern Hemisphere, where it forages for vegetation (figure 6.41). Although the moose generally has a quiet, peaceful nature, it becomes a fierce and unpredictable beast when provoked or frightened. The moose moves between habitats, usually sticking to its preferred cold, snowy environments. It is generally solitary other than the period after birth when the young calf sticks with its mother. The most common natural predators of the moose are the grizzly and wolf, but these predators usually go after calves. The Siberian tiger, which only inhabits a small portion of the moose's range, is the adult moose's only regular predator.

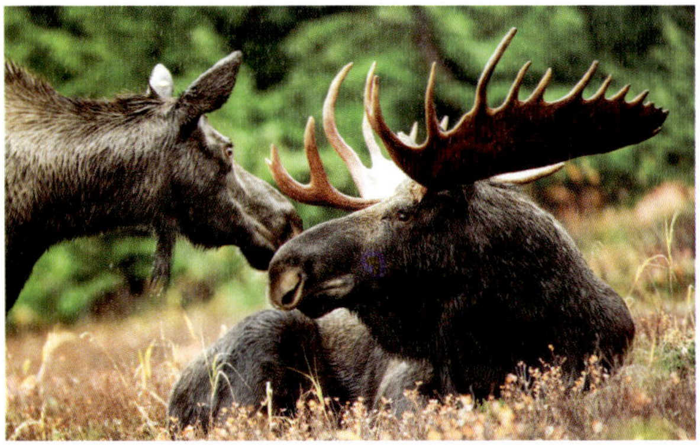

Figure 6.40 Alaskan moose pair. Ryan Hagerty, 2007, courtesy of US Fish and Wildlife Service.

While the moose is not an endangered species, it has experienced a significant shrinking of its range. Like the cougar, the arrival of Europeans to North America increased hunting and habitat loss, causing the moose population to drop drastically. By the 1870s, it was almost completely extinct from the eastern US. While Canadian conservation efforts helped its numbers bounce back over the following century, it is still not found in the eastern US. Since the 1990s, the North American population has experienced another rapid decline, likely due to increased competition and predation as other animals have encroached on its southernmost habitats.

Tyree painted a majestic moose wading through a shallow river using oil paints on velvet (figure 6.42). The peaceful giant pauses to glance in the viewer's direction, with the twisting river disappearing into the darkness of the trees in the background. While the moose's body mostly blends into the brown of the riverbank, its spectacular antlers, for which hunters target it, stand out against the deep green of the forest.

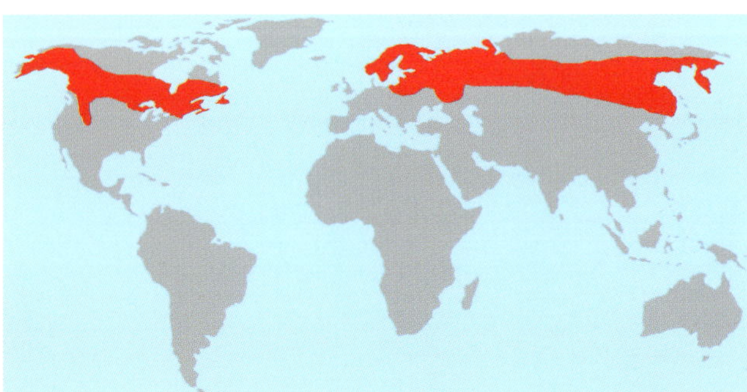

Figure 6.41 Moose distribution. Jürgen Gbruiker, 2007.

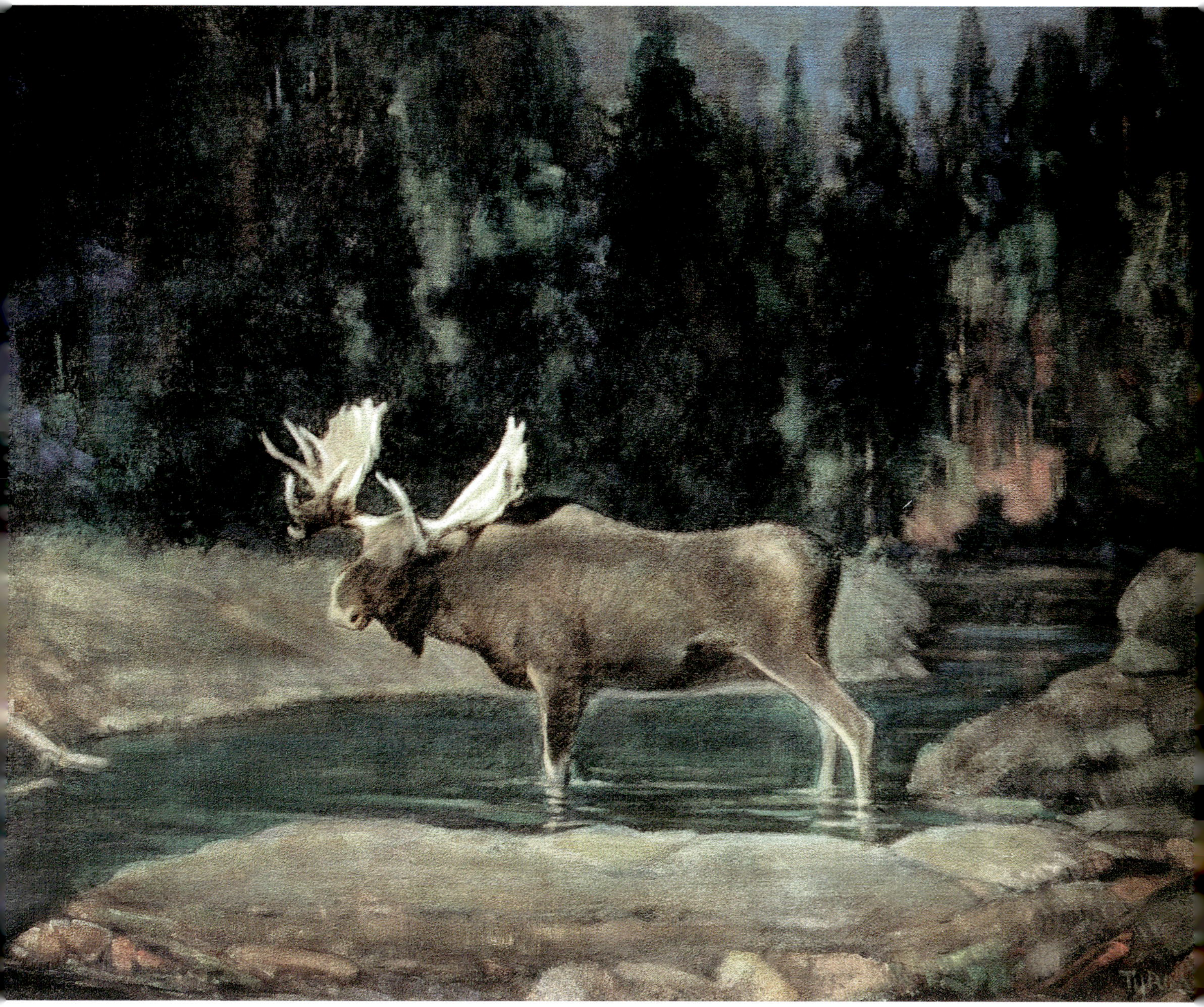

Figure 6.42 *Moose*, Ralph Burke Tyree. Oil on velvet, 16" x 20", Tyree family.

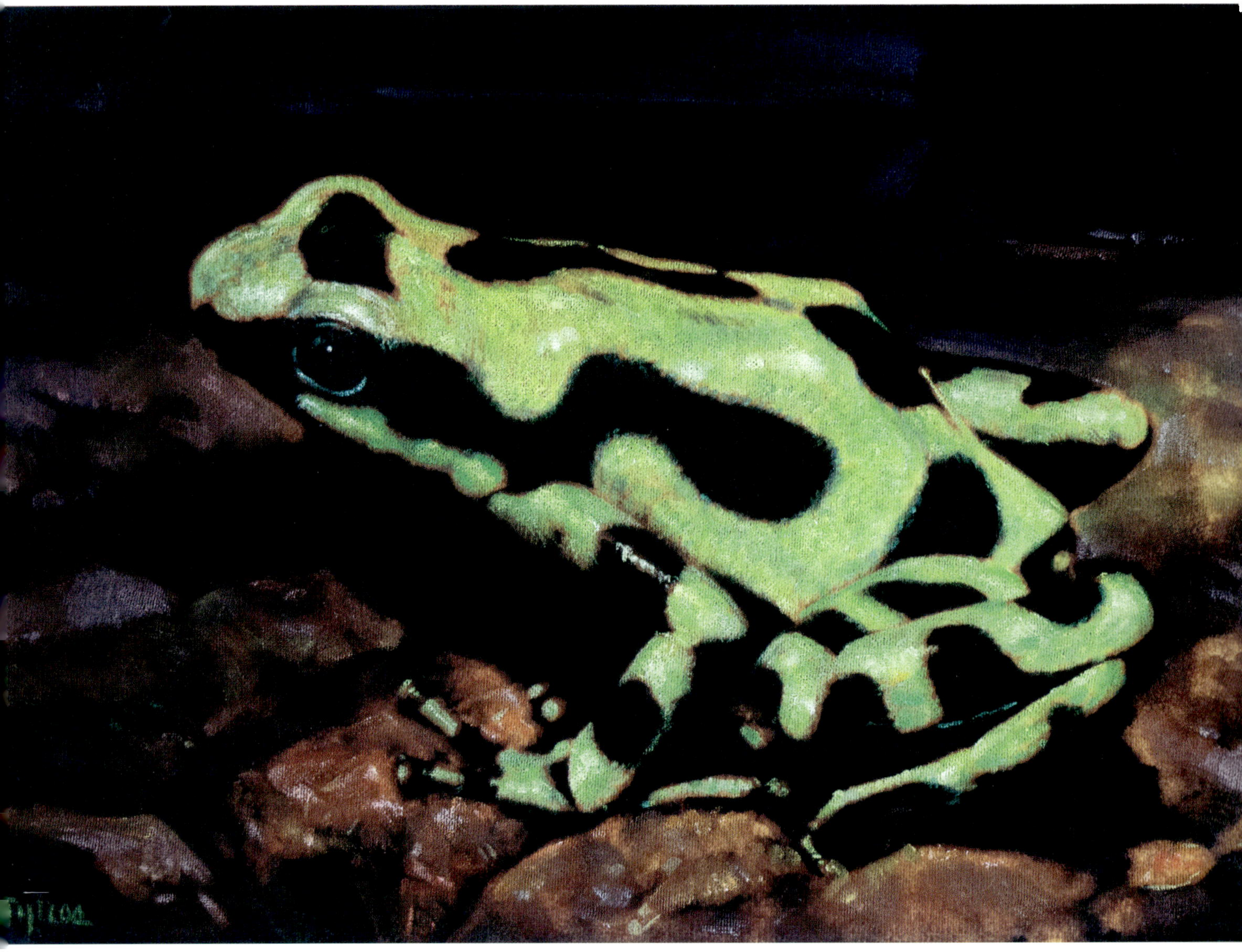

Figure 7.1 *Poison Dart Frog*, Ralph Burke Tyree. 1974, oil on velvet, 12″ x 16″, author's collection.

CHAPTER 7

To the South

"If all the beasts were gone, men would die from a great loneliness of spirit, for whatever happens to the beasts also happens to the man. All things are connected. Whatever befalls the Earth befalls the sons of the Earth."

—CHIEF SEATTLE

While living in the Hawaiian jungle on the island of Maui near Hana, Ralph Burke Tyree took painting inspiration from his surroundings. In 1971, he told reporter Ella Chun,

> I am inspired by nature and animals from Hawaii's lush jungle-like setting. . .It's quiet, and I can hear the wildlife outside—the bullfrogs bellowing or the owls hooting in the night. And there's the constant pounding of the surf.

Despite never traveling to South America, Tyree was fascinated by its natural beauty and unique cultures, and he looked to the colorful and unusual birds, frogs, and monkeys of the South American jungles for further inspiration. These animals made spectacular subjects for his oil paintings on board and velvet.

Poison Dart Frog

Poison dart frogs (*Dendrobatidae*) are a family of about 175 species of frogs native to the rainforests of Central and South America (figure 7.2). These tiny frogs are known for their vividly colored bodies (figure 7.3), which correlate to their toxicity—the greater the frogs' coloration, the more toxic the species. The bright patterning notifies predators of the frog's toxicity, hopefully deterring them and saving the frog. Some species in the family have neutral coloring that helps them blend into their environments, but these species are minimally or non-toxic. They avoid predators thanks to their camouflage appearance.

Some indigenous people of the Amazon used the secretions of poison dart frogs to create a poison called *curare*. This mixture was then used on the tips of blow darts or arrows when hunting. Ironically, the entire species is named for this practice even though only three species are toxic enough to be used for this purpose. Because of this, curare was more often made by boiling the bark of a bush rope vine to create a paste. Weapons dipped into this poisonous paste quickly paralyze prey. Today, anesthesiologists sometimes use a variation of this drug during surgery to relax the patient's muscles.

About 25% of frogs worldwide are endangered, and as such, it should not be a surprise that the IUCN categorizes many species of poison dart frogs as threatened or endangered. Frogs face many threats, including habitat loss, infectious disease, and capture for sale on the pet market. In recent years, poison dart frogs, in particular, have become popular exotic pets.

Due to the bright colors of poison dart frogs, they are a frequent subject of artists. Tyree painted the green and black poison dart frog native to Costa Rica on velvet, seen in figure 7.1. The neon frog glows in the dark composition, with the lime green standing out beautifully against the deep blackness of the velvet. Tyree expertly uses the velvet to highlight this tiny frog's vibrancy.

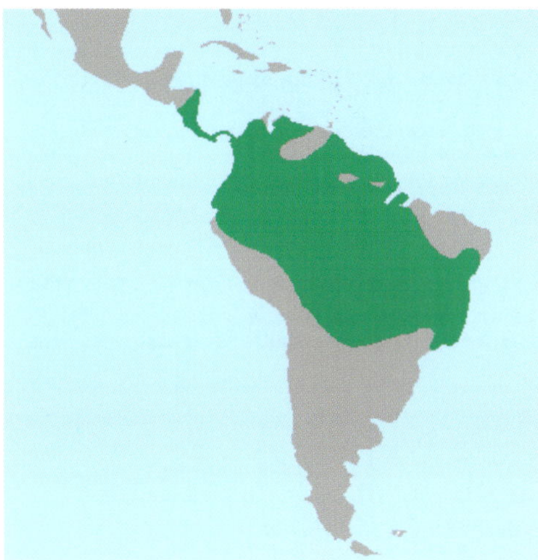

Figure 7.2 Poison dart frog distribution. Tnarg 12345, 2006.

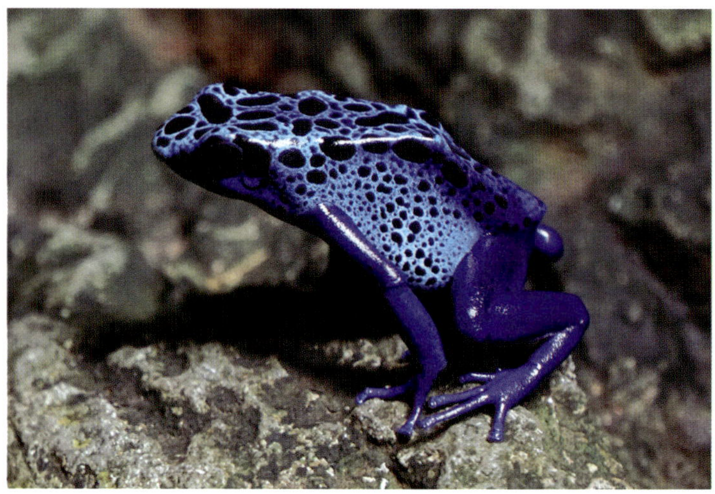

Figure 7.3 Blue poison dart frog. Zachary Spears on Unsplash.

Tree Frog

Tree frog is a general term for the many species of tiny frogs in the order *Anura*. These frogs spend their lives in trees and other moist vegetation, hiding and hunting for insects, rarely descending. With more than 800 species worldwide, some are more terrestrial, whereas others prefer a watery environment. Most of these tiny amphibians grasp foliage using suction cups on their digits. Because there are so many species of tree frogs, their risk statuses vary, but like all frogs, many species are threatened or endangered.

Tyree depicted two classic tree frogs, both gracefully resting on foliage. The frogs are both neutral shades of green, with round toe pads and bulging eyes. In figure 7.4, the frog stands against a black and red background while perched atop an upright piece of wood. This setting interestingly resembles a portrait studio. The same year, Tyree also painted figure 7.5, which shows a striped pale green frog upon deep green foliage, eyeing a nearby moth, likely about to strike.

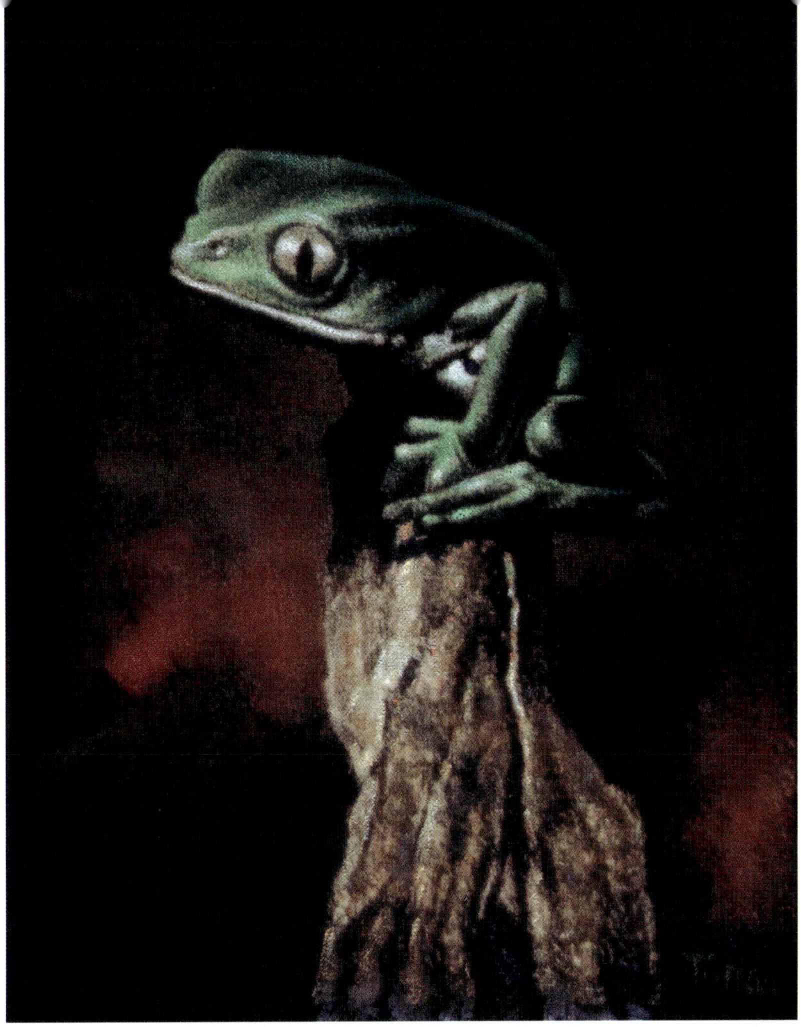

Figure 7.4 *Green Tree Frog*, Ralph Burke Tyree. 1978, oil on velvet, 9" x 7", author's collection.

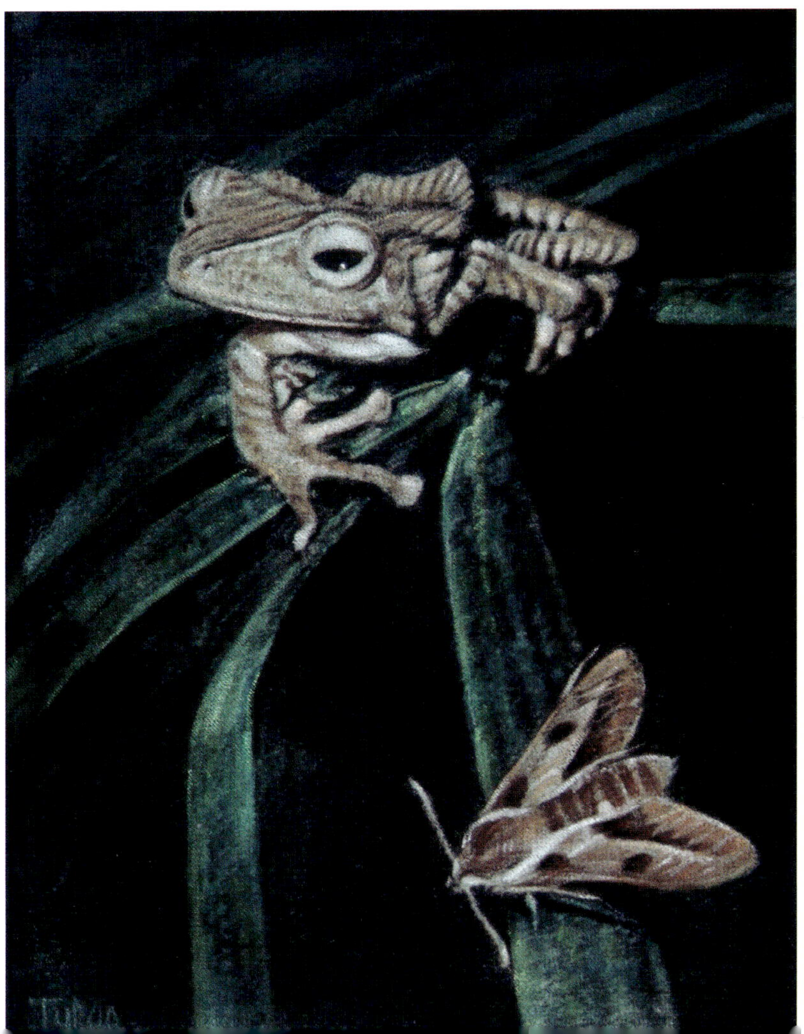

Figure 7.5 *Tree Frog and Moth*, Ralph Burke Tyree. 1978, oil on velvet, 9" x 7", author's collection.

Toucan—The Bills Have It

The **toco toucan** (*Ramphastos toco*), also called the common toucan or giant toucan, is the most famous of the over 40 species in the *Ramphastidae* family of toucans, some of which are shown in figure 7.6. This large toucan is recognizable for its graphic white, black, and red plumage and massive orange and black bill (figure 7.7). This bill helps it reach fruits and raid the nests of smaller birds, but it also provides thermoregulation by moderating blood flow. Like many raptors, it has zygodactyl feet, backward-facing first and fourth toes. This iconic toucan has famously featured in advertising for Guinness beer, Fruit Loops cereal, and cartoons like Toucan Tecs, Tuca & Bertie, and Dora the Explorer.

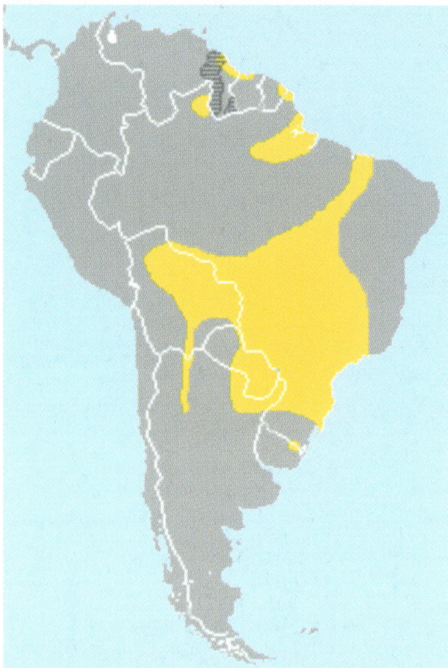

Figure 7.8 Toco toucan distribution. Webbotwill, 2014.

Native throughout South America (figure 7.8), the toco toucan is the only species of the family that doesn't prefer a forest habitat, instead favoring somewhat open habitats like woodlands and savannas. Because of this preference and its wide range, the widespread deforestation occurring in South America has not put it at risk like it has the other toucan species. In fact, not only does the IUCN not consider it threatened, but because of this preference for open environments, it may be one of the few animals that benefits from deforestation.

Tyree painted a toco toucan on board, achieving stunningly lifelike textures even when not working on velvet (figure 7.9), compared to oil on velvet in the painting featured under Einstein's epigraph in the front of the book. He shows the iconic bird in profile, with smooth, soft feathers that shine in the light. Perched on a highly realistic branch in front of a vague grey background, it inhabits its favored open environment.

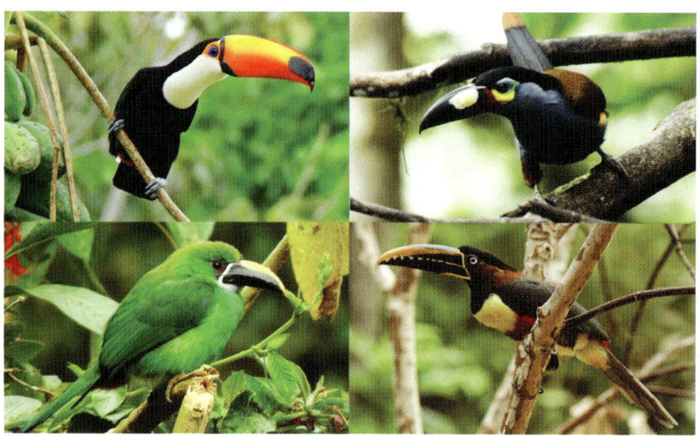

Figure 7.6 Four different toucan species (*clockwise from top left:* toco toucan, plate-billed mountain toucan, white-throated toucanet, and chestnut-eared aracari). Laura Serra Dourada, Bernard Dupont, Alejandro Bayer Tamayo, and Ben Tavener.

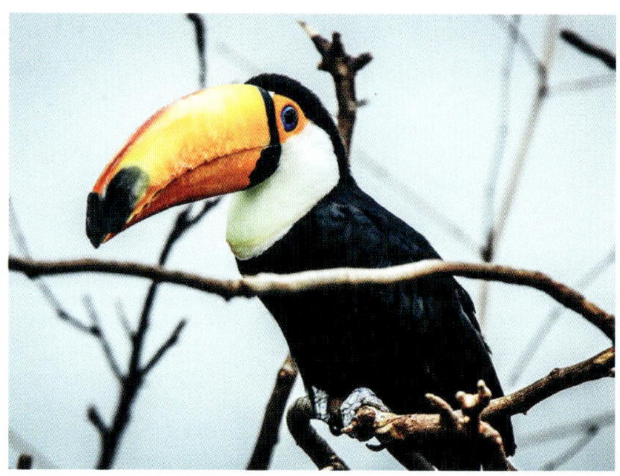

Figure 7.7 Toco Toucan. Nick Karvounis on Unsplash.

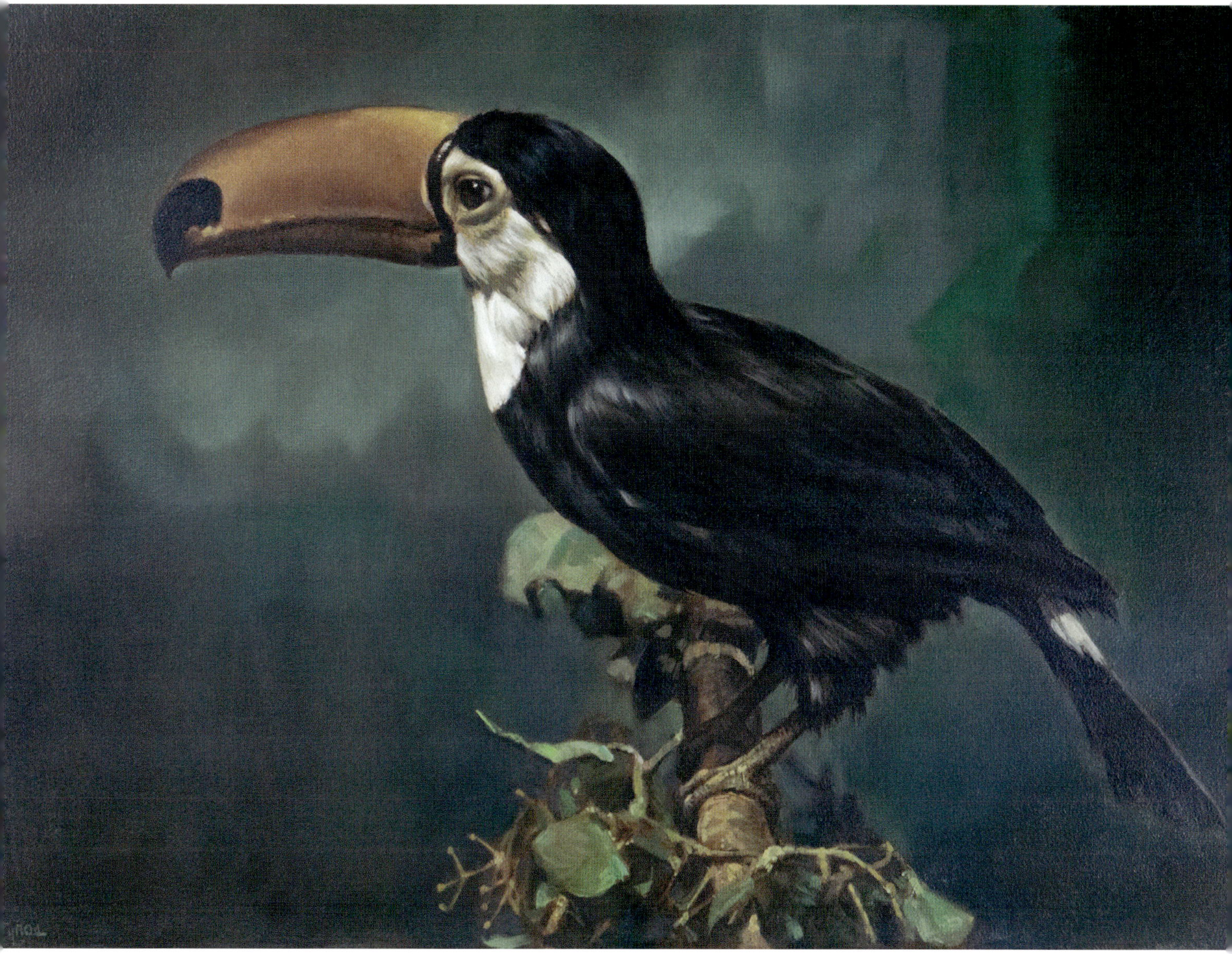

Figure 7.9 *Toco Toucan*, Ralph Burke Tyree. 1976, oil on board, 18" x 24", author's collection.

Scarlet Macaw

The **scarlet macaw** (*Ara macao*) is part of the *Psittacidae* family of parrots that includes six genera of macaws, one of which is shown in figure 7.10. Macaws are a group of parrots characterized by their long tails, large beaks, light facial area, and zygodactyl feet. The sizeable scarlet macaw is about 32 inches long and bright red, blue, and yellow. It is known to make a variety of loud squawks and squeaks. Like most macaws, it is native to the humid forests of Central America, South America, and parts of Mexico and the Caribbean (figure 7.11).

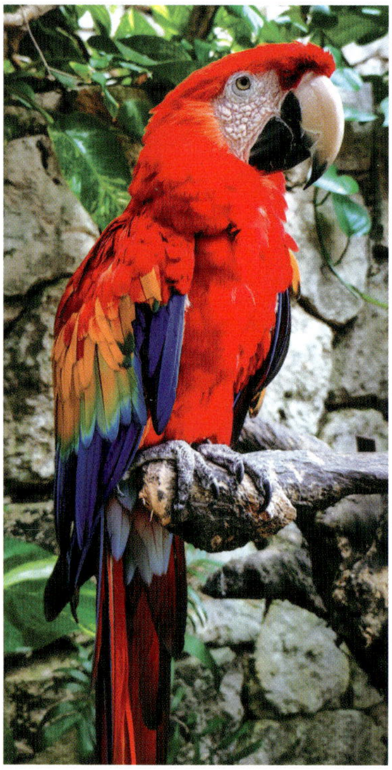

Figure 7.10 Two macaws: scarlet macaw (*right*) and blue-and-gold macaw. Yorman Tamayo on Unsplash.

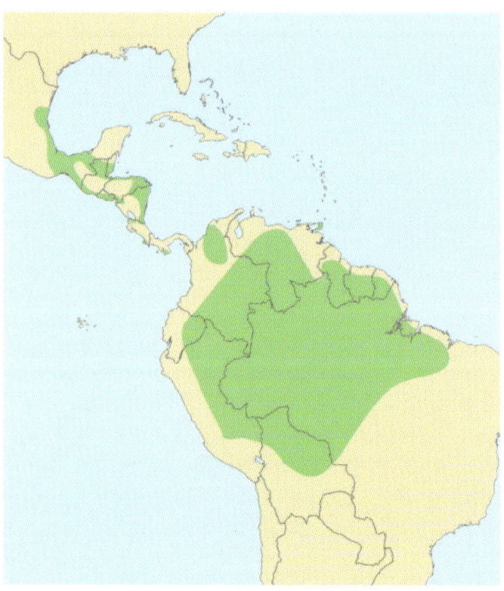

Figure 7.11 Scarlet macaw distribution. Concerto, 2011.

Humans have sought out the scarlet macaw through the ages. Some indigenous pre-Columbian civilizations of South America adorned their clothing and headdresses with this bird's brightly colored feathers. Today, this beautiful macaw is the national bird of Honduras. A popular pet, the scarlet macaw has been domesticated and traded for centuries, most famously by pirates and street vendors.

In 1975, many countries banned international commercial trade of the scarlet macaw, although domestic breeding of captive scarlet macaws continues with much success. This undercuts the need to capture wild scarlet macaws for their continued trading. Still, poaching and habitat loss due to deforestation remain significant threats to this macaw. Some agencies consider both subspecies endangered, although the ICUN does not classify the species as threatened.

Tyree depicts a pair of scarlet macaws on velvet perched next to each other on a branch in front of an abstract jungle background (figure 7.12). He seems to explore the richness of color he can achieve on this unique surface while painting these two prototypical macaws. Playing with the bird's colors, he paints one of the macaws in shades of purple and green, sitting upright. The other macaw, shown in the usual shades of red and blue, leans forward over the branch.

Figure 7.12 *Scarlet Macaw*, Ralph Burke Tyree. 1973, oil on velvet, 20″ x 16″, author's collection.

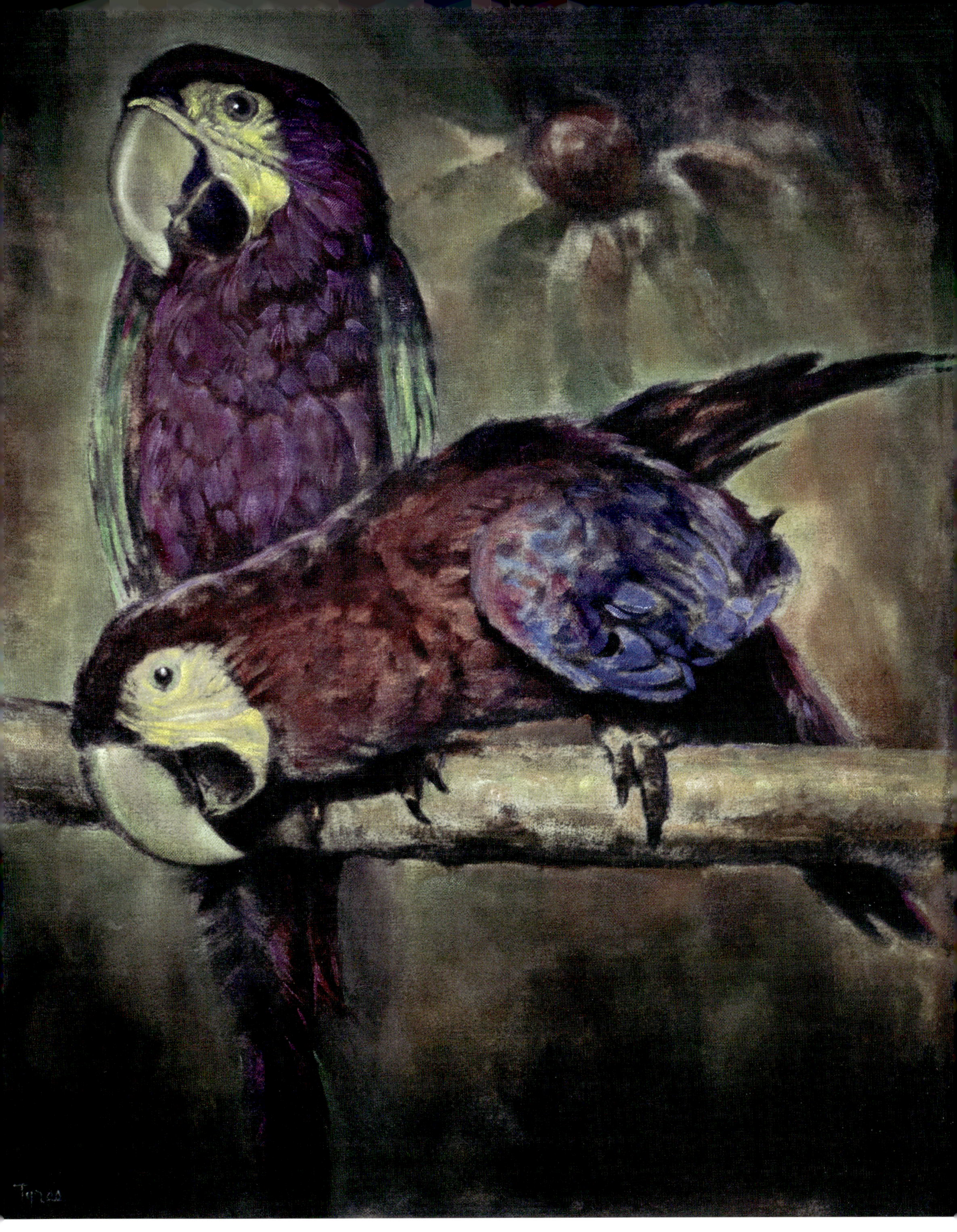

Imperial (Shag) Cormorant

The **imperial shag** (*Leucocarbo atriceps*), or blue-eyed cormorant, is a species from the *Phalacrocoracidae* family of cormorants native to the southern tip of South America (figure 7.13). As seen in figure 7.14, this seabird forms colonies that can number in the hundreds along the rocky coasts of this cold and hostile environment. This black and white cormorant has a distinguishing blue ring around its eyes and is slightly larger than its cousin, the pelagic cormorant. Like other cormorants, it primarily eats fish and other seafood, diving deep into the ocean. Shag cormorants can dive as deep as 200 feet to forage on the ocean floor. The species is relatively common despite its small range, and the IUCN does not consider it threatened.

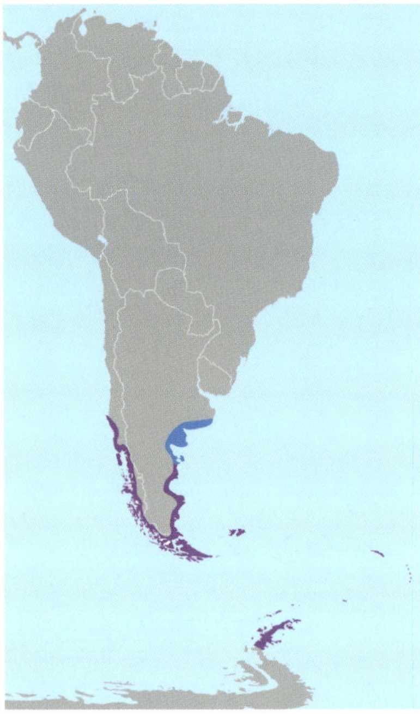

Figure 7.13 Imperial shag distribution. Cephas, 2021.
Blue=non-breeding
Purple=year-round

Tyree depicts a trio of imperial cormorants in their native rocky, coastal setting using oil paints on board (figure 7.15). Massive grey-brown boulders fill the frame as white waves crash against them in the background. In effect, the rugged coastline of this far south region becomes the primary subject of the painting. The overall grey tone in this painting further evokes the imperial shag's cold, remote, and often stormy habitat. The beautiful black and white birds, protected from the waves by the wall of rocks, stand in a line looking out over calmer waters, perhaps preparing to go diving for their next meal.

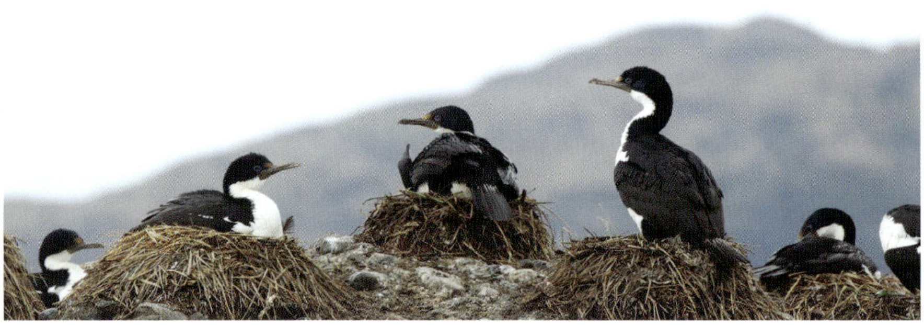

Figure 7.14 Imperial shags nesting. Eowes, Tucker Islets, Patagonia, 2018.

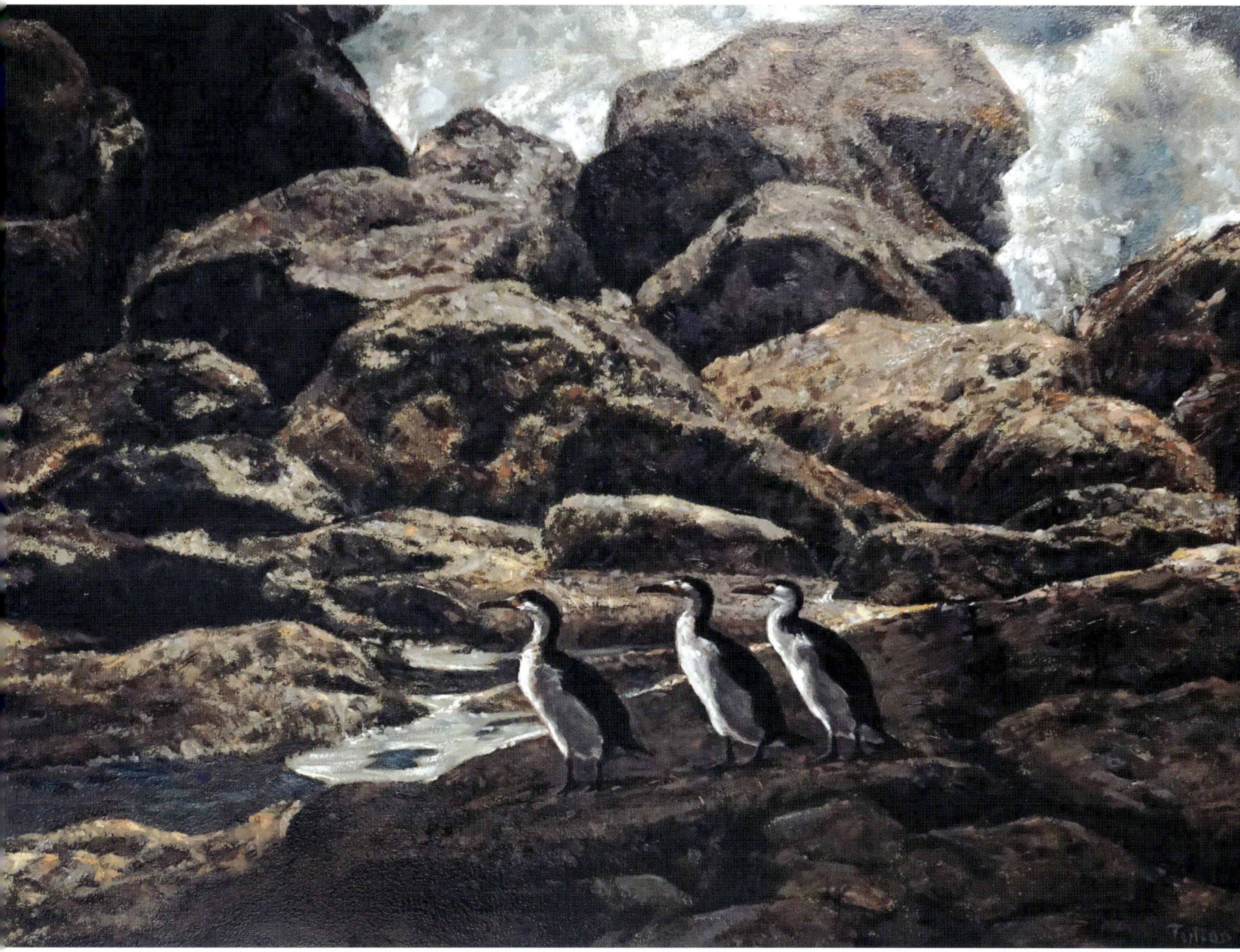

Figure 7.15 *Imperial (Shag) Cormorants*, Ralph Burke Tyree. Oil on board, 16" x 20", Tyree family.

Ornate Hawk-Eagle

The **ornate hawk-eagle** (*Spizaetus ornatus*), also called the crested hawk-eagle, is a large raptor in the *Accipitridae* family. Still, it is small for an eagle at about two feet long with a four-foot wingspan. This elegant bird of prey (figure 7.16) has boldly patterned white, brown, and reddish plumage, a long black crest, a hooked beak, and feathered legs and feet, which make it a member of the booted eagle subfamily. The ornate hawk-eagle makes its home in tall trees throughout humid forests from Mexico to South America (figure 7.17). It is a formidable predator that perch-hunts in the forests' canopies, primarily eating birds and mammals. This eagle is known for mid-morning activity when it soars through the sky and emits a series of loud whistles. Usually solitary, it only joins with others for breeding season, during which the female typically lays a single egg—one reason for this eagle's scarcity.

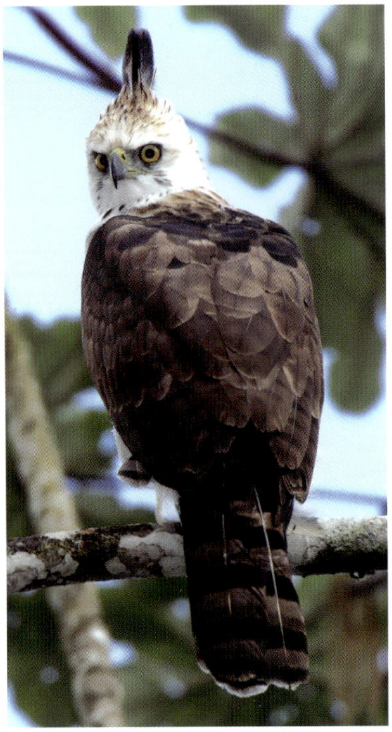

Figure 7.16 Immature ornate hawk-eagle. Mdf, Parque Nacional Darién, Panama, 2008.

Naturally rare, the ornate hawk-eagle is experiencing a notable decline in population. In 2012, the IUCN updated its categorization to near threatened. This eagle is sensitive to deforestation, requiring a habitat with at least 80% canopy cover. Due to this, it has become locally endangered or extinct in some areas, especially throughout the Amazon rainforest. Even with these warning signs, few formal conservation efforts are underway for this magnificent raptor.

Tyree painted a striking portrait of an ornate-hawk eagle on velvet (figure 7.18). Seen from the breast up, the eagle turns its head in profile, showing off its pointed crest and hooked beak. The artist paints the feathers with astounding detail and naturalism, and the back of the bird's head appears particularly downy. Its golden eye sparkles from below a heavy brow. Tyree leaves the deep black velvet of the background bare and unpainted, allowing the bird's stark plumage to stand out and hold all the focus.

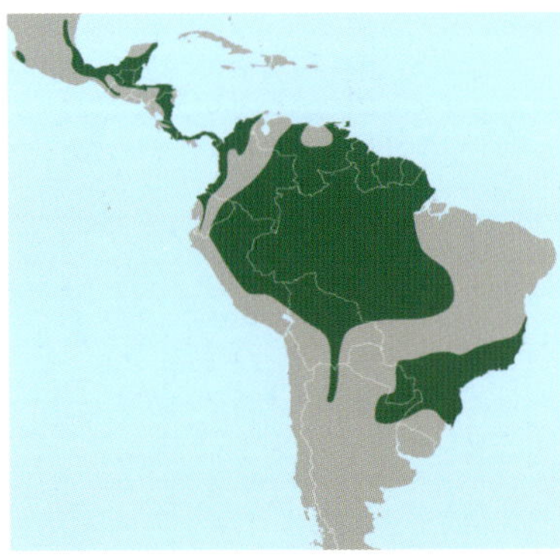

Figure 7.17 Ornate hawk-eagle distribution. Cephas, 2020.

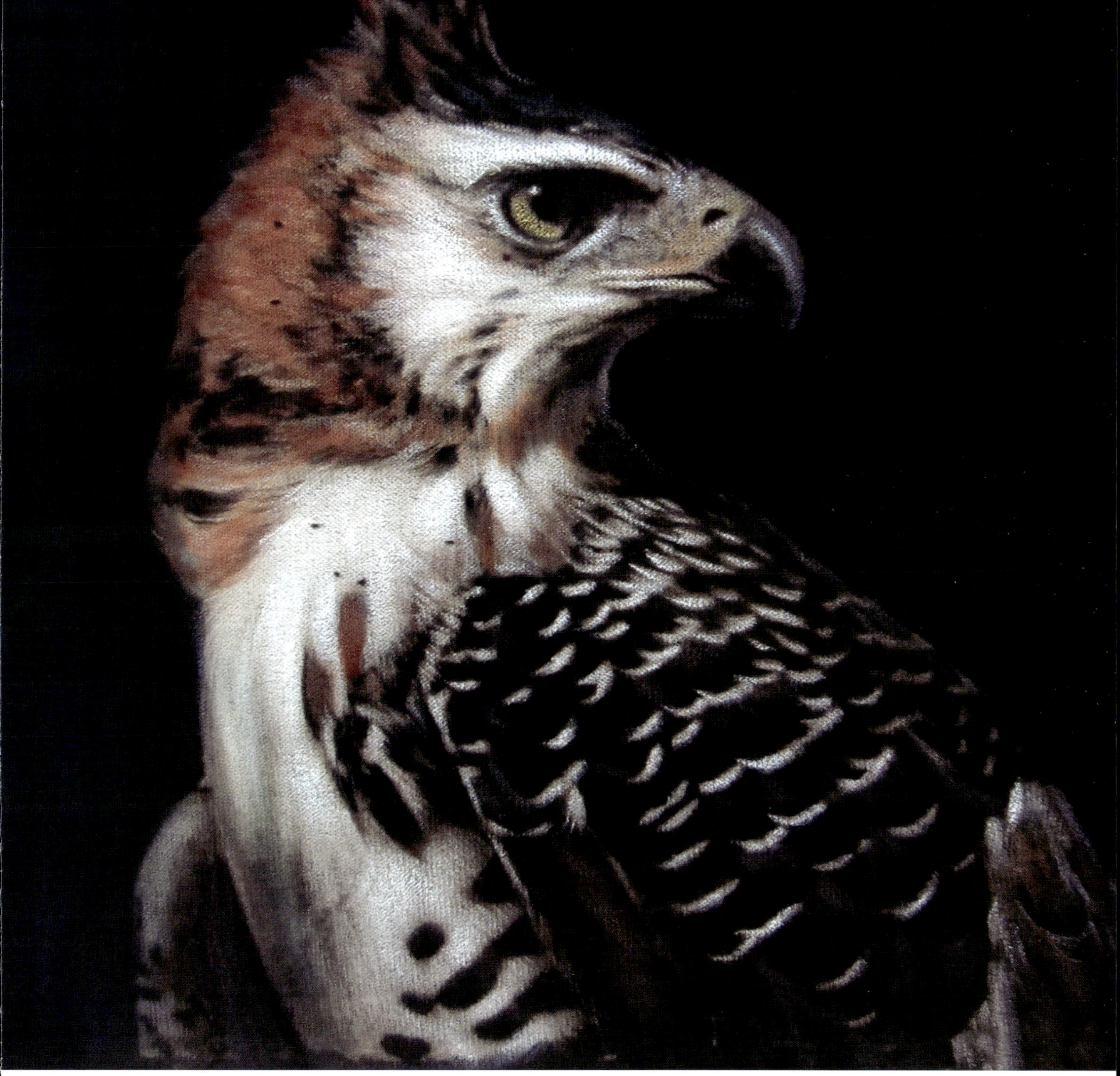

Figure 7.18 *Ornate Hawk-Eagle*, Ralph Burke Tyree. Oil on velvet, 20″ x 16″, private collection.

Cotton-Top Tamarin, a Rare Primate

The **cotton-top tamarin** (*Saguinus Oedipus*) is a monkey from the *Callitrichidae* family of small tree-dwelling primates that is found exclusively in the tropical forests of northwestern Columbia, shown in figure 7.19. At only a pound, it is one of the smallest primates. Its unique name comes from its notable long, white, Einstein-like hairdo that extends from its forehead to its shoulders (figure 7.20). This arboreal monkey sticks to the lower levels of the forest canopies. It spends its days foraging for insects, sap, fruits, and reptiles and acts as a seed disperser in its environment. The cotton-top tamarin is incredibly social with complicated and intriguing communication systems. Researchers have described 40 distinct sounds that this tamarin makes. The family groups also have clear hierarchies and practice cooperative breeding, working together to care for the dominant pair's offspring.

During the late 1960s and early 1970s, medical researchers captured thousands of cotton-top tamarins for biomedical research before this was prohibited in 1976. Unfortunately, because of its prior exportation for medical research and this deforestation, more cotton-top tamarins exist in captivity today than in the wild. This monkey has also experienced dramatic habitat loss, with deforestation taking 95% of its historical range. Today, with only 6,000 remaining in the wild, it is one of the world's scarcest primates, leading the IUCN to categorize it as critically endangered.

The cotton-top tamarin is the rarest South American animal Tyree depicted (figure 7.21) and a perfect example of the endangered species he intended to bring to the public's attention. Two of these tamarins sit on a branch with an epiphyte hanging from it. The pair inhabit a dark, dense jungle environment. He painted these distinctive primates staring straight at the viewer with almost humanoid faces as if to force us to confront how our species has impacted our tiny relative.

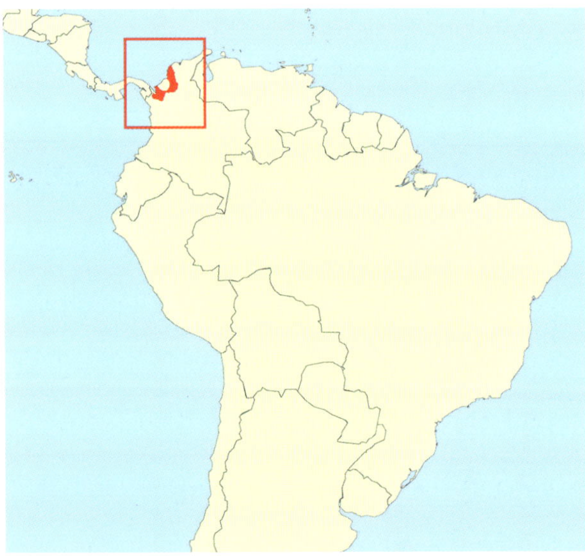

Figure 7.19 Cotton-top tamarin distribution. Oona Räisänen, 2010.

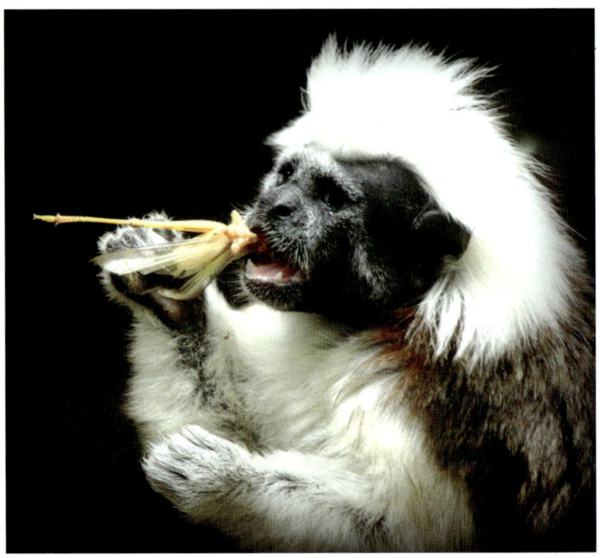

Figure 7.20 Cotton-top tamarin eating a grasshopper. Mickey Samuni-Blank, 2012.

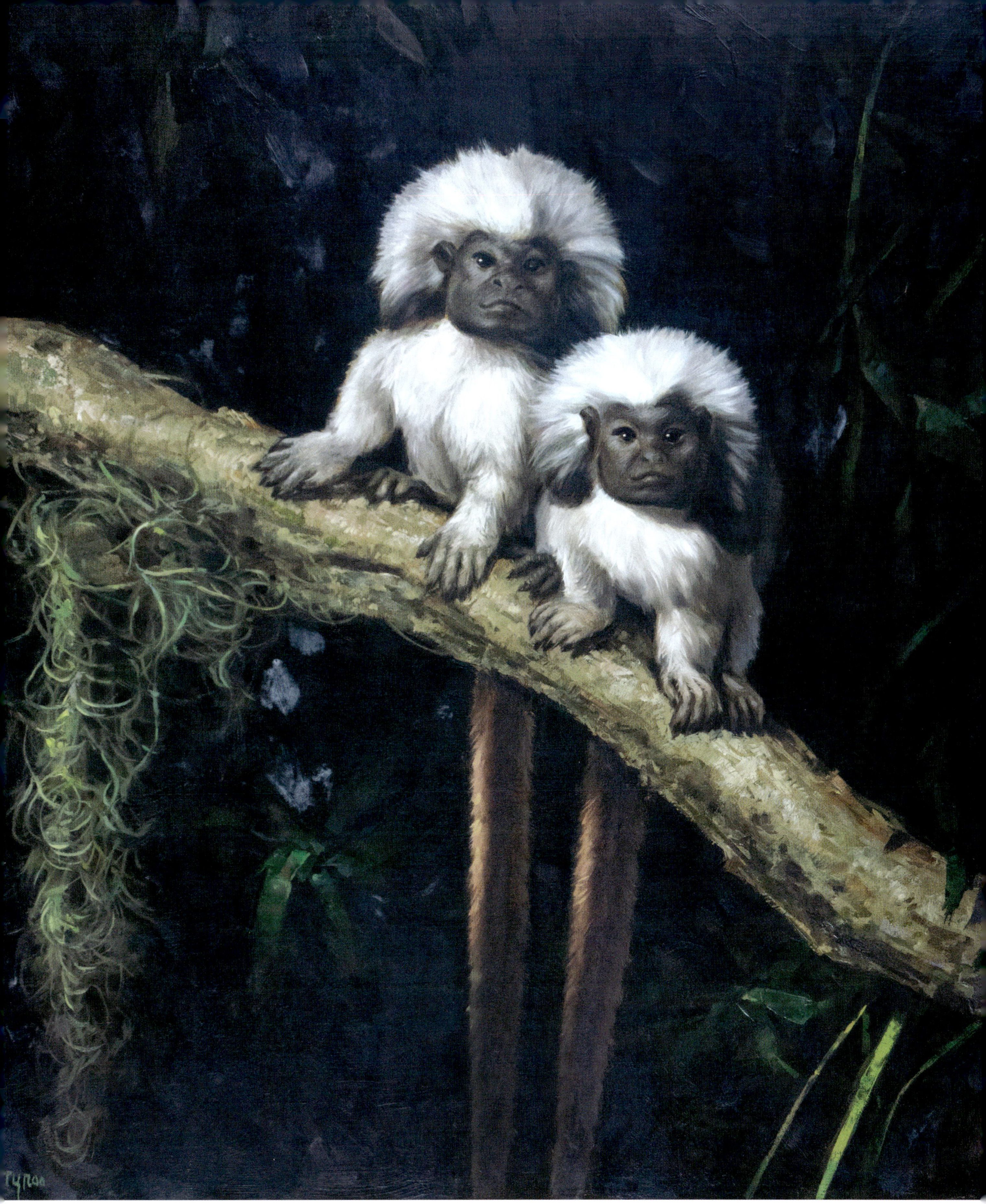

Figure 7.21 *Cotton-Top Tamarin*, Ralph Burke Tyree. Circa 1975, oil on board, 24" x 20", author's collection.

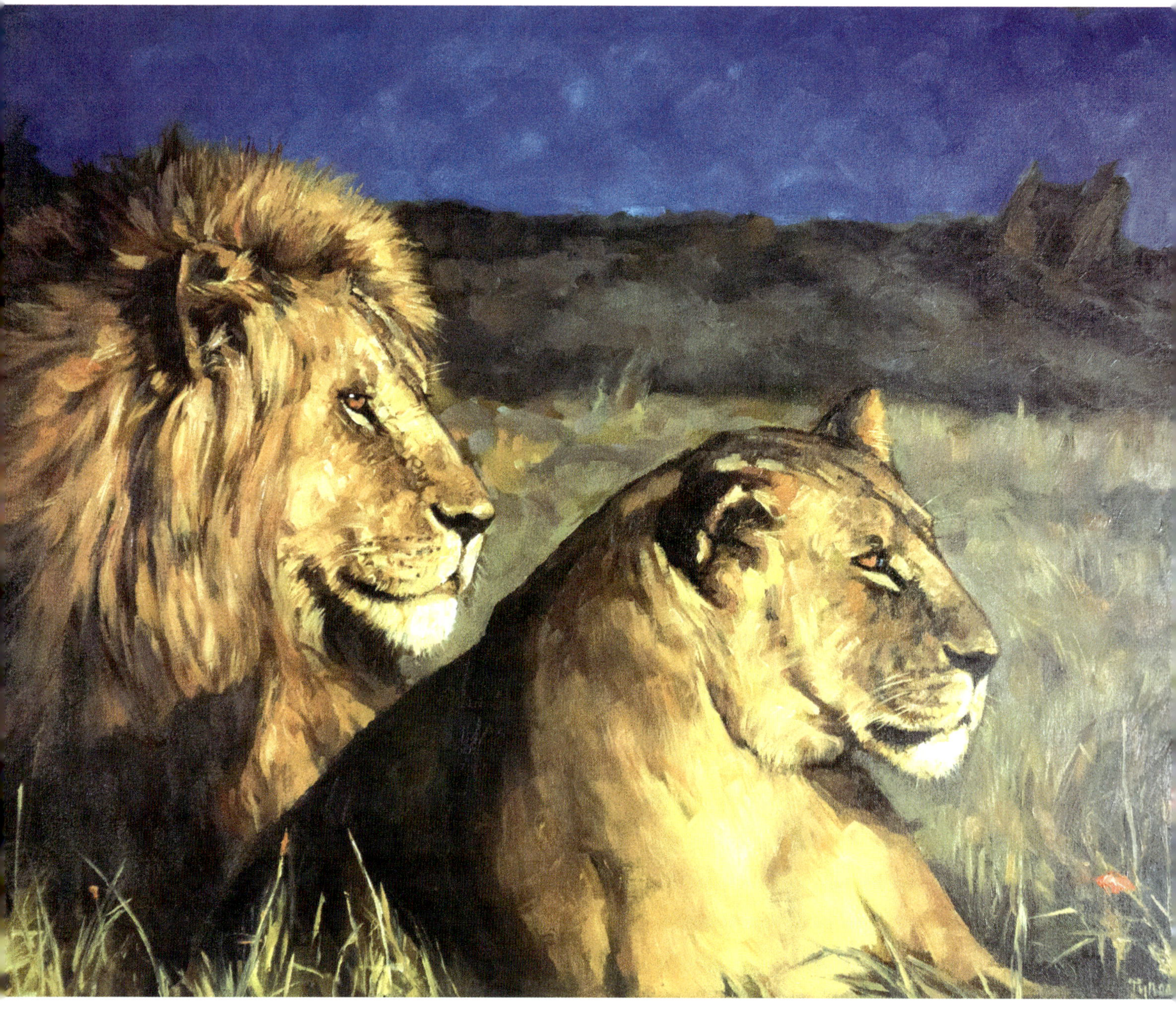

Figure 8.1 *Lion Pair*, Ralph Burke Tyree. 1972, oil on board, 24" x 30", private collection.

CHAPTER 8

Amazing Africa

"Do not try to fight a lion if you are not one yourself."

—AFRICAN PROVERB

Ralph Burke Tyree dreamed of visiting Africa and its wildlife reserves, especially during the 1970s. Although this expedition was out of his reach, he did get the next best thing, helping to construct a safari park in the US when he worked for Remont Studios drafting landscape plans for a drive-thru animal park near Winston, Oregon. Through this work, Tyree's realized his African safari dream. In 1972, this park, called Wildlife Safari, opened under the leadership of WWII army vet Frank Hart, whose love of animals equaled Tyree's. Hart wanted to save rare and endangered species while bringing these threatened species to the public's attention. The over 600-acre reserve has been hugely successful and is now home to about 600 animals across 100 species. The park's most significant achievement is its cheetah breeding program, which has produced over 200 offspring since its inception.

In 1975, the Tyree family moved to Crows Landing, California, a small-town west of Turlock. By now, he and Margo had seven children, and the two youngest sons, Jeff and Marc, accompanied them

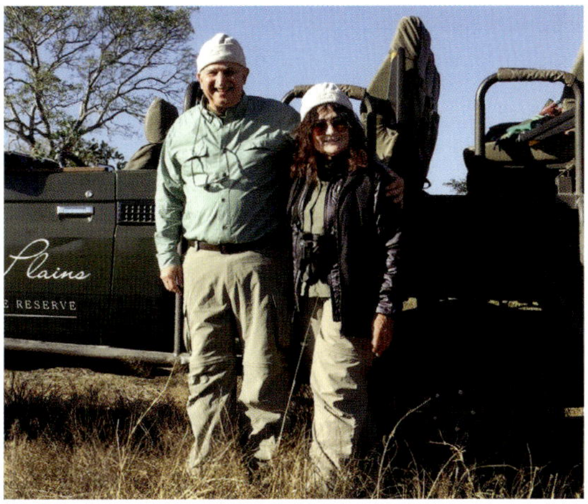

Figure 8.2 My wife and I exploring South African animal parks. 2017, author's photo.

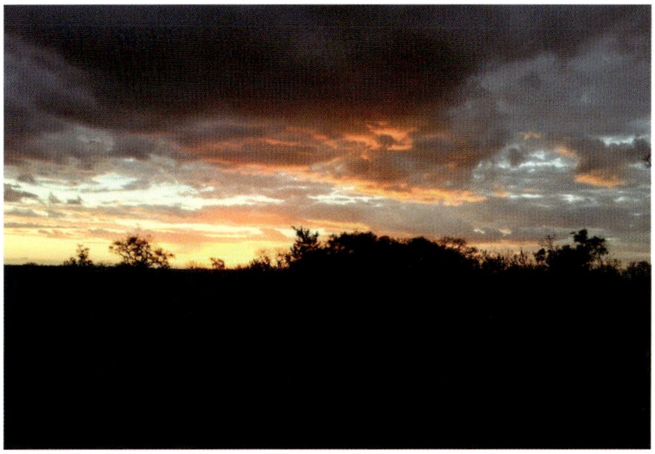

Figure 8.3 South African sunset, Kruger Park. 2017, author's photo.

on this final move. Over the next four years, Tyree mainly painted animals, many hailing from Africa. He traveled to art shows around California and the West, selling to individuals and galleries. Unfortunately, he never fulfilled his dream of visiting Africa. On January 15, 1979, while painting an African elephant at home, Tyree died unexpectedly of a heart attack at only 57 years old.

Like Tyree, I yearned to visit Africa and its wildlife reserves. In the summer of 1982, with a trip planned, a severe illness forced me to cancel this dream vacation. It wasn't until 2017, 35 years later, that my wife and I finally went on our African safari. Some of the photos in this chapter are from that sojourn (figures 8.2 and 8.3).

BURKE'S BIRDS OF AFRICA

Lilac-Breasted Roller Bird

The **lilac-breasted roller** (*Coracias caudatus*), also called the fork-tailed roller or the lilac-throated roller (figure 8.4), is a member of the *Coraciidae* family of birds. It is most well-known for its colorful appearance and aerial acrobatics. During the breeding season, the male rises high in the sky before descending in swoops and plunges while uttering shrill cries to impress his lady and frighten off competing suitors. This roller is about the size of a crow and frequents the woodlands and savannas of the southern half of Africa (figure 8.5), where it can often be spotted perched in the treetops. This perspective allows the roller to spot insects, small reptiles, and other prey on the ground before swooping down to grab their game. Common throughout its range, the IUCN does not consider the lilac-breasted roller threatened.

Impressed by this beautiful and athletic flyer, Tyree painted the lilac-breasted roller using vibrant colors (figure 8.6). The bird perches on a branch with its profile facing the viewer. By painting it against an atmospheric blue-green background, all the focus is on the roller's striking plumage, especially the red that frames its face. The velvet base allows Tyree to achieve these brilliant colors.

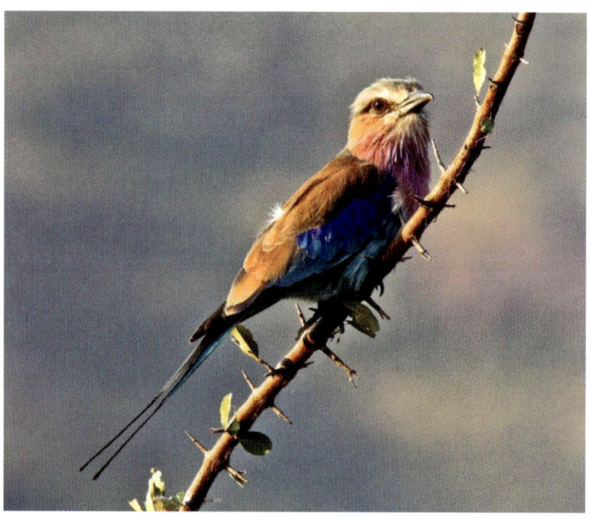

Figure 8.4 Lilac-breasted roller. Katelyn Deelefs (@bushbabyk8), South Africa, 2019, courtesy of the photographer.

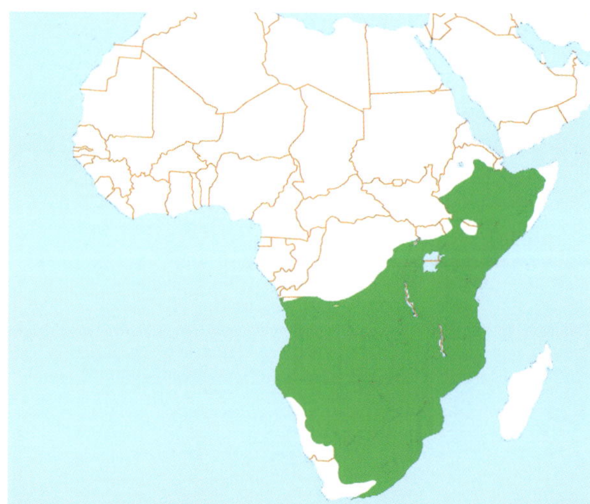

Figure 8.5 Lilac-breasted roller bird distribution. Стефанко1982, 2020.

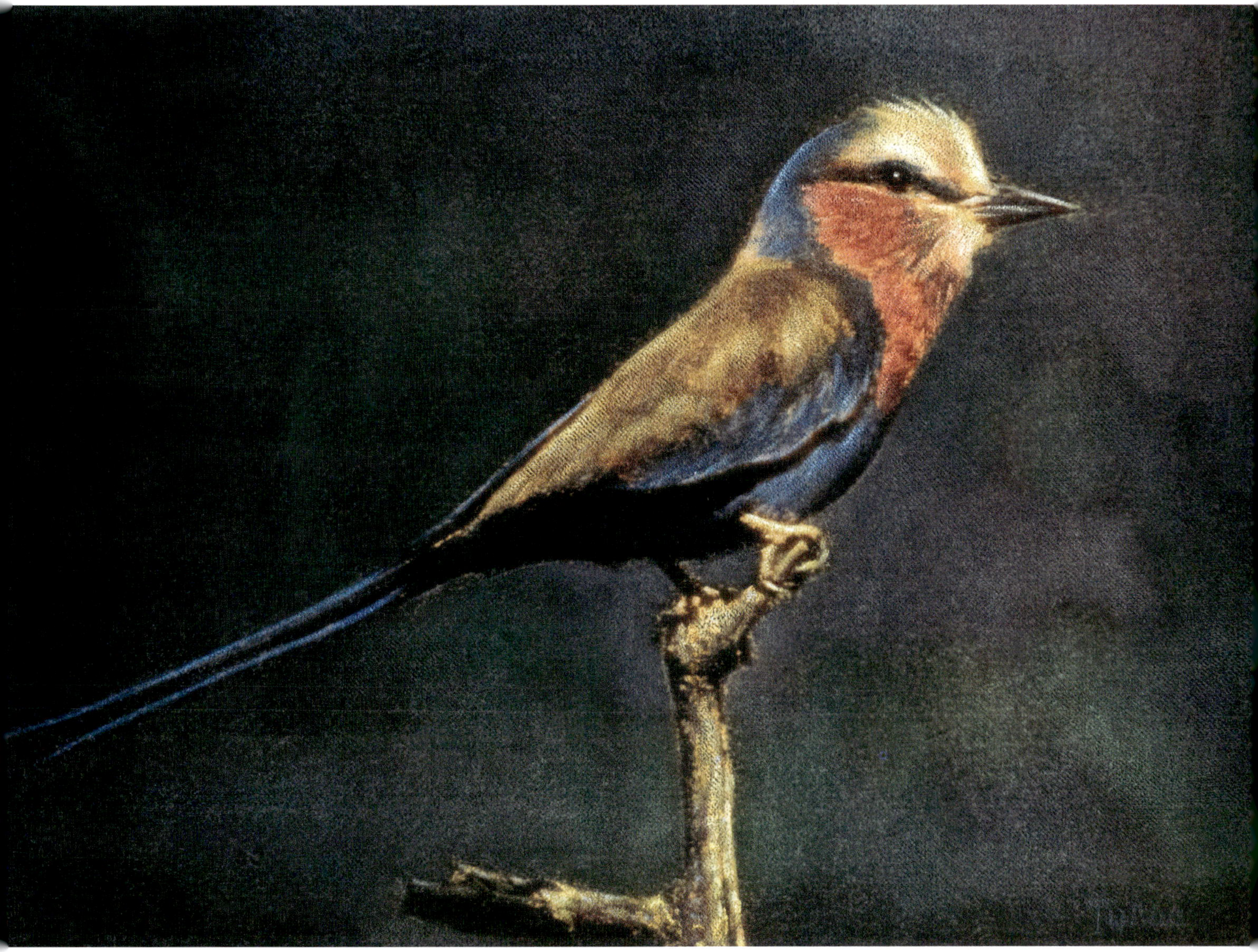

Figure 8.6 *Lilac-Breasted Roller Bird*, Ralph Burke Tyree. 1978, oil on velvet, 11" x 8", author's collection.

Pied Avocet

The **pied avocet** (*Recurvirostra avosetta*) is a striking black and white bird from the *Recurvirostridae* family of stilts and avocets. With its graphic markings, this beachcomber is rather remarkable in appearance (figure 8.7). It also has a wingspan of about 30 inches and a unique upturned bill about three inches long, which it uses to forage for crustaceans and insects in shallow waters and mud flats. This wader spends summers in Europe, Central Asia, and Russia, migrating to Africa and southern Asia for the winter (figure 8.8).

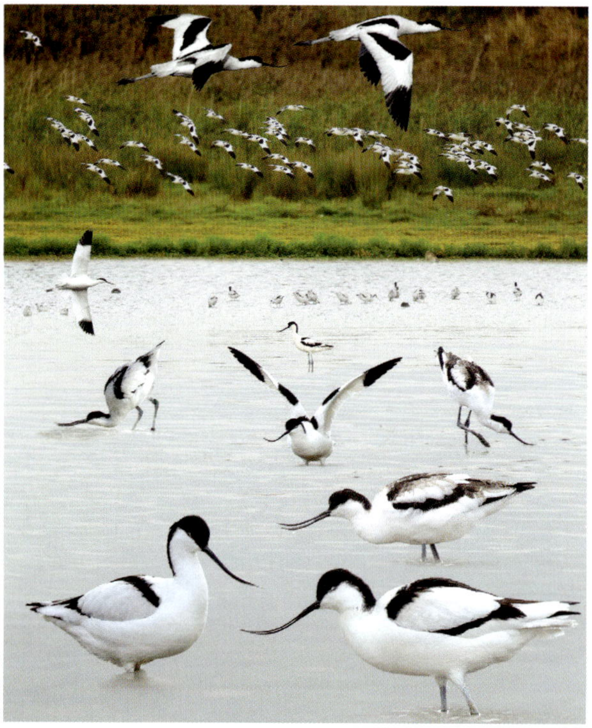

Figure 8.7 Pied avocet. Richard Crossley, 2011.

While the IUCN does not consider it at risk, habitat loss is the pied avocet's greatest threat due to the sensitive nature of its preferred wetlands. Luckily, the Agreement on the Conservation of African-Eurasian Migratory Waterbirds (AEWA) has protected the pied avocet and its habitats since 1999. The UN established this international treaty to preserve bird species that migrate between Europe and Asia and rely on wetland habitats. Today, the AEWA covers over 250 species of waterbirds.

Tyree painted the pied avocet swimming in dark green water (figure 8.9). The artist shows the pied avocet in profile, cropped off slightly at the tail end, suggesting its movement and speed as it glides across the water. The captivating bird's reflection creates a distorted double image on the water, highlighting its distinctive bill.

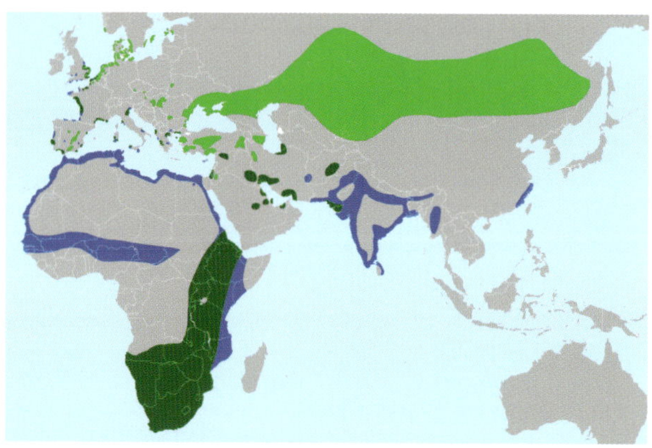

Figure 8.8 Pied avocet distribution. SanoAK: Alexander Kürthy, 2019.

Light green=breeding
Dark green=resident
Blue=non-breeding

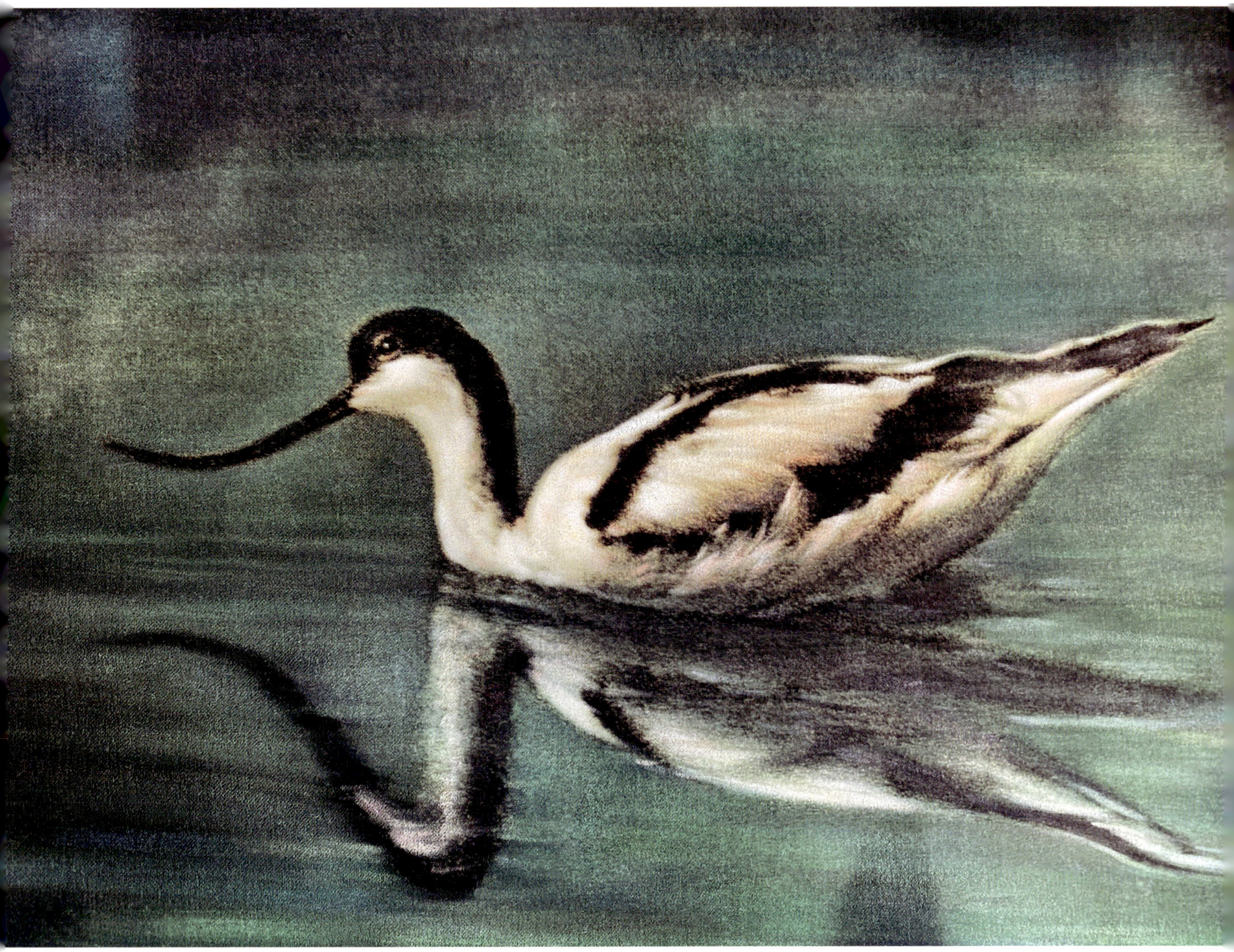

Figure 8.9 *Pied Avocet*, Ralph Burke Tyree. 1978, oil on velvet, 16" x 20", private collection.

Helmeted Guineafowl

The **helmeted guineafowl** (*Numida meleagris*) is a species in the *Numididae* family of guineafowl that is endemic to the African plains south of the Sahara. Weighing about three pounds, it has a distinctive appearance with a large black, gray, and white body and a small head topped with a bony knob (figure 8.10). This guineafowl follows herds of antelope, zebras, and elephants, eating parasites off the animals and worms from their manure. First domesticated in the fifth century BCE, it is still sold today in North America, Australia, and Europe as a food source, a form of insect control, and a pet. Due to this, it is common beyond its native distribution and is not in any peril.

Interestingly, when painting the helmeted guineafowl on black velvet, Tyree chose to show a close-up of the bird's head, omitting its unique round body (figure 8.11). This choice highlights the incredible colors and characteristic "helmet" of the guineafowl's head. Painting it against a dark background further focuses attention on its striking colors.

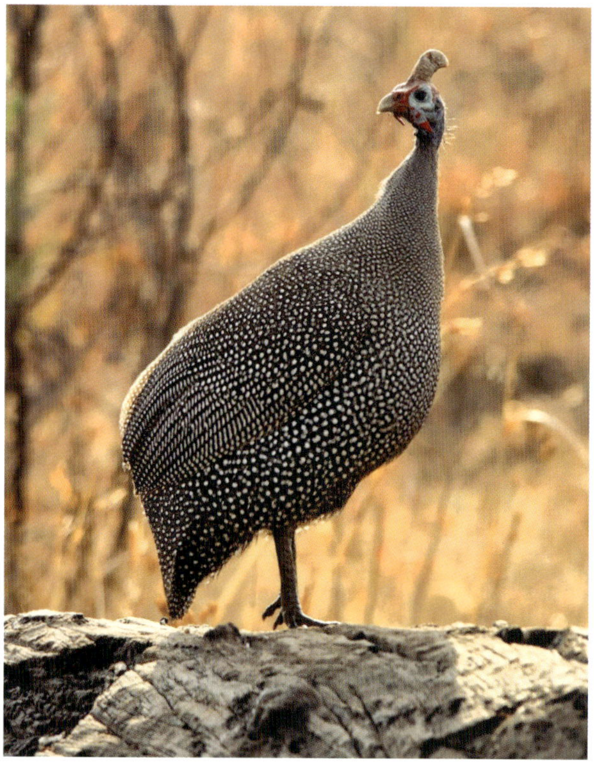

Figure 8.10 Helmeted guineafowl. Katelyn Deelefs (@bushbabyk8), South Africa, 2019, courtesy of the photographer.

Figure 8.11 *Helmeted Guineafowl*, Ralph Burke Tyree. 1974, oil on velvet, 16" x 12", author's collection.

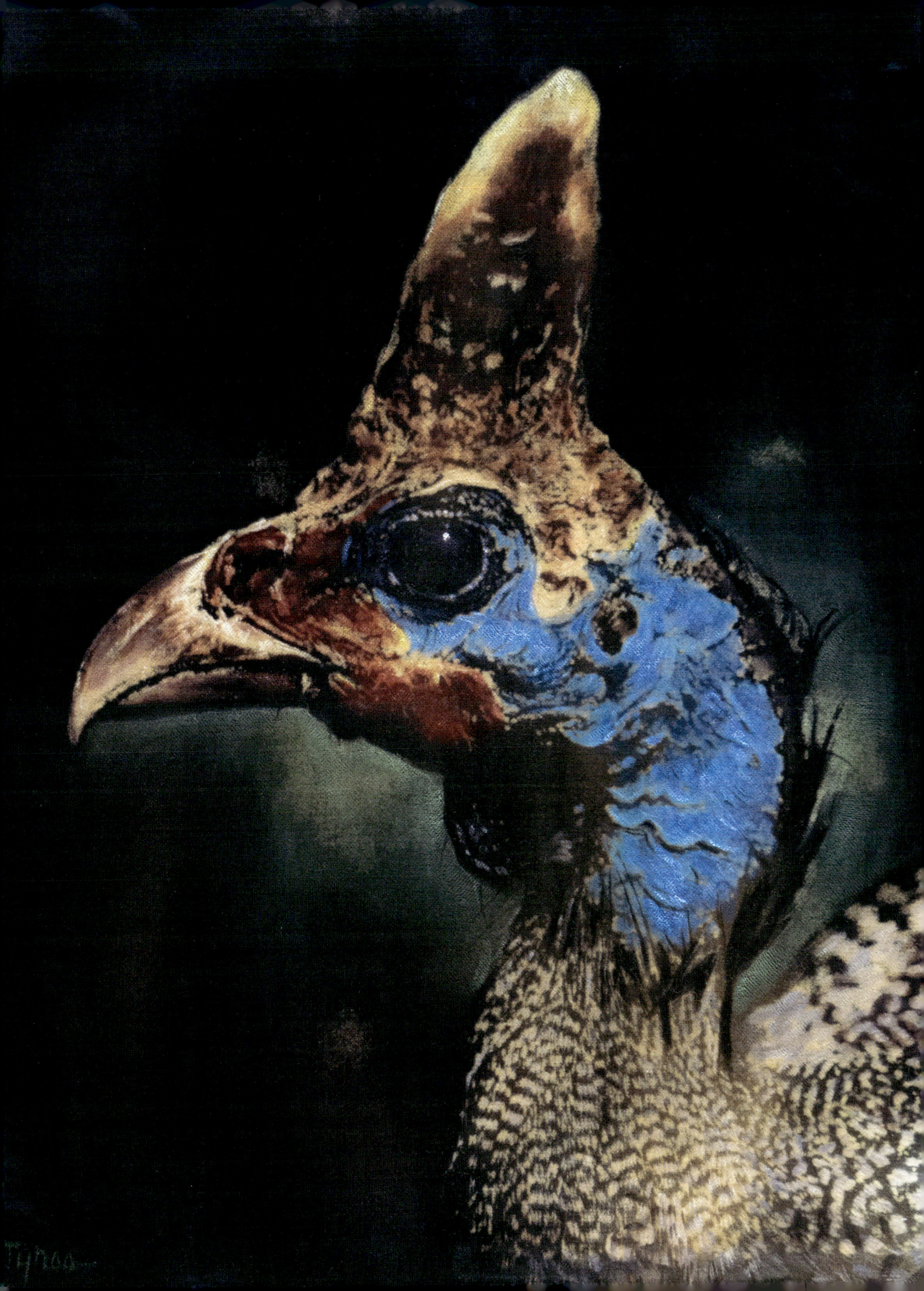

Grey Crowned Crane

The **grey crowned crane** (*Balearica regulorum*) has many names, including the African crowned crane, golden crested crane, and crested crane (figure 8.12). This member of the *Gruidae* crane family is native to eastern and southern Africa. Named for its crown of stiff golden feathers, it stands over three feet tall and weighs almost eight pounds. It has a primarily grey body, white wings, a black and white head, and a bright red inflatable throat patch. Although this crane spends most of its time in savannas, marshes, and grasslands, it builds a large nest in the wetlands during the breeding season. It is known for its unique hunting technique, during which it stamps its feet, driving insects out of the earth for it to eat. The crane also enjoys eating plants, frogs, and small fish. As Uganda's national bird, the grey crowned crane features on the country's flag and coat of arms.

Although the grey crowned crane was once plentiful across its range, habitat loss has reduced its numbers to around 20,000 individuals. This decline occurred rapidly over the last 45 years, and in 2012, the IUCN Red List listed it as endangered. Conservation efforts primarily focus on rehabilitating and protecting wetlands in Kenya, South Africa, Uganda, and Zimbabwe.

Tyree painted a flock of grey crowned cranes mid-flight against a cloudy grey-brown sky (figure 8.13). The painting has a dynamic sense of movement as the birds flap their huge wings and fly out of the frame. The work is almost monochromatic, with the grey, white, and black birds shown against an overcast grey sky. The cranes' dark silhouettes create a graphic composition full of movement.

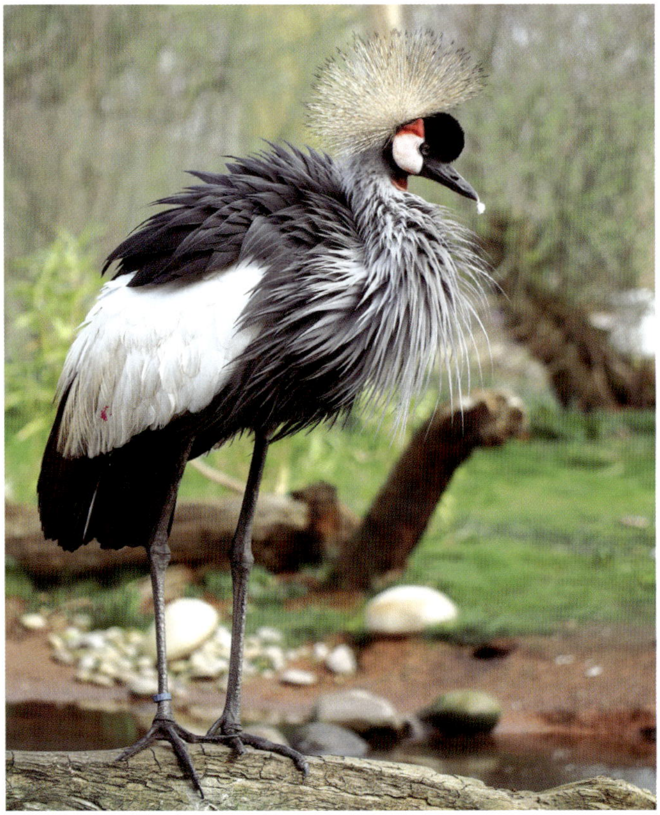

Figure 8.12 Grey crowned crane. Francis C. Franklin, United Kingdom, 2014.

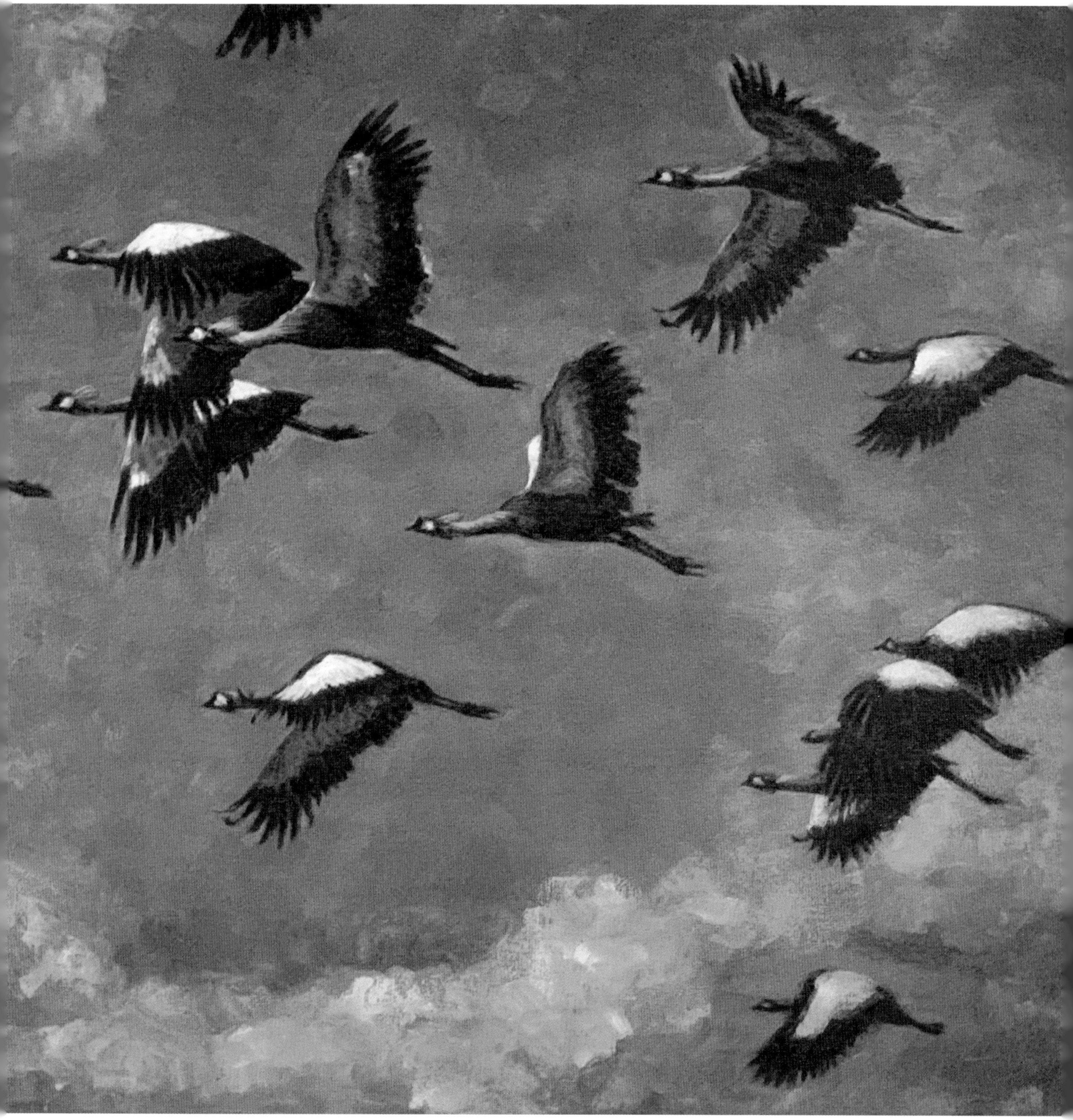

Figure 8.13 *Crested Crane Flock*, Ralph Burke Tyree. 1975, oil on board, 20" x 24", private collection.

NOCTURNAL PRIMATES

Bush Baby

Bush babies (*Galagidae*), also called galagos, are a family of small nocturnal primates similar in size to rabbits (figure 8.14). The family is part of the *Strepsirrhini* suborder, which also includes lemurs from the island of Madagascar and lorises from India and Southeast Asia. All members in this order have wet noses, like domestic cats and dogs. Bush babies are the most abundant members of the order, both in terms of actual numbers and variety of species.

In Afrikaans, the 20 species of this family are referred to as *nagapies*, meaning night monkeys. These primates use strong hind legs, human-like fingers and hands, and tails to move quickly through the trees, escaping danger and capturing prey. Bush babies' characteristic large round eyes give them strong night vision. They also have superb hearing that allows them to track insects in the dark and swipe them right out of the night sky. Bush babies are also fantastic jumpers and can leap up to seven feet. Due to their small size and nocturnal lifestyle, they are elusive and seldom seen during African safaris.

Bush babies generally are not considered at risk, but they face several threats from humans. Deforestation, rising temperatures, and the pet trade are the most significant threats to their populations. However, they are hard to domesticate and produce an unpleasant cry, similar to a baby's. The IUCN categorizes three species as near threatened, one as vulnerable, and one as endangered.

Tyree depicts two bush babies clinging to a branch with humanoid fingers and nails (figure 8.15). The two primates stare straight at the viewer with massive, saucer-like eyes. Their golden-hued fur stands out against the dark forest. Both have their ears upright, with their right ear turned outward as if reacting in real time to a rustling sound. Altogether, the composition emphasizes the large ears and eyes bush babies use to track prey at night.

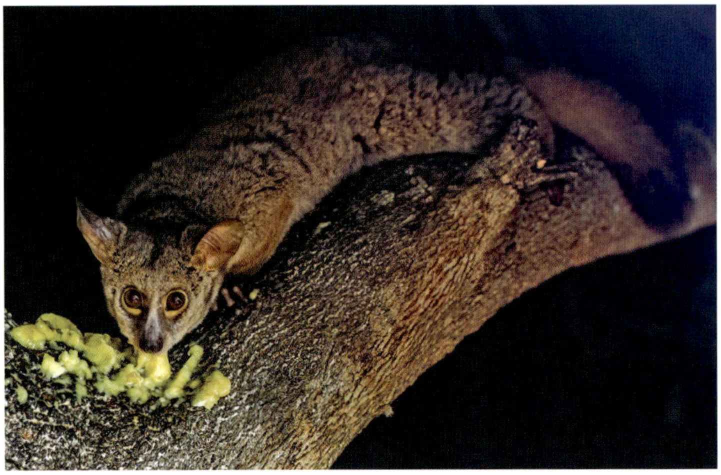

Figure 8.14 A bush baby eating a banana. Jacobmacmillan, South Africa, 2014.

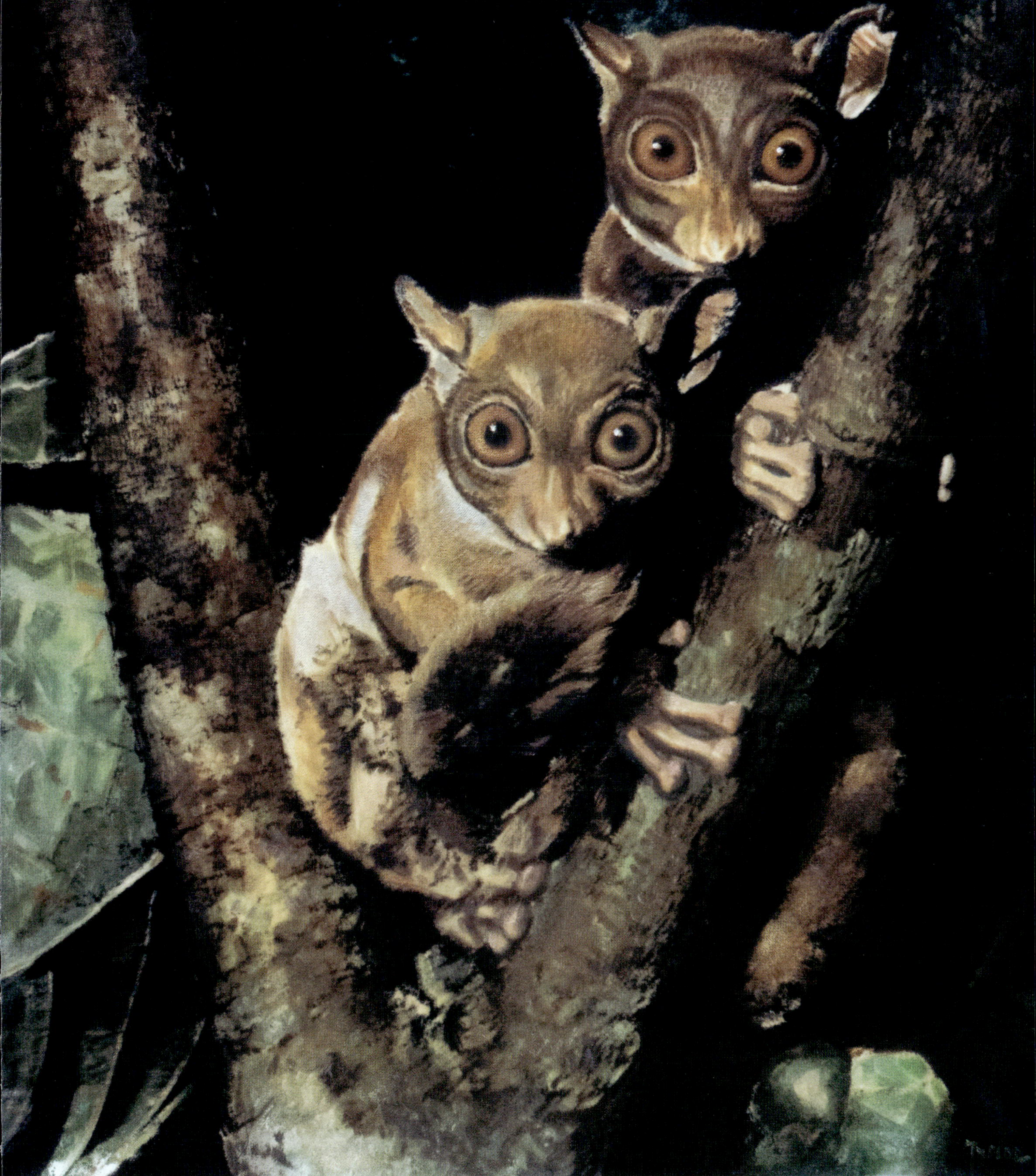

Figure 8.15 *Bush Babies*, Ralph Burke Tyree. 1974, oil on velvet, 24" x 20", private collection.

Dwarf Lemur

Like bush babies, **dwarf lemurs** (*Cheirogaleus*) of the *Lemuroide* superfamily of lemurs are a genus of primates within the *Strepsirrhini* suborder (figure 8.16). This genus consists of nine known species of small lemurs with long tails. During the winter, they enter a state of inactivity called torpor and use their tails to store fat. Dwarf lemurs have round heads with little pointed muzzles and a grip similar to the monkeys of South America. With nimble fingers and widely abducted great toes, they can firmly grasp branches between their second and third fingers. A lemur's tail, although long, is not able to hold onto trees like its monkey relatives. Native throughout the eastern coastal forests of Madagascar, they nest in tree holes alone or in pairs and primarily eat fruits and flowers.

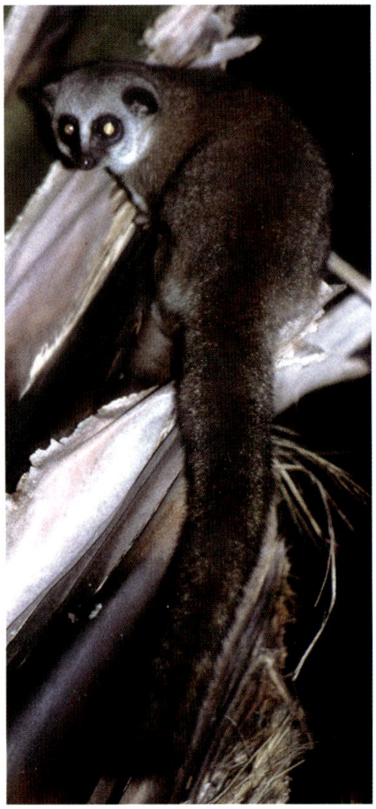

Figure 8.16 Greater dwarf lemur. Adam Britt, Madagascar, 2004.

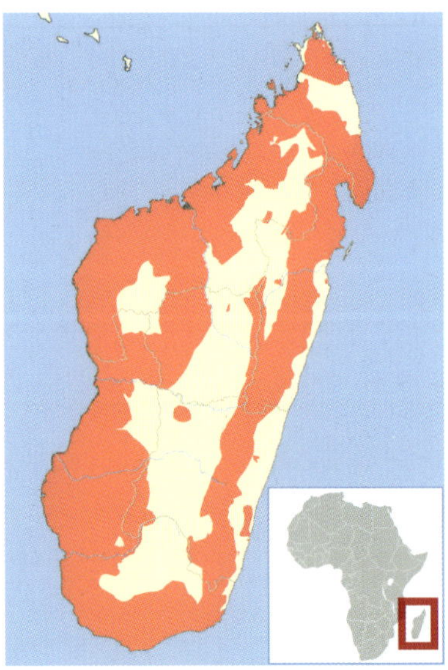

Figure 8.17 Lemur distribution. Alex Dunkel (Maky), 2015.

All of the over 100 different species of lemurs are exclusively native to Madagascar (figure 8.17). In the last decade, scientists have chronicled many new species that vary significantly in size. The most petite is Madame Berthe's mouse lemur, weighing about one ounce, and the largest is the indri at about 20 pounds. This variance in size was even more significant when the giant sloth lemur, which went extinct around 350 BCE, walked the earth. This lemur weighed around 400 pounds and was the size of a male gorilla.

Lemurs are the world's most endangered mammals, with 90% of the species listed as threatened on the IUCN Red List. Most lemur species face habitat loss from the deforestation of their island home, as well as climate change, capture for the pet trade, and hunting for their meat. Since the arrival of humans on Madagascar, over a dozen species of lemurs, eight genera, and three families have gone extinct. Without significant conservation efforts, many more may soon follow. Fortunately, documentaries and cartoon movies such as *Madagascar* are helping raise public awareness of Madagascar's unique beast, the lemur.

Tyree's ability to masterfully work with velvet to portray the soft textures of nature with incredible realism is on full display in his painting of a dwarf lemur (figure 8.18). The lemur's fur appears soft and fluffy to the touch, in direct contrast to the roughness of the bare tree post upon which it balances. While this choice highlights the artist's skill with his chosen medium, it also places the lemur in a precarious position that calls attention to the literal danger facing these animals.

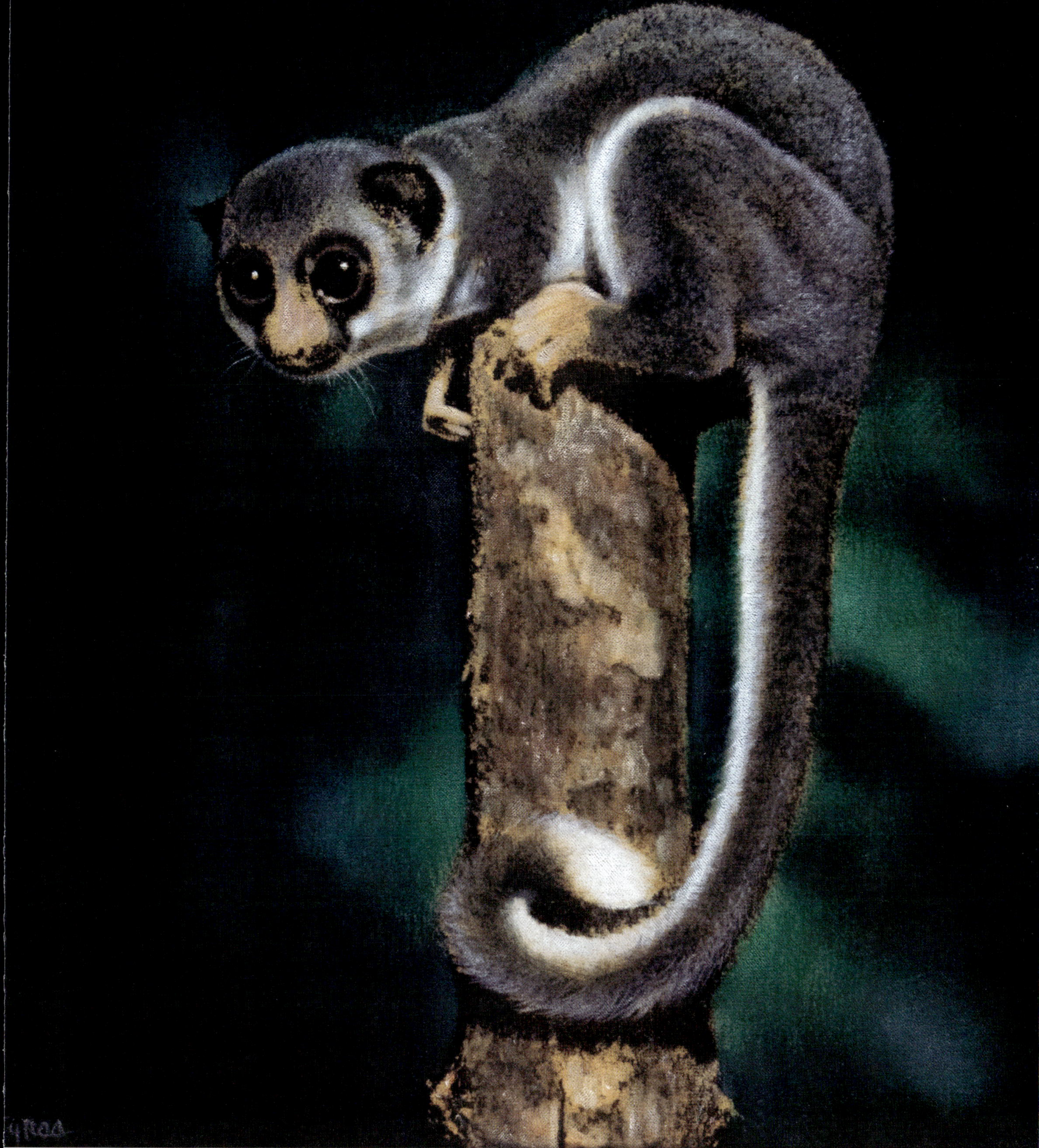

Figure 8.18 *Dwarf Lemur*, Ralph Burke Tyree. 1972, oil on velvet, 20" x 16", author's collection.

AFRICAN HERBIVORES, FOOD FOR THOUGHT

Thomson's Gazelle

The **Thomson's gazelle** (*Eudorcas thomsonii*) of the *Bovidae* family is one of the most recognizable gazelle species (figure 8.19). It is named for Joseph Thomson, a peaceable nineteenth-century British explorer with the personal slogan, "He who goes gently, goes safely; he who goes safely, goes far." This gazelle is also sometimes referred to as a tommie. It has a light brown coat, a distinctive black stripe down each side, a white spot on the rump, and ridged horns. At less than three feet tall, it is the smallest species of gazelle. Thomson's gazelle can reach an impressive 50 miles per hour, making it the fourth-fastest land animal, but not quite fast enough to outrun its primary predator, the cheetah.

This gazelle is found across the grasslands and savannas of the Serengeti region of Tanzania and Kenya and is the most common gazelle in East Africa (figure 8.20). As such, the IUCN does not classify it as at risk. Despite this, its numbers have declined sharply since the 1970s, mostly due to hunting and habitat loss.

Tyree painted a pair of Thomson's gazelles, a mother and a young fawn, in oil on velvet (figure 8.21). The two stand together amongst rocks, in a position that appears both relaxed and protective. While the mother observes their environment, watching for predators, the fawn stares at the viewer almost knowingly as if it has spotted the true threat. Together, the glowing quality of the light and the abstracted, natural background produce a peaceful, almost mystical effect while also highlighting the fawn. All of this creates an almost otherworldly moment of connection and understanding between the viewer and the fawn, the future of this species.

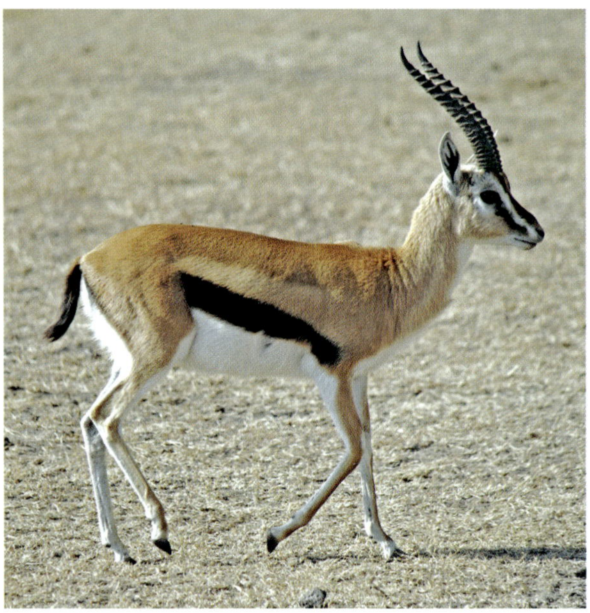

Figure 8.19 Thomson's Gazelle.
Jaliya Rasaputra on Unsplash.

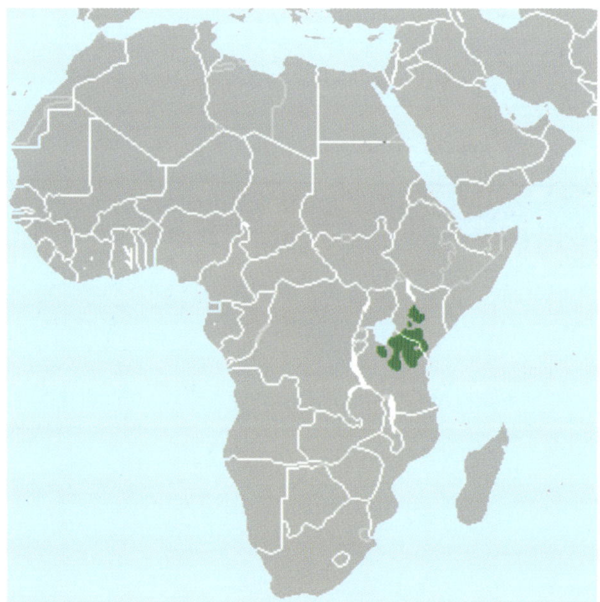

Figure 8.20 Thomson's gazelle distribution.
Vardion, 2015.

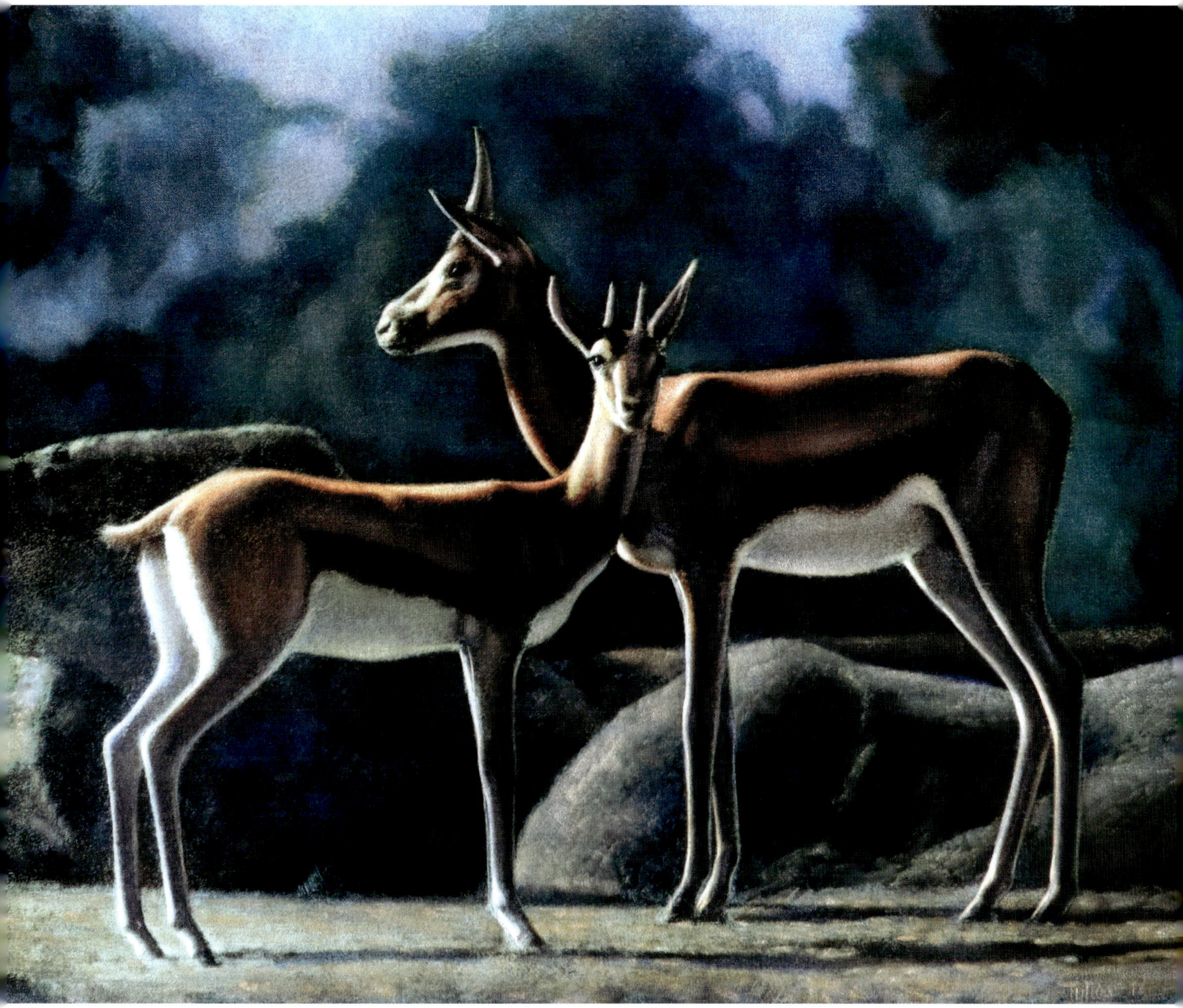

Figure 8.21 *Thomson's Gazelle, Mother, and Youngster*, Ralph Burke Tyree. 1978, oil on velvet, 20" x 24", private collection.

Impala

The **impala** (*Aepyceros melampus*), an antelope from the *Bovidae* family, is usually found foraging grass in the woodlands and savannas of southern Africa (figure 8.22). Medium-sized with a slender build, it has a glossy red-brown coat with a white underbelly and black markings. The male has long, curved horns that interlock with its opponent's when fighting, protecting its skull. Impala is an important food source for cheetahs, leopards, and lions, but its quickness and agility enable it to escape most cat ambushes. Not only can it jump almost ten feet high, but it also has the unique ability to leap in different directions on its forelegs, confusing predators.

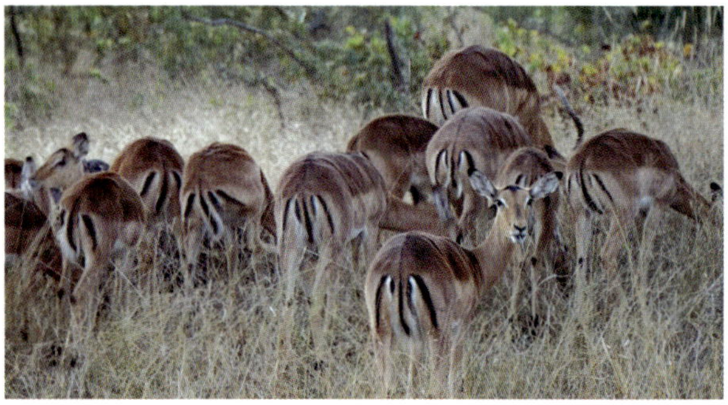

Figure 8.22 Impalas. Note the "M" for McDonald's (African humor) on their rears. 2017, author's photo.

Not only is the impala one of the most hunted antelopes, but it also faces habitat loss and fragmentation. Luckily, its numbers are plentiful, and the IUCN does not consider the common impala at risk. This is not the case for the rare, black-faced subspecies, which the IUCN classifies as vulnerable. Only about 3,000 black-faced impalas remain in the wild, with 90% residing in Namibia, where they are specially protected.

Zebra

Zebras (*Hippotigris*) are an African subgenus of the *Equidae* horse family, known for their characteristic black-and-white striped hides (figures 8.23 and 8.24). These iconic markings are unique to each zebra, like a fingerprint.

Due to the blur these black and white striped beauties create when running across the savannah, a herd of zebras is called a dazzle.

Zebras inhabit various habitats in southern and eastern Africa, where they graze on vegetation and are prey for lions and other large carnivores. Like the wild horse, a dominant zebra stallion usually breeds with many females in the dazzle.

The IUCN classifies all three zebra species as threatened to some extent. This is primarily due to hunting and habitat destruction, although their food supply has also diminished. Since the invasive mesquite plant was introduced to Ethiopia about 30 years ago, it has been replacing the grasses they rely on for nourishment.

The zebra subgenus includes three species: the Grévy's zebra, the plains zebra, and the mountain zebra. Also known as the imperial zebra, **Grévy's zebra** (*Equus grevyi*) is not only the largest of the three but also the largest living wild horse. It was discovered in 1882 by Émile Oustalet and named after Jules Grévy, then President of France. This zebra's distinguishing features are big ears and a prominent mane. Sadly, it is the most threatened zebra species, partly because the mesquite plant's takeover

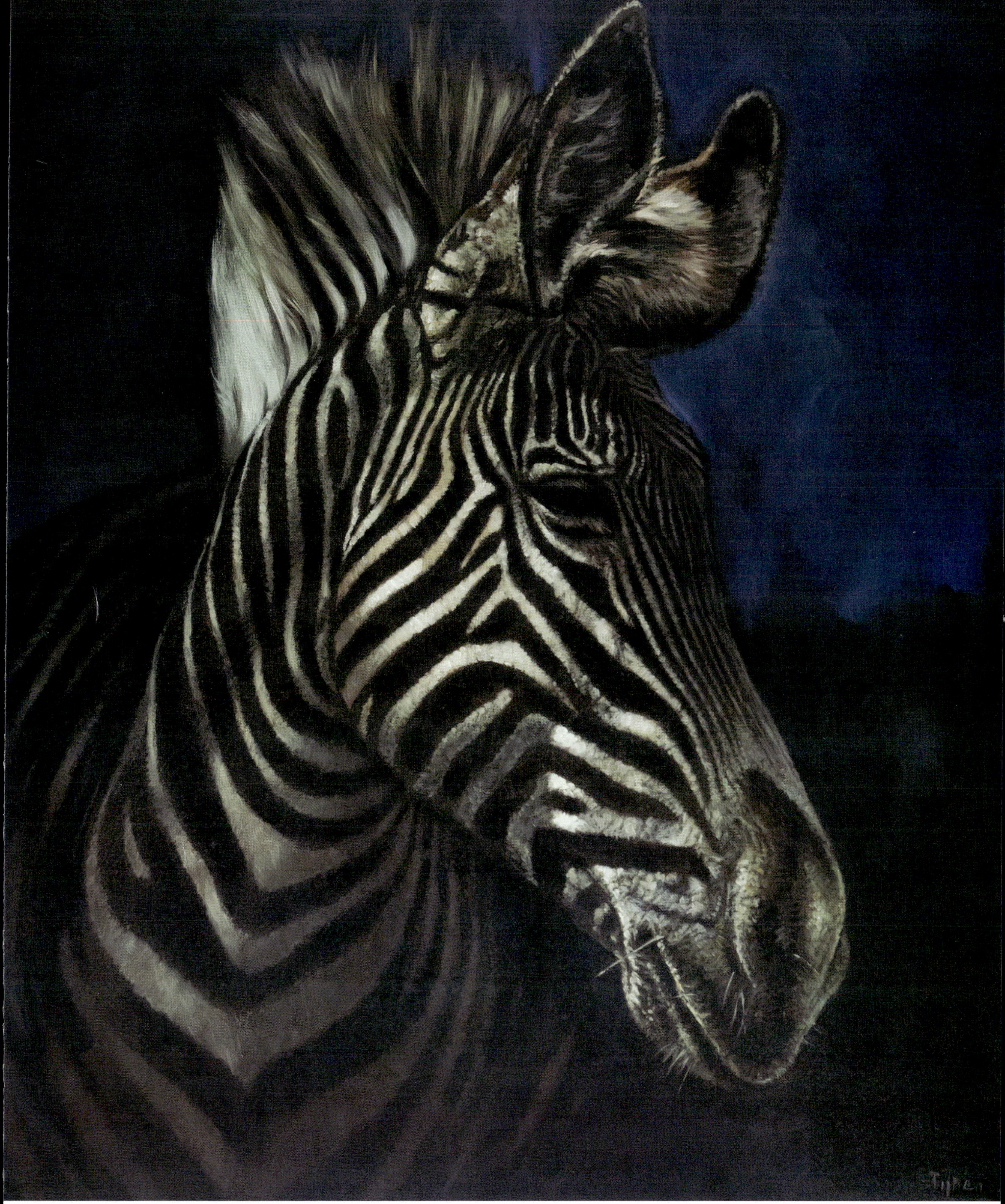

Figure 8.25 *Grevy's Zebra*, Ralph Burke Tyree. 1975, oil on board, 24" x 20", author's collection.

in Ethiopia has particularly impacted it. Since the 1970s, its total population has declined from 15,000 zebras to less than 3,000, leading the IUCN to list it as endangered.

With their stunning stripes, zebras are a frequent subject of photographers and other artists. Tyree was no exception, painting a portrait of a Grévy's zebra in oils on board (figure 8.25). He highlights the qualities that set the Grévy's zebra apart from the other, less seriously threatened species of zebras by painting a close-up of its face. These qualities include elongated nostrils, large conical ears, an ash-colored muzzle, whiskers, and a tall, stiff mane. Overall, it has a more mule-like appearance than other zebra species. The bright blue sky in the background allows the zebra's bold black and white coat to stand out beautifully.

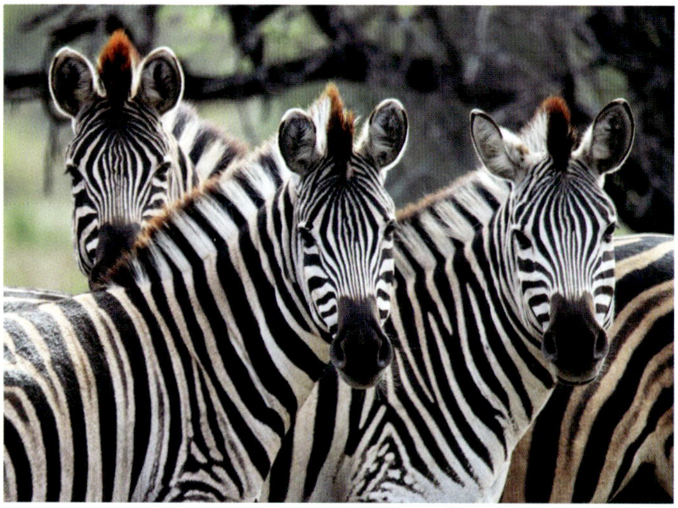

Figure 8.23 A dazzle of zebras. Katelyn Deelefs (@bushbabyk8), South Africa, 2019, courtesy of the photographer.

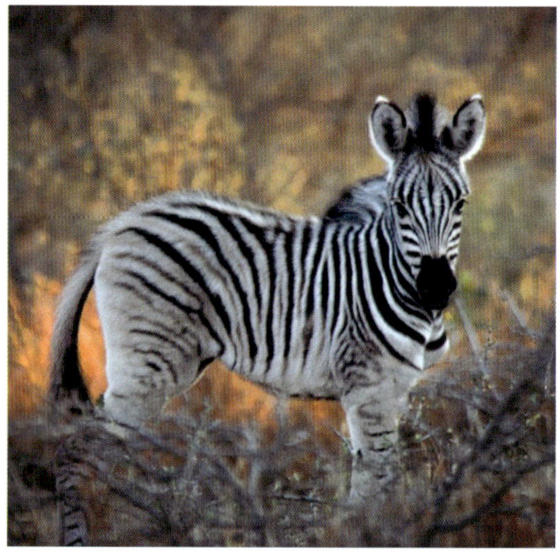

Figure 8.24 Baby zebra at golden hour, Katelyn Deelefs (@bushbabyk8), South Africa, 2019, courtesy of the photographer.

SMALLER CATS OF AFRICA

Two smaller African cats that were both tamed long ago by humans merit mention—the cheetah and the caracal. The cheetah featured prominently in Egyptian religion as early as the First Dynasty, around the thirtieth century BCE. By the New Kingdom, which began in the sixteenth century BCE, pharaohs commonly kept cheetahs as pets. Similarly, the caracal was also religiously significant for ancient Egyptians, but it wasn't until the thirteenth century in India, under the Delhi Sultanate, that the caracal was tamed for hunting. The caracal became highly valued and was often given as a gift or traded for gold, silver, and silk. Both cats are uniquely athletic, with the cheetah famously the fastest land animal and the caracal a skilled vertical jumper. In Africa, they both endure despite competing for much of the same prey with the dominant lions who hunt in their large prides, in contrast to these solitary hunters.

Cheetah

"A cheetah who chases two impalas catches neither."

—CHINESE PROVERB

The **cheetah** (*Acinonyx jubatus*) is one of the most famous members of the *Felidae* family (figures 8.26–8.28). Native to Africa and central Iran, this cat can run up to 80 miles per hour. With a slender, flexible body, strong legs, and a long tail, it is built to run, and it's no wonder it is the fastest land animal. While the cheetah is most recognizable for its spotted coat, it also has black tear marks and a rounded head just under the size of a leopard's. Because its head is built similarly to that of smaller cats like bobcats, cougars, and even house cats, it is the only big cat that can purr. Female cheetahs lead nomadic lifestyles, usually accompanied by their cubs, searching for prey across a much broader region than the territory solitary males cover. The cheetah is active at dawn and dusk when it hunts for ungulates like impala and gazelle. Like other cats, it stalks its prey until within range before pouncing. Then, with incredible speed, it knocks its target off-balance and grabs the throat, suffocating its prey.

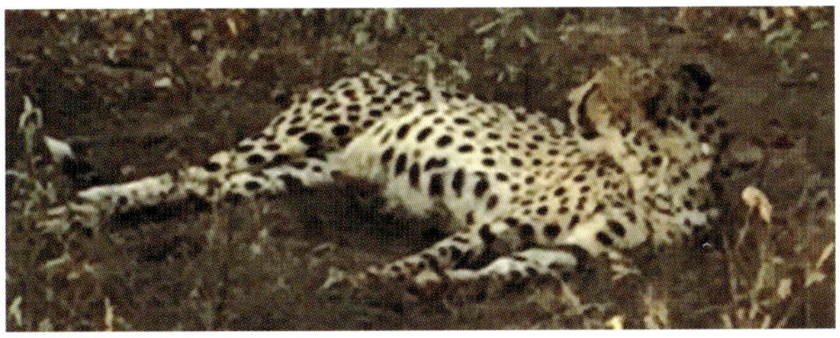

Figure 8.26 Very satiated cheetah resting after consuming most of an impala. South Africa, 2017, author's photo.

Habitat loss particularly impacts the cheetah because of the extensive range of land it requires to roam. Its iconic appearance and constant threat to livestock also make it a common target of hunters and traffickers. As a result, the IUCN Red List categorizes the cheetah as vulnerable. Cheetah breeding programs at some prominent zoos have had notable successes. Wildlife Safari, the animal park in Oregon that Tyree drafted landscape plans for, is home to America's most successful cheetah breeding center, with 235 cheetah offspring born since 1972. Effective conservation programs are underway in Africa and Iran, and India has even reintroduced this legendary cat.

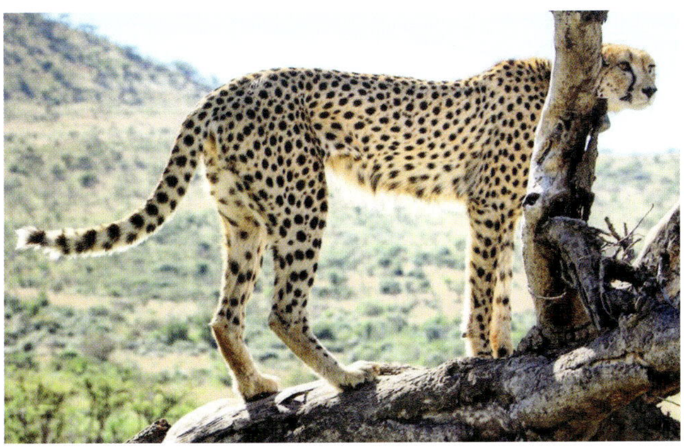

Figure 8.27 Cheetah on the lookout. Katelyn Deelefs (@bushbabyk8), South Africa, 2019, courtesy of the photographer.

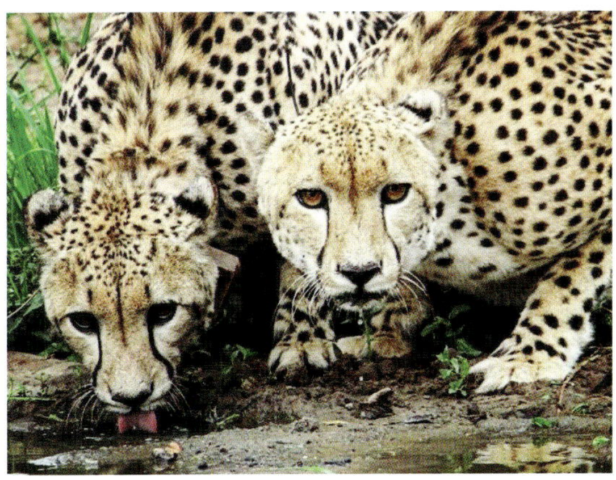

Figure 8.28 Cheetah brothers, Katelyn Deelefs (@bushbabyk8), Mkhuze Game Reserve, South Africa, 2019, courtesy of the photographer.

Caracal

The **caracal** (*Caracal caracal*) is a mid-sized member of the *Felidae* family from Africa, the Middle East, and parts of India (figure 8.29). Also known as the desert lynx and the Persian lynx, the caracal is not actually a lynx but one of its cousins. As seen in figures 8.30 and 8.31, this confusion makes sense with its lynx-like long-ear tufts, but the caracal is a solid tan color, in contrast to the spotted lynx. Primarily a solitary nocturnal hunter, the caracal stalks small mammals, rodents, and birds with great agility, even taking down larger prey like antelopes, sheep, and goats. Its powerful hind legs allow it to leap more than ten feet to catch prey, even plucking birds straight from the air.

The IUCN does not list the caracal as threatened, as it is found throughout an array of habitats across more than 50 countries. Still, it is considered locally endangered in certain regions like North Africa, Pakistan, and Jordan. This is primarily due to habitat loss and hunting, despite the widespread prohibition of this practice.

Like Tyree's painting of a Grévy's zebra, he highlights the caracal's distinctive facial markings by painting it in the style of a traditional portrait (figure 8.32). He masterfully works with the velvet, achieving astonishing realism and depicting fur that appears as soft as any housecat's. Tyree positions the caracal in a three-quarter profile against yellow-green grasses, which mimic the color of its glowing eyes. As it surveys its domain, it gives off a distinct regal air, befitting its ancient association with the Egyptian pharaohs.

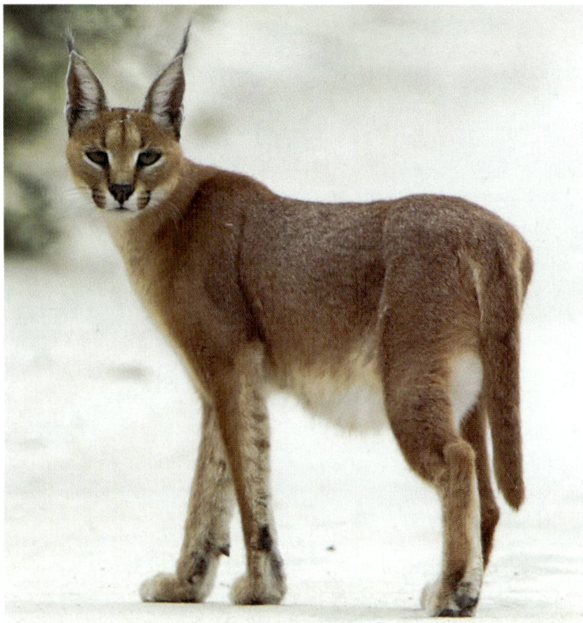

Figure 8.30 Caracal on the road in the early morning. Derek Keats, Kgalagadi, South Africa, 2017.

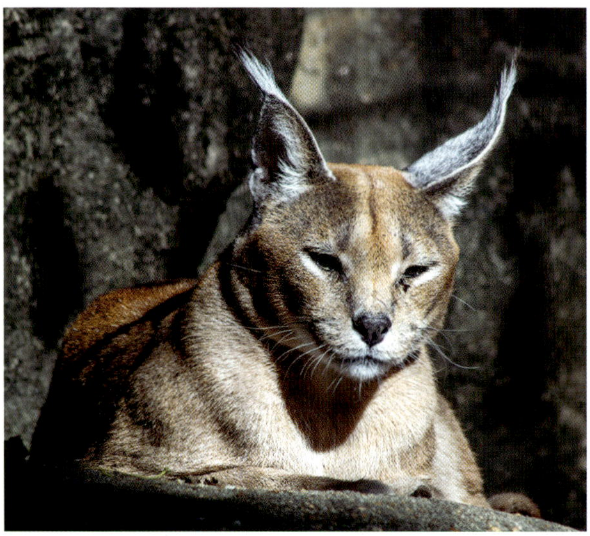

Figure 8.31 Caracal, Joshua J. Cotten on Unsplash.

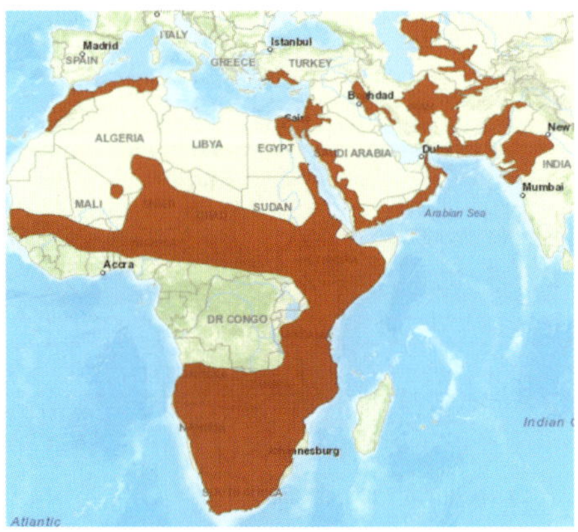

Figure 8.29 Caracal distribution. BhagyaMani, 2019.

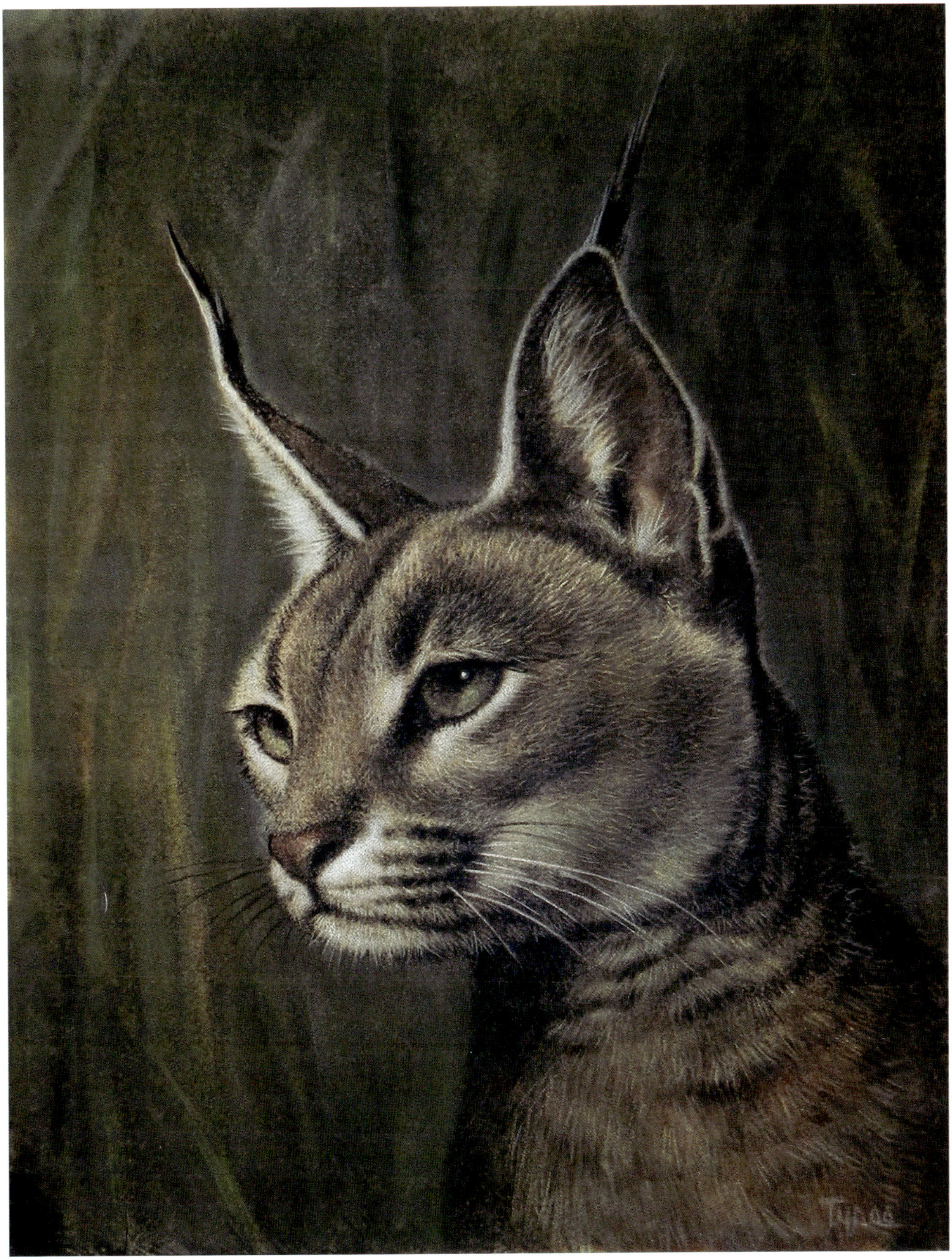

Figure 8.32 *Caracal*, Ralph Burke Tyree. 1979, oil on velvet, 16" x 12", Tyree family.

THE BIG FIVE GAME ANIMALS OF AFRICA

The Big Five consists of some of Africa's most famous megafauna: the leopard, lion, rhinoceros, Cape buffalo, and African elephant (figure 8.33). During the late 1800s, as big-game hunting grew in popularity, hunters coined the phrase "Big Five" to refer to the five African animals considered the most challenging and dangerous to hunt on foot. By the middle of the nineteenth-century, big game hunting in Africa was a favorite sport for Europe's wealthy elite. Popularity peaked in the early twentieth century when several notable Americans, like Ernest Hemingway (figure 8.34) and at least one former president, ventured into Africa for these hunts.

A year removed from his US presidency, Teddy Roosevelt famously embarked on an African safari in 1909 with his son Kermit (figures 8.35 and 8.36). Between the two, they killed over 500 animals during this year-long hunt, including all of the Big Five, specifically 20 rhinoceroses, 10 buffalo, 11 elephants, 17 lions, and 3 leopards. By no means did they limit themselves to these five, also killing 7 cheetahs, 8 hippopotamuses, 29 zebras, 47 gazelles, and 9 giraffes, to name a few other beasts they targeted. Roosevelt downplayed these shocking numbers in his 1910 book, *The Great Roosevelt African Hunt and the Wild Animals of Africa*, stating, "Kermit and I kept about a dozen trophies for ourselves; otherwise, we shot nothing that was not used either as a museum specimen or for meat."

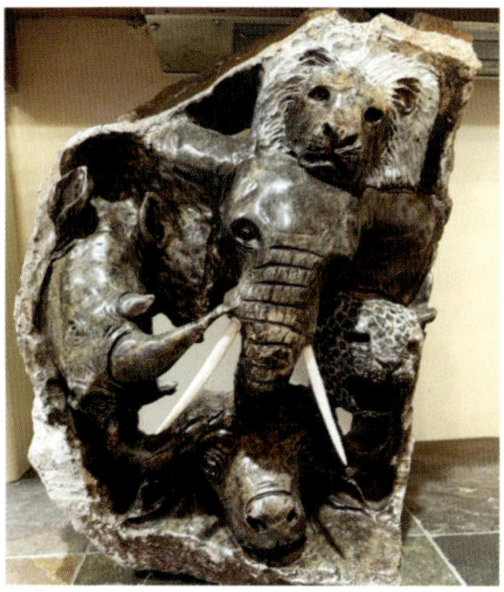

Figure 8.33 Large carved granite South African sculpture of Africa's Big Five (top-down: lion, rhino, elephant, leopard, and buffalo), author's collection.

While much less common and accepted today, big game hunting does sadly continue. Anyone can still hunt the Big Five for the right price as trophy kills. Despite this, the term Big Five is now more often used by safari operators challenging tourists to shoot with their cameras, not weapons.

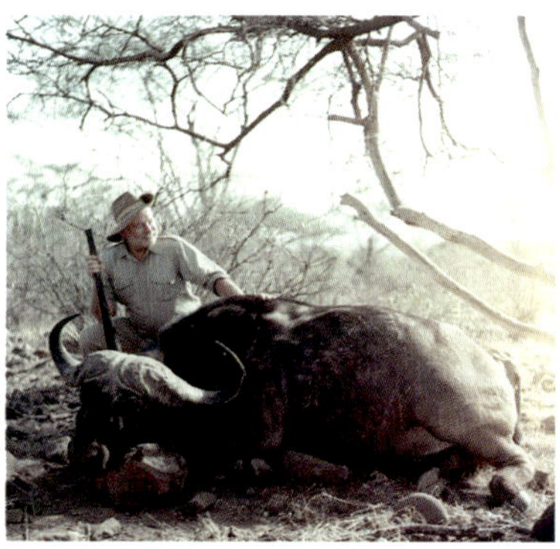

Figure 8.34 Ernest Hemingway poses with a Cape buffalo he killed while on safari in Africa, 1953-1954. Ernest Hemingway Photograph Collection, courtesy of John F. Kennedy Presidential Library.

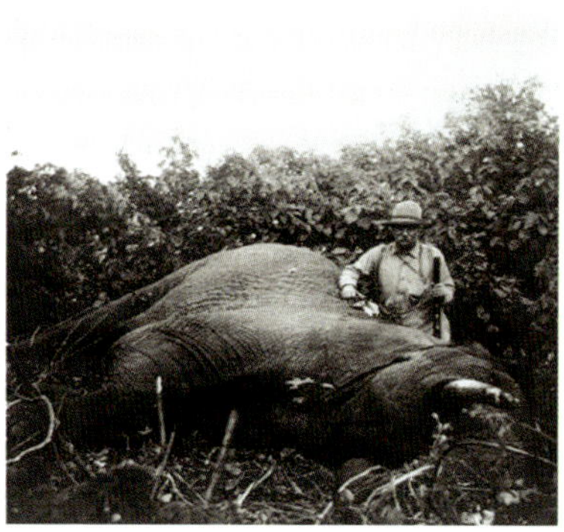

Figure 8.35 Teddy Roosevelt with one of the 11 elephants he and his son killed during his African Safari, 1909-1911. Courtesy of the Smithsonian-Roosevelt African Expedition.

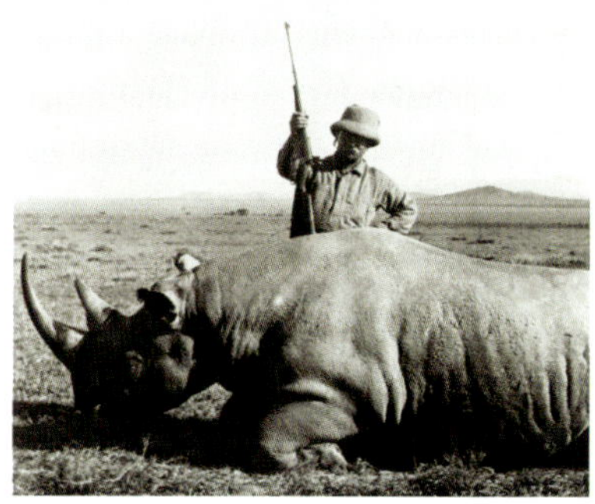

Figure 8.36 Teddy Roosevelt and one of the 20 rhinos he killed during his African Safari, 1909–1911. Courtesy of the Smithsonian-Roosevelt African Expedition.

Leopard

The **leopard** (*Panthera pardus*) is a member of the Panthera genus from the *Felidae* family. Although it has a less muscular physique than its American cousin, the jaguar, it is spotted with similar rosettes (figure 8.37). The leopard can stalk prey with near-perfect ambush concealment thanks to this uniquely patterned tan and black fur (figures 8.38–8.39). With shrewd hearing and vision, it is a fierce predator, primarily targeting ungulates, primates, and smaller carnivores.

The nine subspecies of leopards occupy the broadest range of all wild cats, extending across Africa, the Middle East, central and southern Asia, and into southern Russia (figure 8.40). It is a solitary, territorial cat that prefers undisturbed savannas, forests, and woodlands where it can avoid interaction

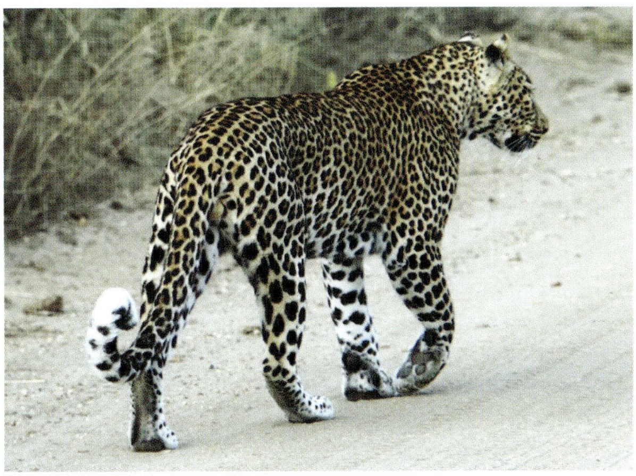

Figure 8.37 Leopard. Kruger National Park, South Africa, 2017, author's photo.

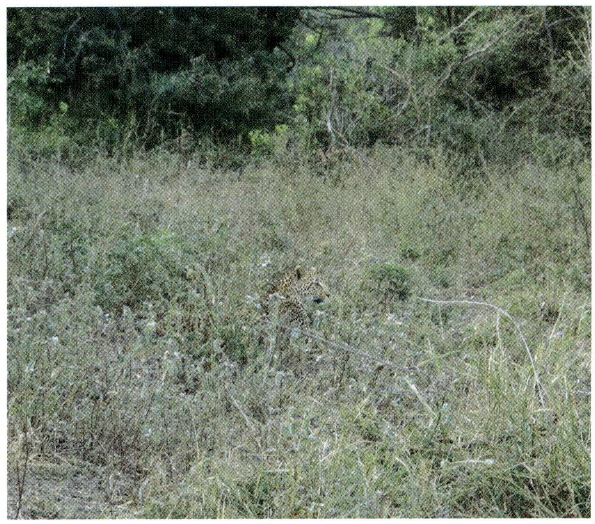

Figure 8.38a The leopard, a camouflage queen. Kruger National Park, South Africa, 2017, author's photo.

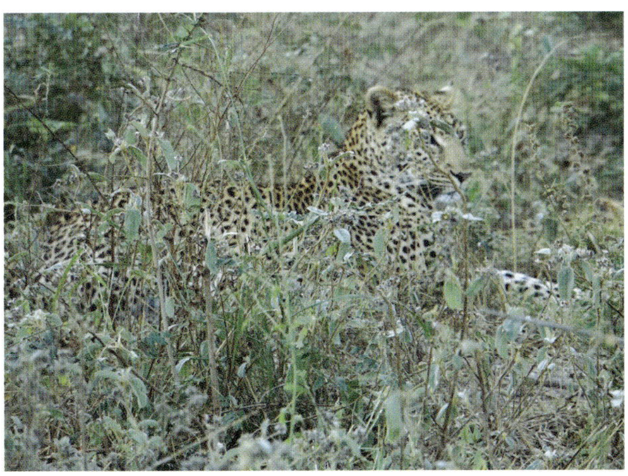

Figure 8.38b Close-up of the camouflage queen. Kruger National Park, South Africa, 2017, author's photo.

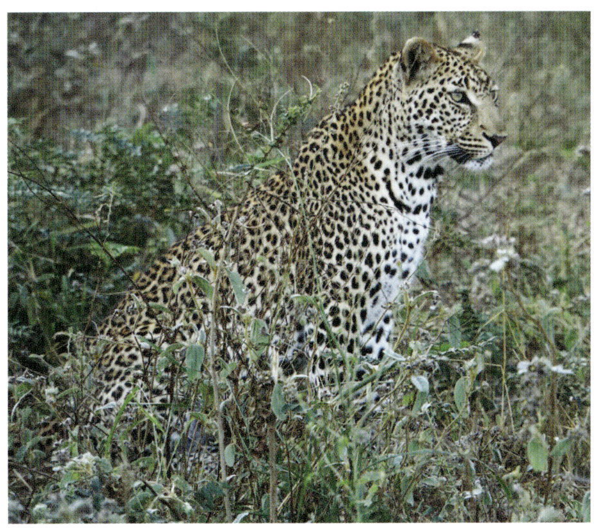

Figure 8.38c The camouflage queen sits up. Kruger National Park, South Africa, 2017, author's photo.

with humans and other leopards. Due to its wary and nocturnal tendencies, this cat is considered the most difficult of the Big Five to hunt.

The leopard faces multiple threats to its population, and the IUCN Red List has thus classified it as vulnerable. Habitat loss and fragmentation have caused a decrease in available prey while increasing this cat's interactions with humans. It is also a common target of illegal hunting and poaching, often sought out as a trophy kill.

As seen in figure 8.41, Tyree painted a leopard in oil paint on board. The leopard stands in a tree, observing its surroundings. Similar to previously discussed works like the *Great Blue Heron* and *Squirrel*, Tyree chose a monochromatic palette of black, white, and grey for this smaller piece. Again, this allows him to focus on light and shadow and demonstrates his keen eye for details. In this case, it also draws attention to the leopard's striking spot

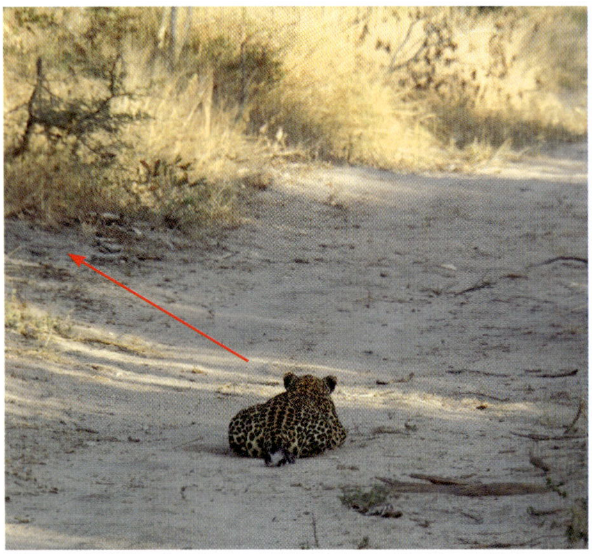

Figure 8.39a Female leopard stalks a young impala buck. Arrow shows where she will hide in next photo. Kruger National Park, South Africa, 2017, author's photo.

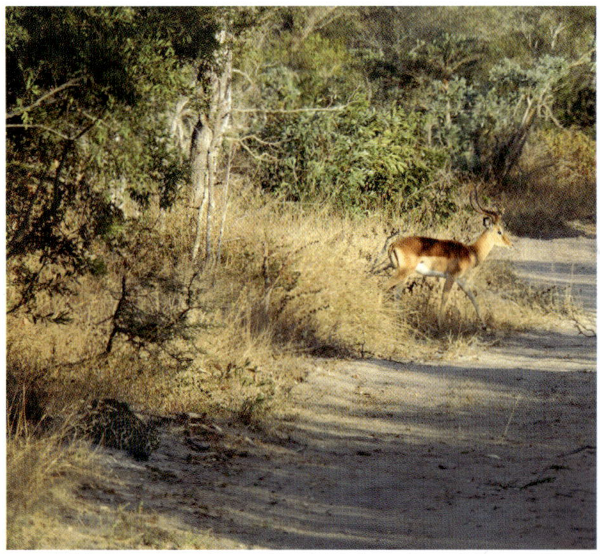

Figure 8.39b The leopard hides and waits to pounce. Kruger National Park, South Africa, 2017, author's photo.

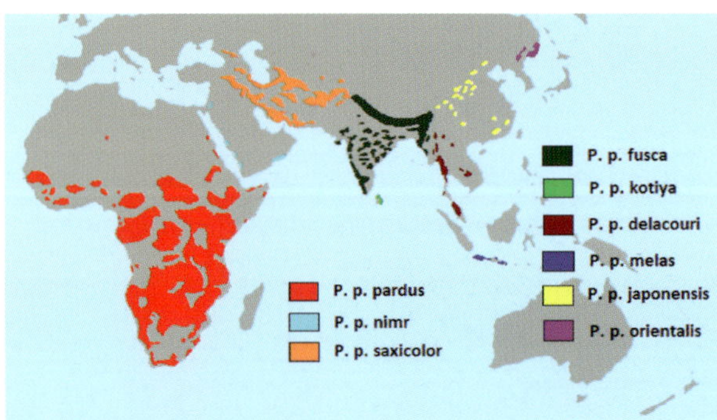

Figure 8.40 Leopard subspecies distribution. Whitesachem, 2016.

Figure 8.41 *Leopard*, Ralph Burke Tyree. 1975, oil on board, 14" x 11", author's collection.

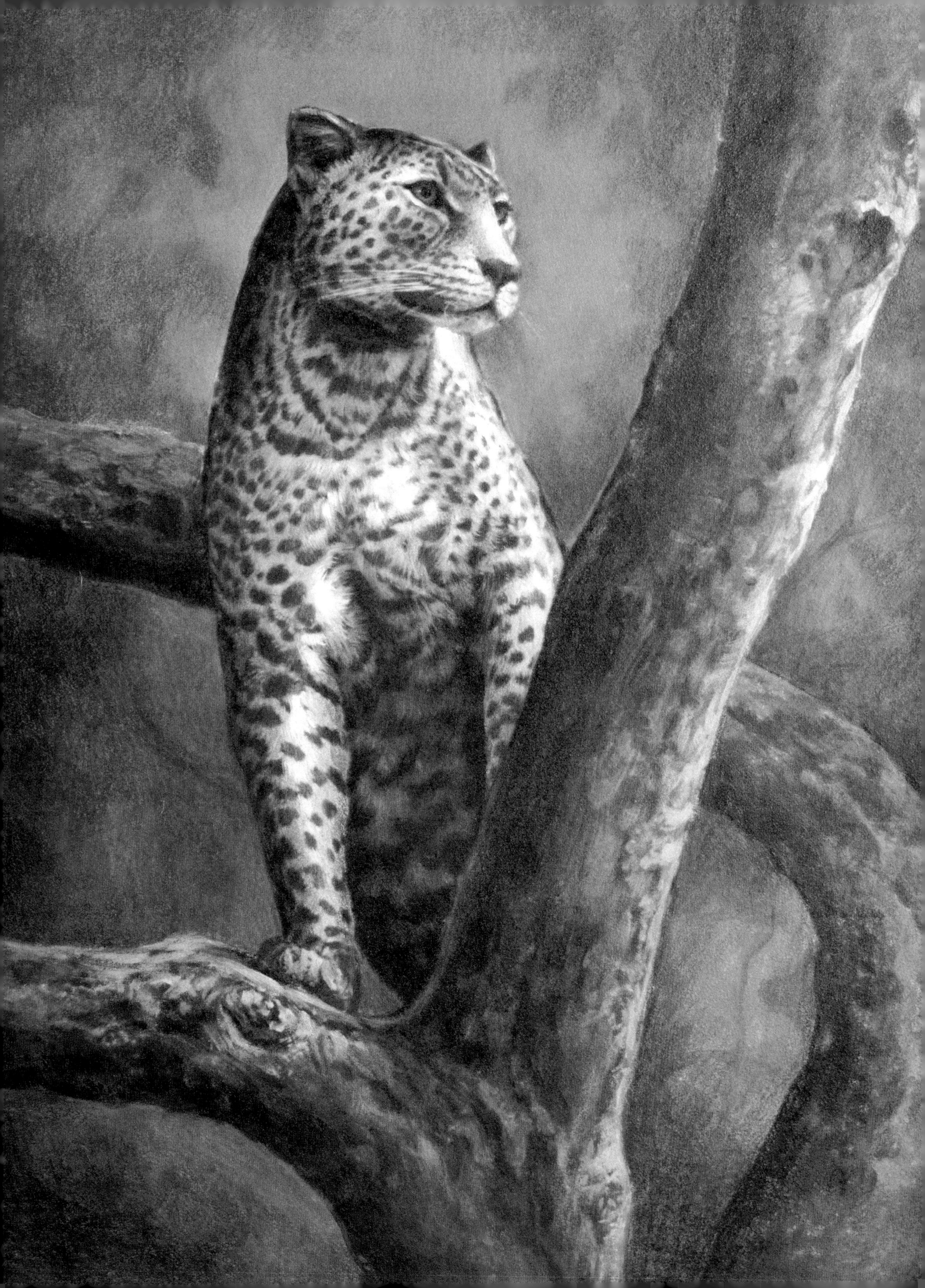

Lion

"Lions make leopards tame."

—WILLIAM SHAKESPEARE

The **lion** (*Panthera leo*) is another member of the *Panthera* genus that hunters have long sought out for thrill and glory because of the genuine danger it poses. This cat is the largest in Africa and has a muscular body, round ears, and a long tail tipped with a hairy tuft (figure 8.42). The adult male lion is larger than his female counterparts at up to 490 pounds and is recognizable by the iconic, prominent mane (figure 8.43). This social creature is the only cat that lives in groups, forming family units called prides, which consist of related lions and their cubs. As an apex predator of African savannas, this fearsome cat (figure 8.44) mainly preys on large herbivores like wildebeests, zebras, buffalo, and giraffes. Groups of females usually hunt together for the pride, conducting fast, powerful ambushes on their prey.

As one of the most famed animals, the lion has been depicted in an array of contexts (figures 8.45 and 8.46) and included in the folklore of many cultures since antiquity. Befitting its place at the top of the food chain, the lion has become a symbol of nobility and royalty. It is even referred to as the "king of the jungle." This popularity continues into the modern day with films like *The Lion King* (1994). The IUCN Red List has categorized the lion as vulnerable since 1996. Despite widespread protection, this cat has disappeared from 94% of its historic range (figure 8.47). As we encroach on the lion's habitat, it is pushed out and forced to interact with humans as it roams and searches for food. This increases human-lion conflict, especially when farmers retaliate against the cat for attacking their livestock. It also continues to be a target of poachers, hunted as a trophy and for use in traditional medicine.

In 1960, when Tyree painted one of his earliest animal paintings, his subject of choice was a lion

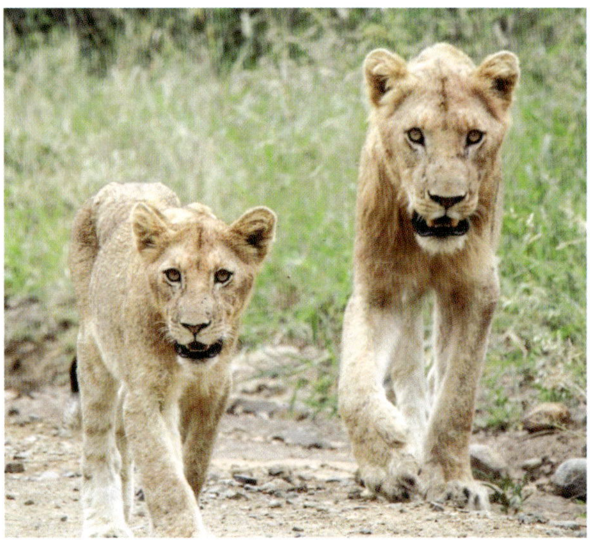

Figure 8.42 Scary but beautiful lionesses. Katelyn Deelefs (@bushbabyk8), South Africa, 2019, courtesy of the photographer.

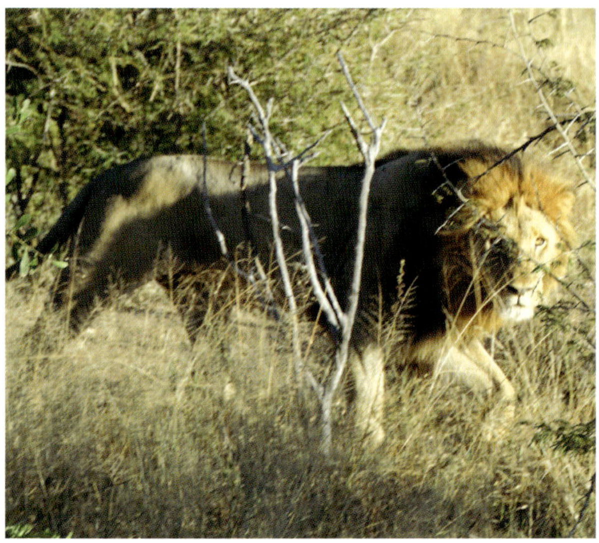

Figure 8.43 Stalking lion. Kruger National Park, South Africa, 2017, author's photo.

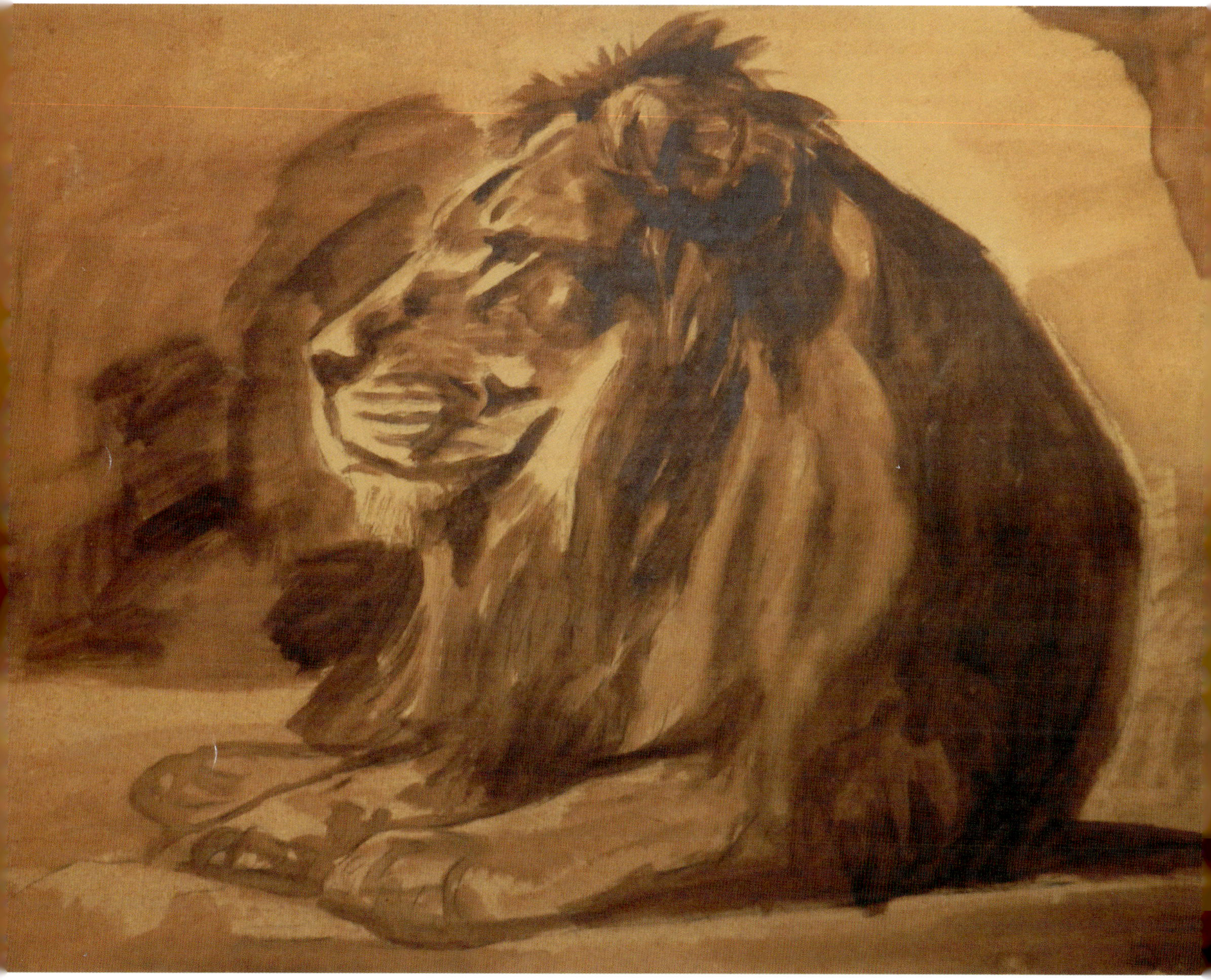

Figure 8.48 *Resting Lion*, Ralph Burke Tyree. 1960, oil on board, 14" x 18", Tyree family.

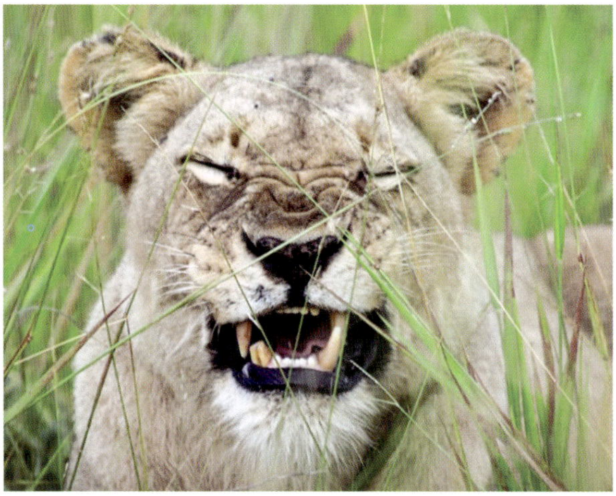

Figure 8.44 A look at a female lion's teeth—does anyone know a good dentist? Katelyn Deelefs(@bushbabyk8), South Africa, 2019, courtesy of the photographer.

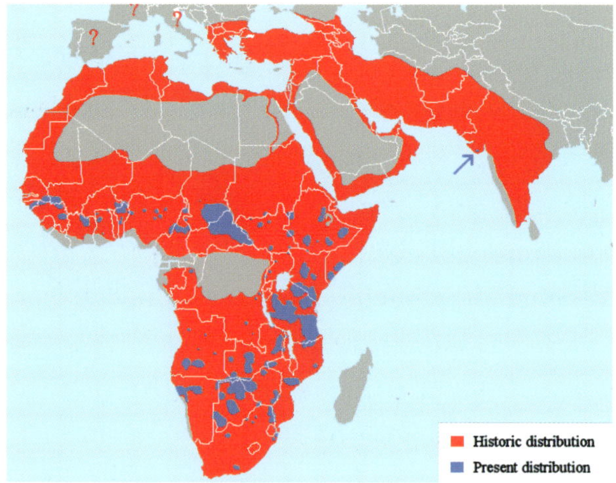

Figure 8.47 Lion distribution. Tommyknocker, 2009.
Red (and blue)=historic range
Blue=current range

(figure 8.48). He painted the male lion in profile with his great mane on display. The cat reclines while passively observing his surroundings. Tyree utilizes large, expressive brushstrokes and shades of brown on a brown board, a departure from his usual style. He created this unique work for his children, never intending to sell or publicly display the piece.

Later, in the 1970s, Tyree returned to this magnificent beast, creating more detailed compositions. One of these, shown in figure 8.1, is a large painting that depicts a resting lion pair. Tyree again paints on board, focusing on the majestic cat's profile, but this time, he shows the pair in stunning realism, a distinct contrast to his 1960 work. Once again, the male lion's mane is on full display as he reclines, but this time he is joined by a female companion who sprawls out alongside him as the two look out over a broad savanna vista and clear, blue sky.

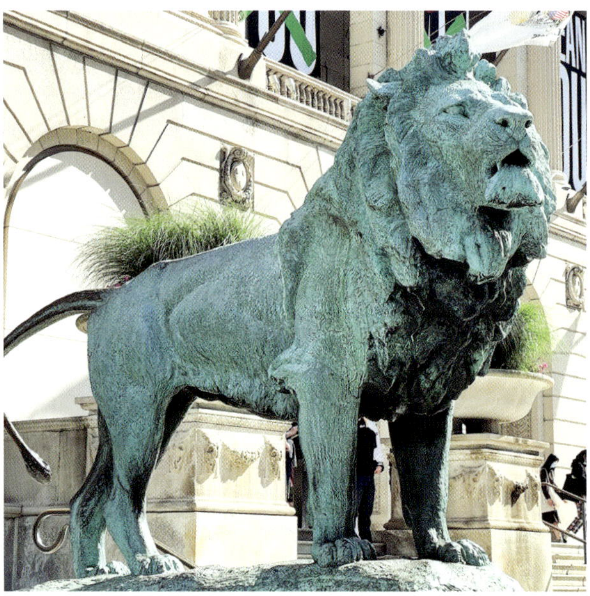

Figure 8.45 *Lion (One of a Pair)*, Edward Kemeys. 1893, Art Institute of Chicago, photographed 2021, author's photo.

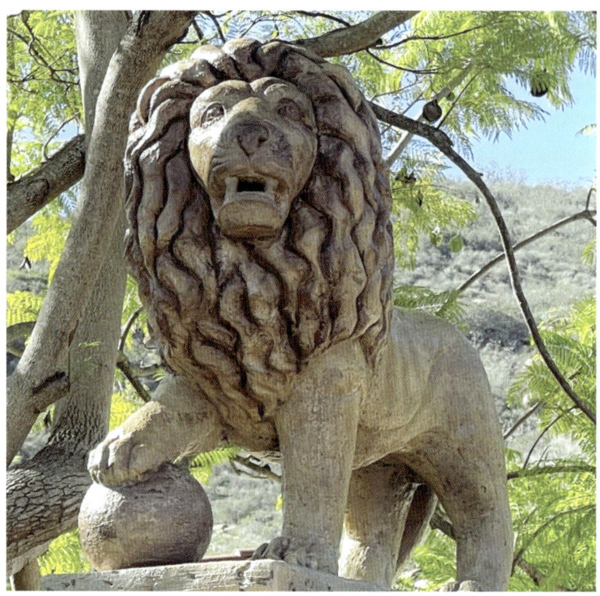

Figure 8.46 Lion guarding a home's entrance. 2022, author's photo.

Rhinoceros

Rhinoceroses (*Rhinocerotidae*) are a family of species in the *Perissodactyla* order of ungulates and one of the largest living megafaunas (figure 8.49). These beasts are found throughout Africa, India, and Southeast Asia. The African varieties—African white and black rhinos—are part of the Big Five. Not distinguishable by color, both species are grey-brown and massive with horns of solid keratin. Although the white rhinoceros is considerably heavier at up to 5,000 pounds, the most apparent difference is the mouth shape. The white rhino has wide, flat lips for grazing on grasses, whereas the black rhino, who prefers leaves and branches, has pointed, hook-shaped lips. Surprisingly fast, rhinos can run over 30 miles per hour.

While young rhinoceroses may be prey to some predators like big cats, crocodiles, and hyenas, the only predator of adult rhinos is humans. The horns have been used across cultures, including in traditional medicines, as an aphrodisiac, for dagger handles, and as an antidote to poison. Although CITES declared the international trade of rhinoceros horns illegal in 1977, protection efforts have been largely ineffective, and poaching increased during the twenty-first century. Various methods to prevent poaching have been tested, including horn removal and poisoning—practices that do not harm the rhino. In Namibia, rhino trophy hunting is federally supported, with profits going toward conservation, an approach other governments have proposed but some conservationists question.

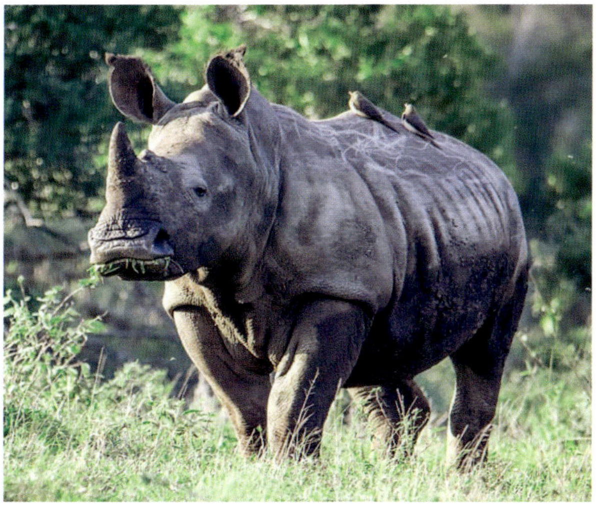

Figure 8.49 White rhinoceros. Katelyn Deelefs (@bushbabyk8), South Africa, 2019, courtesy of the photographer.

Cape Buffalo

The **Cape buffalo** (*Syncerus caffer caffer*) is one of four subspecies of African buffalo (figure 8.50). This member of the *Bovidae* family roams the continent in massive herds, grazing near bodies of water (figures 8.51 and 8.52). Its huge horns fuse across its forehead, forming a shield called a boss, which makes a powerful battering ram against its foes. Other than humans, the buffalo's only natural predator is the lion (figure 8.53). Ill-tempered and weighing around half a ton, it is a formidable and extremely dangerous beast. Unlike its Asian counterpart, the water buffalo, the wild and unpredictable Cape buffalo has never been domesticated.

The Cape buffalo is the only animal of the Big Five that the IUCN does not list as endangered, although they do categorize it as near threatened. Widely considered the most dangerous animal to hunt, wounded buffalo are known to ambush and attack pursuers. It is estimated to kill several hundred people annually, leading to hunters nicknaming it the "black death." Despite this, conflict with humans remains its biggest threat, although habitat fragmentation also impacts this wide-roaming beast.

Tyree's oil painting of an African buffalo shows the great beast in an environment of brown grasses (figure 8.54). He paints the buffalo using the same color scheme as the background. This creates a visual association between the buffalo and its landscape, implying that it is as much apart of the environment as the grasslands that surround it. With a wrinkled and weathered appearance, the buffalo seems to embody the wise old man archetype as it gazes at the viewer.

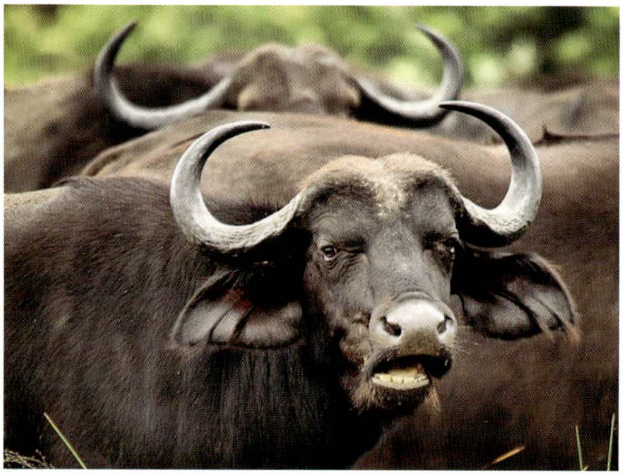

Figure 8.50 Buffalo. Katelyn Deelefs (@bushbabyk8), South Africa, 2019, courtesy of the photographer.

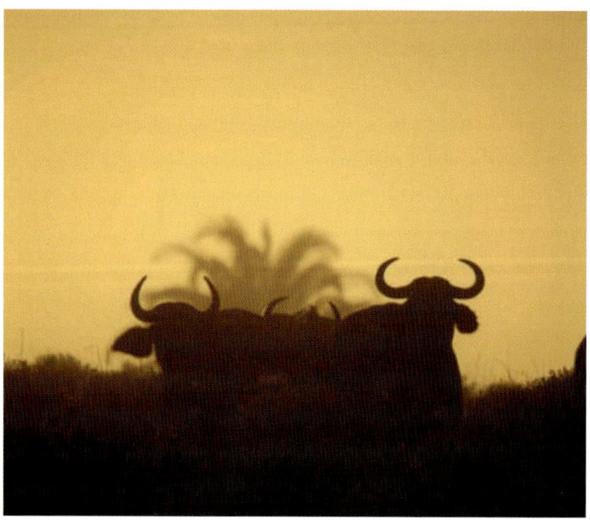

Figure 8.51 Buffalo on a misty day. Katelyn Deelefs (@bushbabyk8), Isimangalisco Wetland Park, South Africa, 2019, courtesy of the photographer.

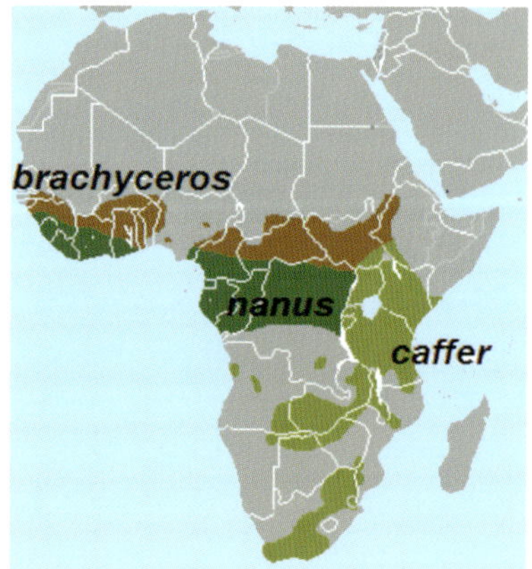

Figure 8.52 African buffalo subspecies distribution. Altaileopard, 2008.

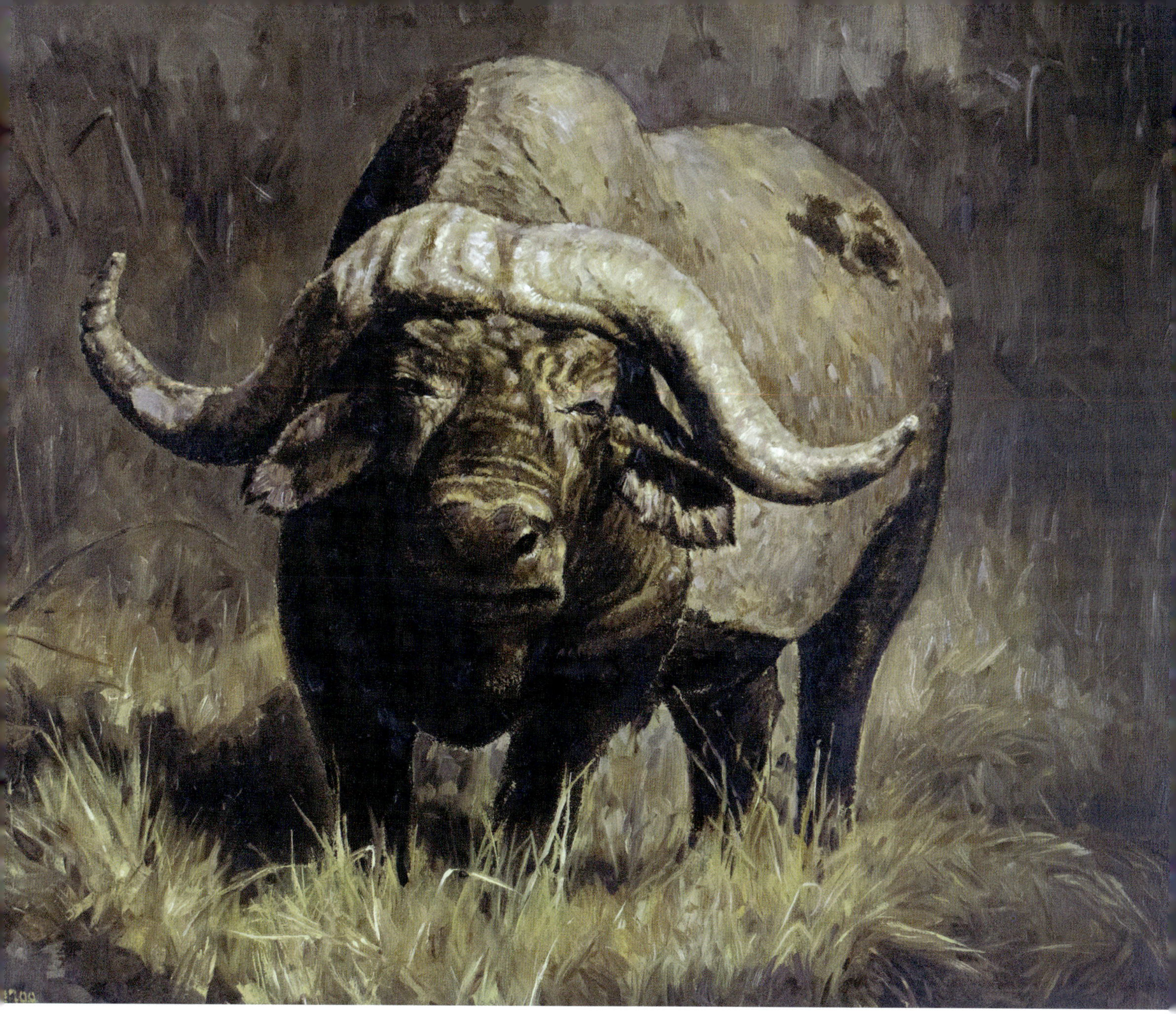

Figure 8.54 *African Buffalo*, Ralph Burke Tyree. 1975, oil on board, 20" x 24", author's collection.

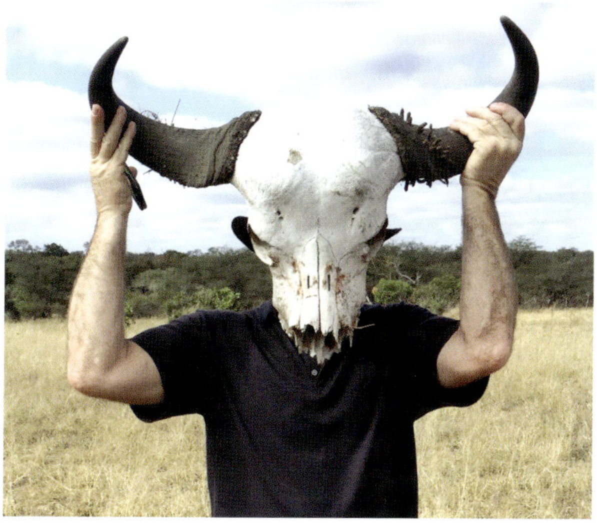

Figure 8.53 Our safari guide shows us the skull of an African buffalo killed by lions. South Africa, 2017, author's photo.

African Elephant

African elephants (*Loxodonta*) are a genus of the *Elephantidae* family made up of the two species of elephant found in Africa—the African bush elephant and the smaller African forest elephant. Both species have the characteristic grey skin with tiny hairs, large ears, long trunks, and tusks. Massive male elephants can be up to 13 feet tall and weigh as much as seven tons (figures 8.55a–c). These social animals form family groups of five to 25 elephants led by a matriarch (figures 8.56 and 8.67). Their brains have as many neurons as our own, making them one of the world's most intelligent animals, likely as smart as dolphins or primates. As nearly the largest living animal, second only to the blue whale, the African bush elephant, also known as the African savanna elephant, is the grand champion beast of our planet.

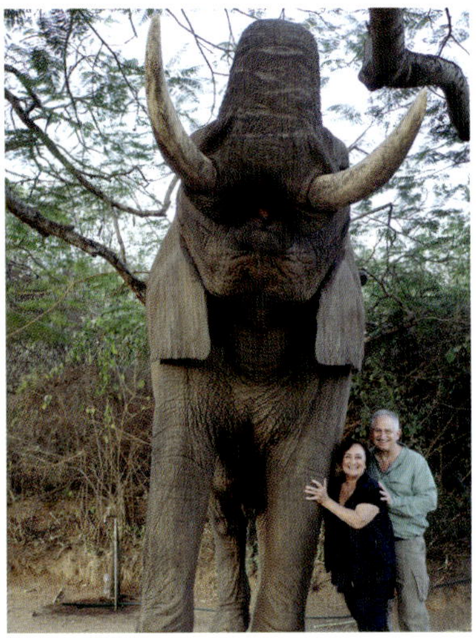

Figure 8.55a My wife and I with a bull African elephant, the largest land animal, near Kruger National Park, South Africa. 2017, author's photo.

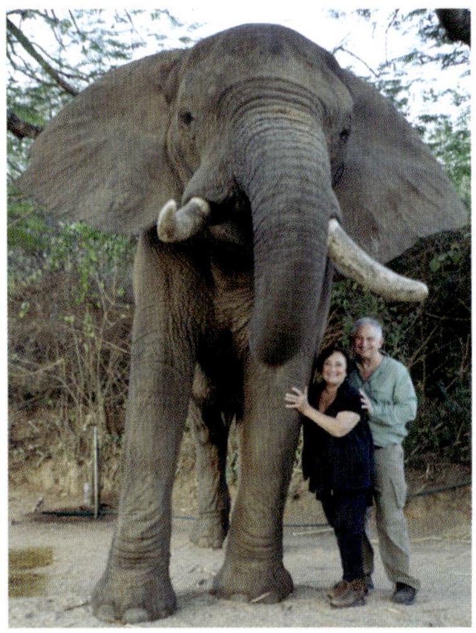

Figure 8.55b Bull elephant lowers his trunk.

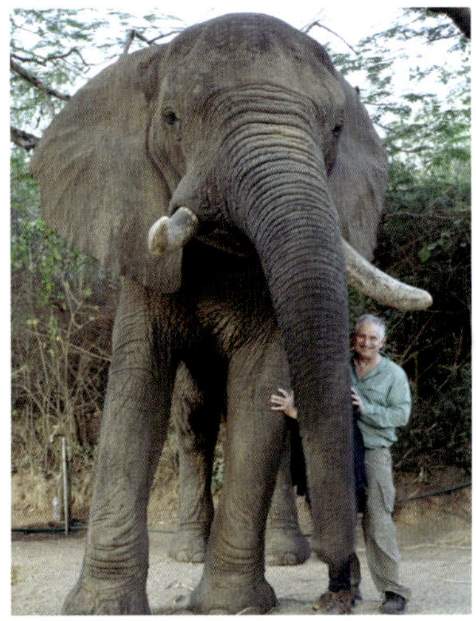

Figure 8.55c His massive trunk covers my wife.

For thousands of years, humans have used elephants as work animals. This practice is primarily associated with India and its neighboring countries, where upwards of 15,000 Asian elephants remain in use today. These elephants carry people, transport supplies, and pull wagons. To a lesser degree, the elephants of Africa were also trained as beasts of burden. Over the past decade, I have had opportunities to ride both species, as seen in figures 8.58 and 8.59—note the larger overall size and ears of the African elephant.

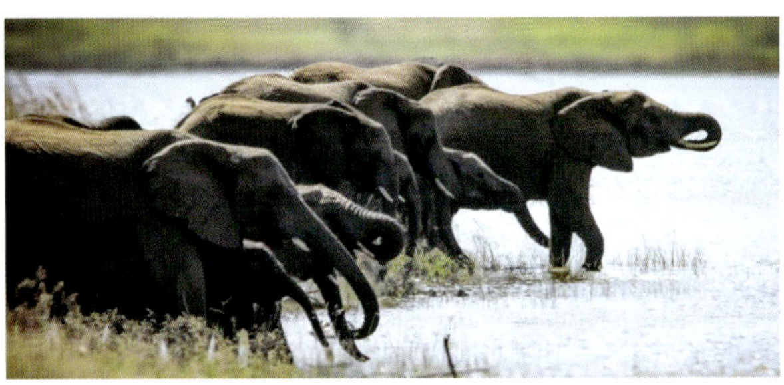

Figure 8.56 Happy Hour at the watering hole for a family of elephants. Katelyn Deelefs (@bushbabyk8), Pilanesberg Park, South Africa, 2019, courtesy of the photographer.

In ancient warfare, elephants played a similar role to today's armored tanks. Armies from Asia to the Mediterranean and into Northern Africa utilized these great beasts to varying degrees until the fifteenth century with the

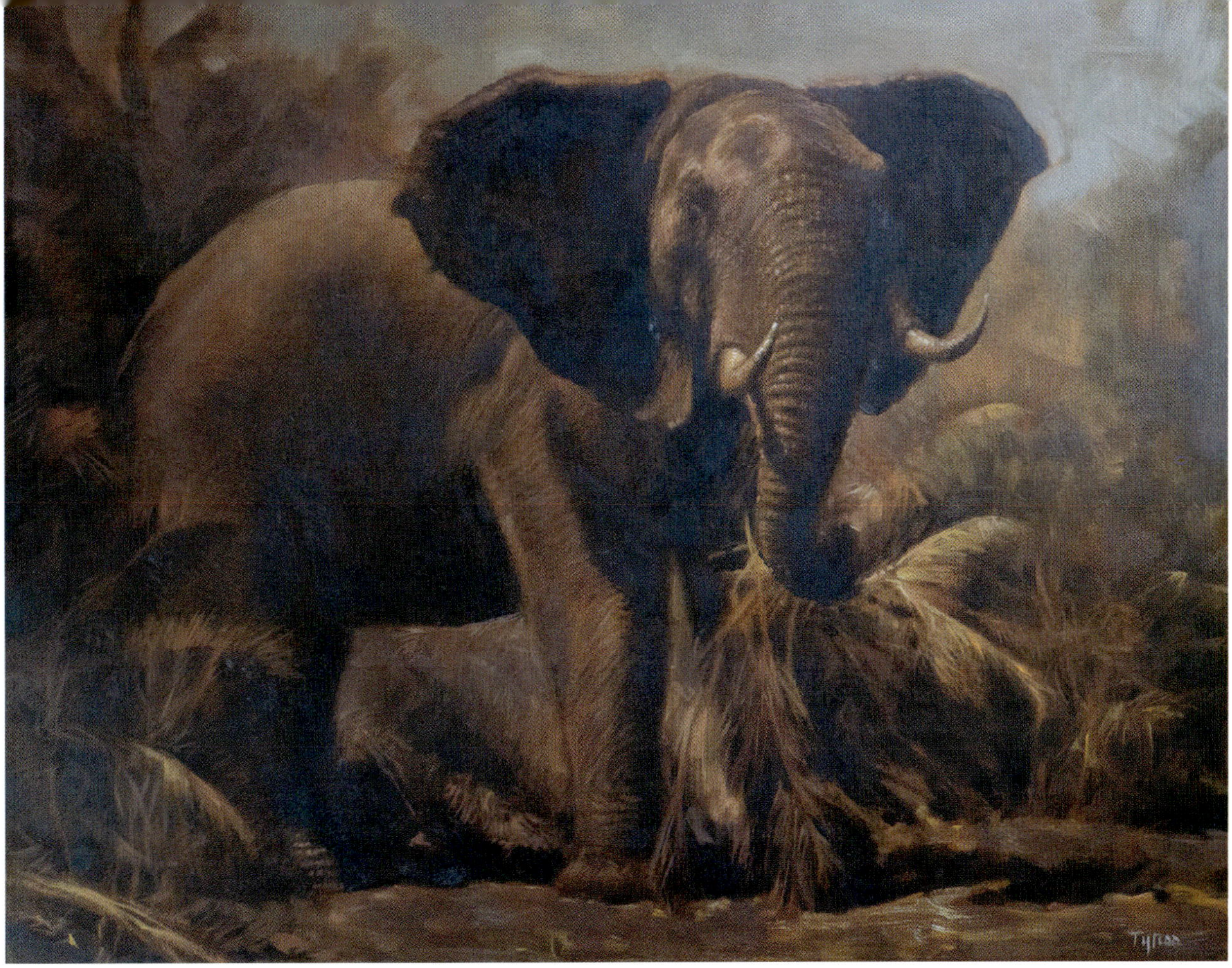

Figure 8.61 *African Elephant*, Ralph Burke Tyree. 1975, oil on board, 20" x 24", private collection.

emergence of modern warfare. However, some armies from Southeast Asia continued this practice into the nineteenth century. While carrying out his conquests throughout the third century BCE, Alexander the Great and his troops encountered stiff resistance from war elephants when facing both the Persian and Indian armies. The following century, when Hannibal and the Carthaginian army invaded the Roman Empire, they marched elephants over the Alps.

Today, elephants are often associated with circuses and zoos, where they are a central attraction and are considered entertainers. In 2016, animal cruelty complaints led Ringling Brothers and Barnum and Bailey circus to retire their elephants. Some advocates also object to these intelligent beasts being held captive in zoos, citing increased physical and mental afflictions compared to those in the wild. Since 1999, almost two dozen zoos across the US have shut down or announced plans to shut down their elephant exhibits due to the growing understanding of these giants' true nature. Notably, the Association of Zoos and Aquariums set new minimum requirements for elephant exhibits in 2011.

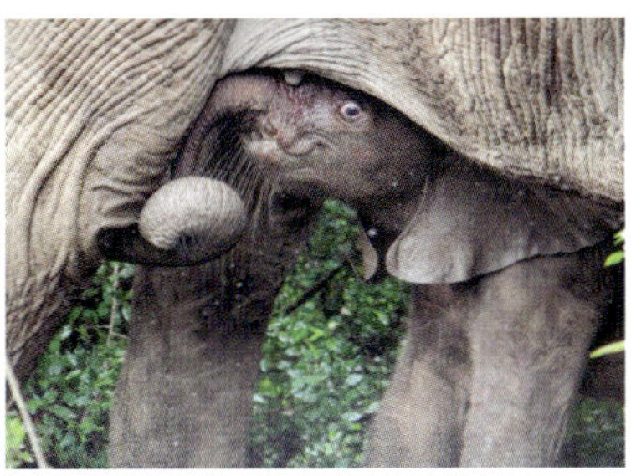

Figure 8.57 Newborn baby elephant hiding under her mother. Newborns can stand within 20 minutes and are three feet tall. Katelyn Deelefs (@bushbabyk8), South Africa, 2019, courtesy of the photographer.

Cultural references to the elephant often emphasize the beast's size and uniqueness. A person with excellent memory is said to have the "memory of an elephant," and the "elephant in the room"

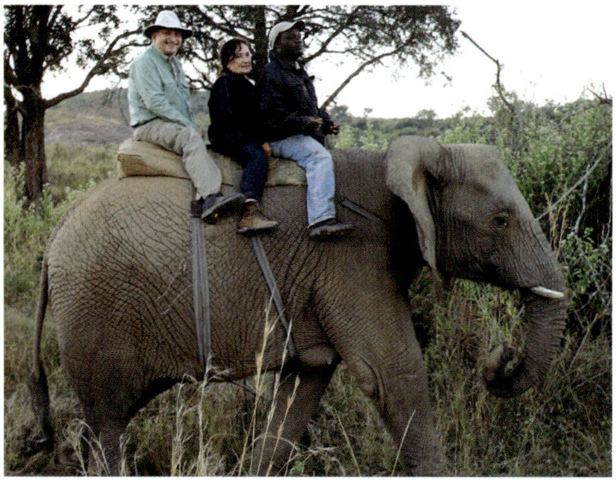

Figure 8.58 My wife and I riding an African elephant near Kruger National Park, South Africa. 2017, author's photo.

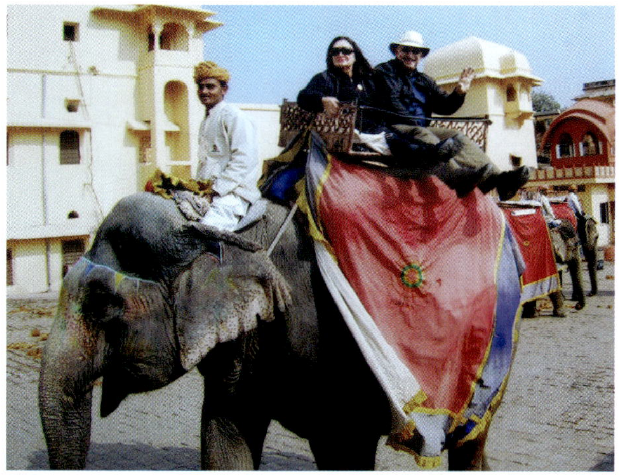

Figure 8.59 My wife and I touring on an Indian elephant. India, 2012, author's photo.

refers to an ignored obvious truth. White elephant gift exchanges are said to go back to the King of Siam, who, instead of killing members of his court who had dissatisfied him, would give them a beautiful and sacred white elephant. This lavish and impractical gift would financially ruin the recipient, who had no choice but to thank the king and find a way to care for it. Today, we no longer seek to bankrupt the recipient in a white elephant exchange, instead typically gifting bizarre, entertaining, and not particularly useful items.

Throughout history, the iconic elephant has been included in countless religions and mythologies, often representing wisdom and strength. Ganesha, one of the most widely worshipped Hindu gods, is recognizable for his elephant head. This is only one example of many, as this beast is integral to many religions and cultures, especially across Asia.

The elephant even holds symbolic significance in modern America. In 1847, the father of the contemporary political cartoon, Thomas Nast, used the elephant to represent "the Republican vote" (figure 8.60), beginning its use as a symbol of the US Republican Party. This stuck, and it has been used ever since. Children's media has featured many elephant cartoon characters—most famously, Disney's *Dumbo* (1941).

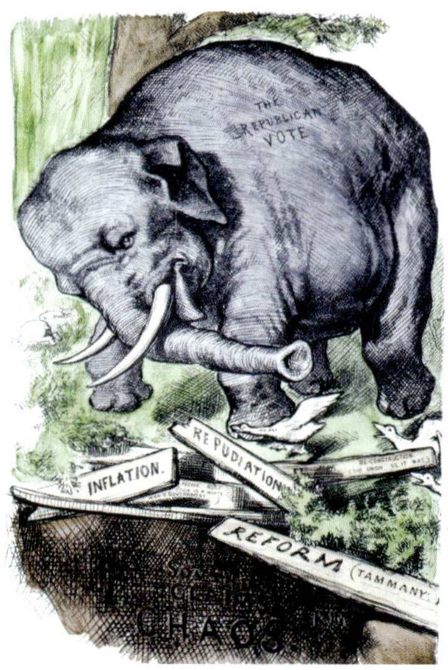

Figure 8.60 Cropped and colorized version of The Third-Term Panic, Thomas Nast. Courtesy of Harper's Weekly, November 7, 1874.

With the elephant's large size, intelligence, and propensity to charge its stalkers, especially if accompanied by calves, it was once a significant challenge to hunt. Today, high-powered rifles, scopes, guides, and fast vehicles have greatly minimized the risk and difficulty for hunters. However, this beast is still known for stampeding gun-toting enemies or, on rare occasions, innocent photographers. The IUCN considers both species of African elephant at risk for extinction due to loss of habitat and poaching for their ivory tusks, with the bush elephant categorized as endangered and the forest elephant as critically endangered. Like the rhino, park rangers are

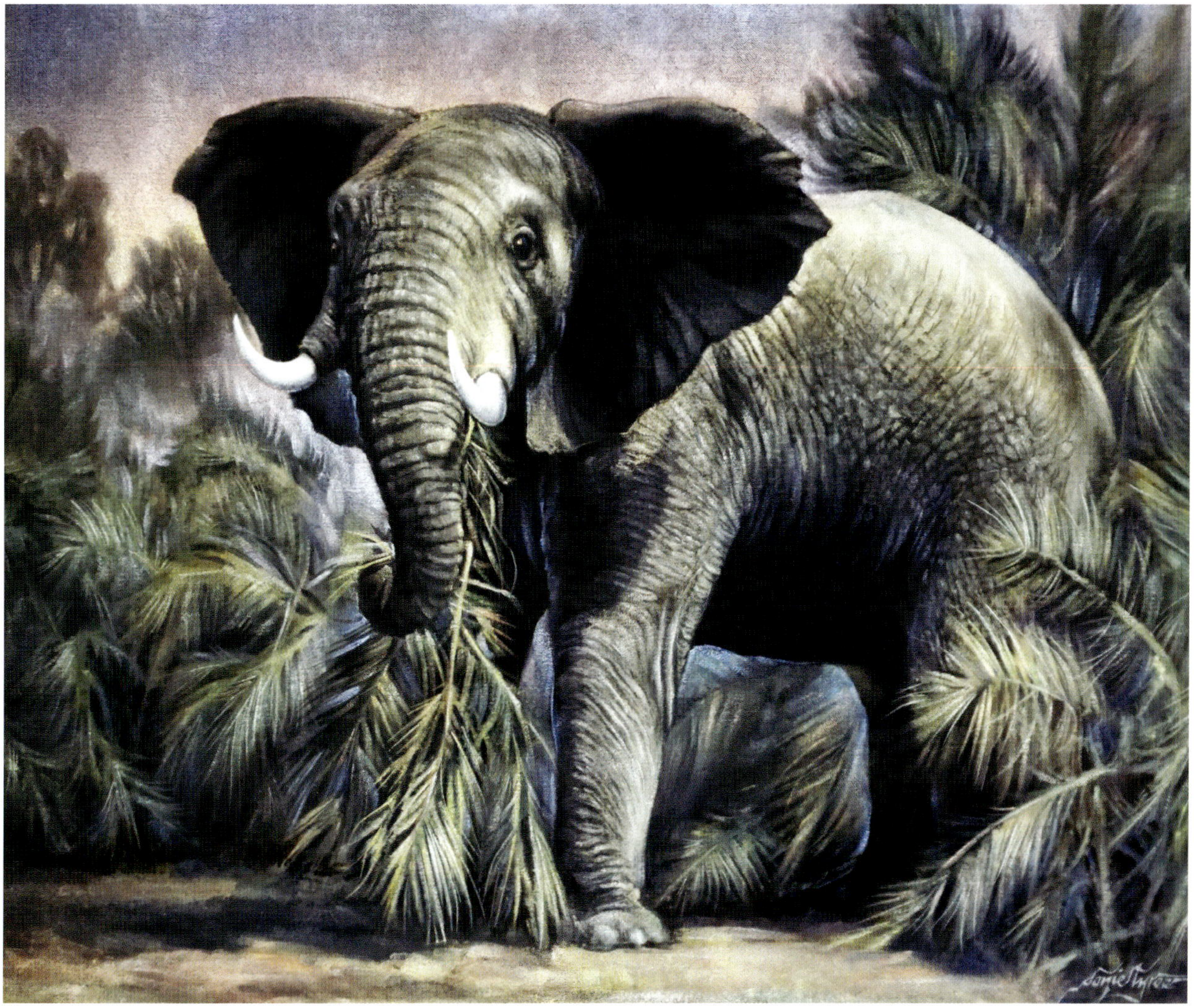

Figure 8.62 *African Elephant*, Ralph Burke Tyree. Tyree's final piece, completed by Danny Tyree. 1979, oil on velvet, 22" x 24". private collection.

at war with poachers who slaughter up to 20,000 of the estimated 415,000 African elephants yearly. They target elephants for their valuable ivory tusks, which, as of 2023, can be sold for about $1,500 a pound. A tusk from an adult male elephant can weigh up to 175 pounds, meaning a single tusk can fetch over a quarter of a million dollars, undoubtedly plenty of incentive for poachers to risk their lives against the well-armed rangers.

In 1975, Tyree painted a solitary African elephant on board (figure 8.61) as the ivory trade and elephant poaching were on the rise. The elephant stands amongst plant life, munching on what appear to be palm fronds, with its next mouthful grasped with its trunk. Like the buffalo, Tyree uses a similar color palette for the elephant and its surroundings. Again, this creates a visual association between the two, showing the elephant at one with its environment and as much a part of the African landscape as the trees and soil. Painted with great care and precision, Tyree displays the magnificence of the massive beast.

Later, in 1979, he began painting another elephant (figure 8.62). This work is a near-mirror image of the 1975 piece but on velvet. Unfortunately, while working on it, he suddenly passed away and could not complete it himself. The following year, his son Danny finished the piece.

Big Five Hunting in Africa Today

"The fate of animals is of far greater importance to me than the fear of appearing ridiculous."

—EMILE ZOLA

Today, the beasts of the African Big Five are somewhat safe as long as they stay within the nature preserves. However, armies of park rangers are still needed to protect the animals, as hunters will pursue all of them for the right price. Wealthy visitors pay tens of thousands of dollars to hunt these big, beautiful animals and bring them home as wall trophies for their dens (figures 8.63 and 8.64). A few companies offer less expensive and less harmful tranquilizer dart gun adventures, but the customer doesn't take home a souvenir.

Hunters can hire local safari companies to find whatever animal they desire. With some animals, like buffalo and elephants, these companies seek out easy targets, pursuing older buffalo and all-male elephant herds. If looking for an elephant, once an individual is selected, the guide perfectly positions the customer within firing range, as close as 20 yards. Then, with skillful maneuvering of the jeep and an easy headshot, the customer gets to take down the largest land animal on the planet and go home with a trophy. When it comes to cat hunting (figure 8.65), the process is not about finding a weak link but baiting the cat into a trap. The safari companies lure lions and leopards with freshly killed impala or zebra placed in a tree with its scent around the base. The guide also sets up a portable blind about 70 yards away, where the customer waits for motion detectors to signal that the cat has taken the bait. Once the guide has verified that the target is male, the hunter takes their shot.

In 2015, the main attraction at the Hwange National Park in Zimbabwe, a 13-year-old male lion named Cecil (figure 8.66), was infamously killed. An American dentist paid $54,000 for safari guides

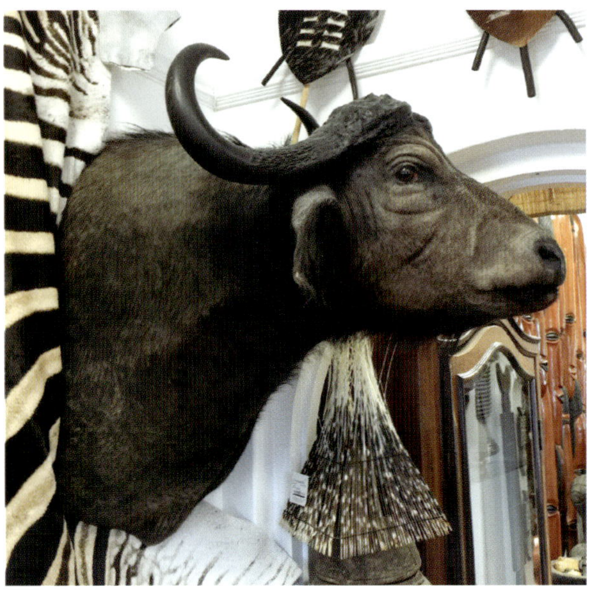

Figure 8.63 Trophy of an African buffalo in a shop. Cape Town, South Africa, 2017, author's photo.

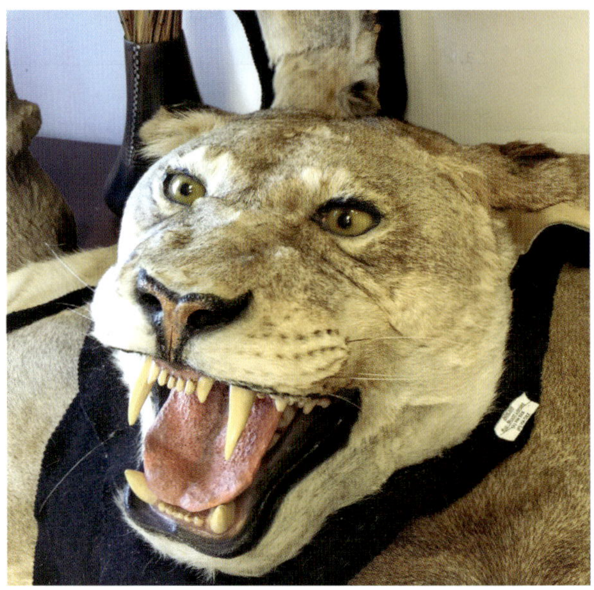

Figure 8.64 Trophy of a lion in a shop. Cape Town, South Africa, 2017, author's photo.

to lure Cecil outside the park's perimeters so he could shoot the famous lion with a compound bow. This brought an international maelstrom of outrage from news sources and social media users. Most people were unaware that in the present day, the rich continued to hunt big game.

The breaking point was the devious nature used to coax the cat out of his protective park.

After this, attitudes toward trophy hunting notably shifted, and the US Fish and Wildlife Service made it more difficult for citizens to hunt lions legally, resulting in a decline in hunters traveling to Zimbabwe.

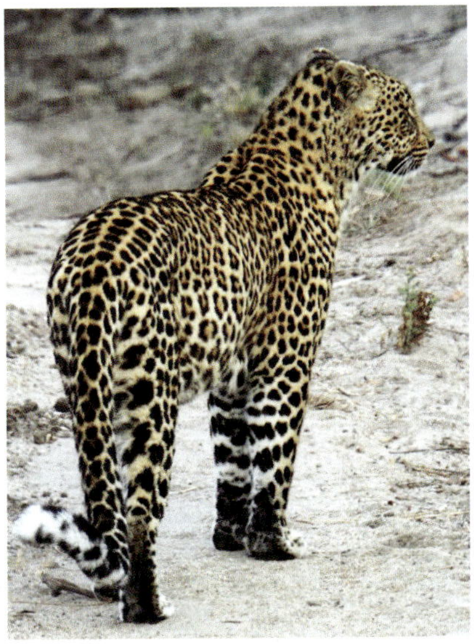

Figure 8.65 Leopard. Sadly, hunters pay to shoot these magnificent cats in Botswana, Namibia, Zimbabwe, and South Africa. 2017, author's photo.

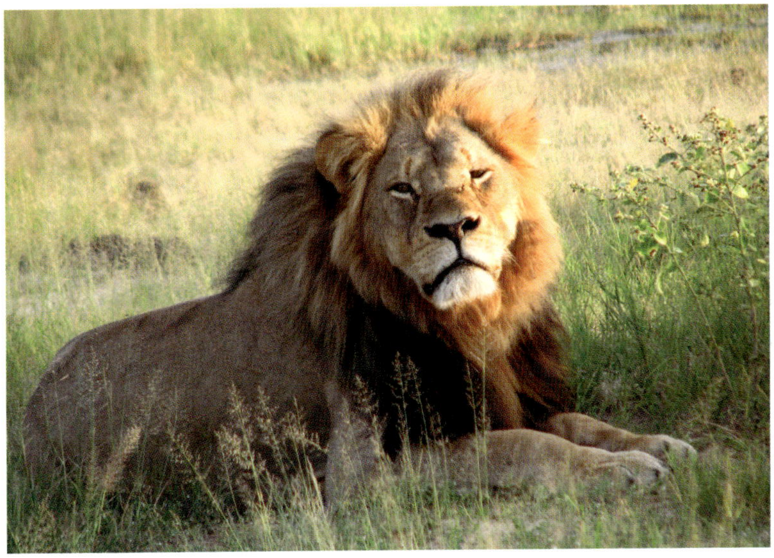

Figure 8.66 Cecil, a major tourist attraction at Zimbabwe's famous Hwange National Park. Daughter #3, 2010.

Park Overpopulation and the Culling Controversy

A side effect of park rangers' protection is that animals can flourish so much their populations outgrow their parks. Elephants are a perfect case study. This keystone species plays a critical role in their ecosystems, dispersing seeds and pulling down trees for shorter animals to eat (figure 8.67). A single elephant eats over 350 pounds of plant matter daily, often from toppled trees. Sometimes these beasts even seem to knock trees over for fun. As such, an excess of elephants will destroy forests and vegetation, causing smaller species to lose their habitats and creating biodiversity loss.

For example, the elephant population in Kruger National Park has become the subject of much debate, with some claiming that it has become overpopulated. In 2021, the park's estimated elephant population was 28,000, far greater than the 7,000 commonly considered the ideal capacity for this South African park. The reserve cannot support such a large population, but how to manage the growing number of elephants (figure 8.68)?

In some instances, park rangers perform the regulated culling of complete elephant families to quickly reduce a park's population. This is a controversial practice, with arguments on both sides. Supporters of legalized elephant culling highlight the profit governments make from selling the resulting meat, skin, bones, and ivory. On the other hand, killing elephants tends to bring bad publicity, tremendous public outcry, and even boycotts of the offending country or park. Many African reserves, including Kruger National Park, have prohibited hunting. Thus, alternative methods are required to manage elephant overpopulation.

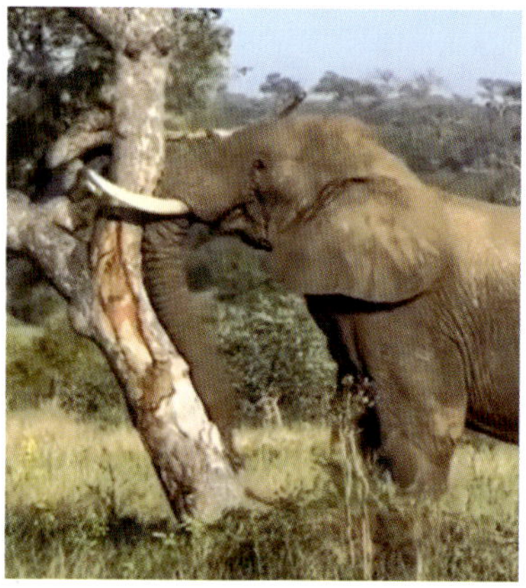

Figure 8.67 Elephant uprooting tree for leaves. TheSecretLifeof Connor (@thsecretlifeofconnor), 2022, courtesy of @africanimals.

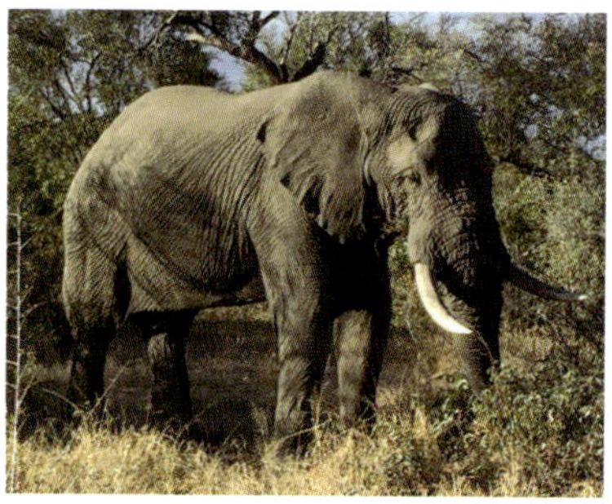

Figure 8.68 African bush elephant. Kruger National Park, South Africa, 2017, author's photo.

The most apparent solution to over-population is the relocation of some elephants to Central African countries with low populations of these giants. However, this is a costly endeavor, and suitable protected habitat is becoming scarce. Similarly imperfect, various forms of birth control are also used for population reduction. Elephant castration is a highly challenging surgery, as the elephant's internal testes are close to its spine. A vasectomy, first performed on an elephant in 2004, is far safer and faster. This procedure requires an innovative anesthesia, a sling, an enormous scope, a scalpel, and a brave veterinary team to complete the vasectomy quickly (figure 8.69–72). The male elephant retains its sexual drive but becomes infertile, which helps to decrease herd size slowly over time. Female birth control is another option. The administration of the sperm-blocking vaccine requires park rangers to shoot females with a dart. Unfortunately, this becomes an enormous task as all fertile females require vaccination, which must be readministered frequently. As such, there is no clear solution, and relocation remains the best practice if a reserve can find the funds.

The South African government has argued for the ivory trade's legalization, hoping to sell off their stockpile and make a profit while flooding the market and undercutting the poachers' business.

There is also the question of what governments are to do with the illegal ivory they have seized from traffickers. Across Africa, millions of tusks are held in storage by governments. While some African governments have argued for the legalization of the ivory trade as a solution to this dilemma, many nations and CITES oppose it, arguing that it is overly simplistic. They recall the 2008 CITES-approved one-time sale of three countries' ivory stockpiles. An elephant poaching crisis soon ensued, likely a result of this sale. Instead, opponents push for the destruction of these stockpiles to prevent the ivory from re-entering the market and ultimately deter its purchase.

As such, it remains today that elephants and the other beasts of the Big Five are somewhat protected while in their African nature preserves, but they still need armies of park rangers to protect them from poachers. Even more, poaching creates new issues with unclear solutions, like the ivory stockpile. Ultimately, difficult decisions and choices remain.

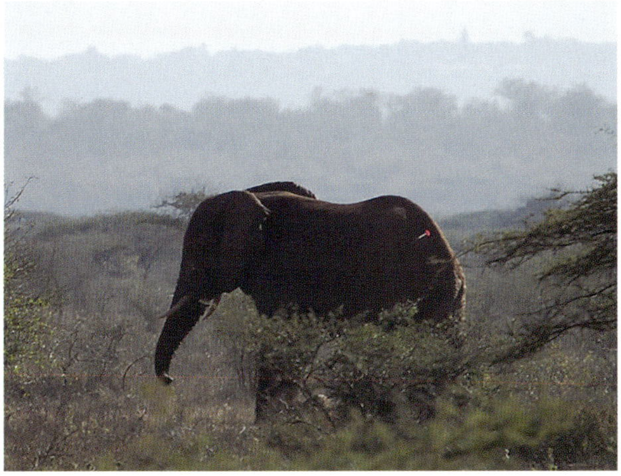

Figure 8.69 Free-ranging bull African elephant in the bush of South Africa darted in the left hip from a helicopter for surgery. 2010, all photos on this page courtesy of Jeffery R. Zuba, DVM, and Elephant Population Management Program (EPMP).

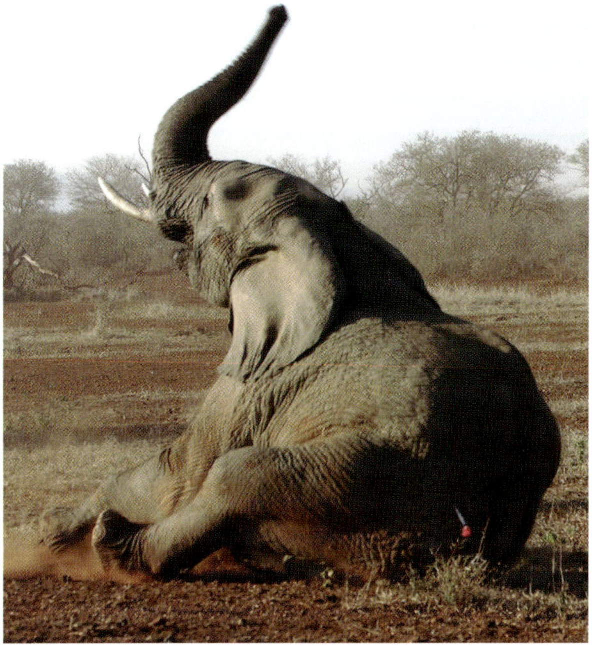

Figure 8.70 The dart causes the elephant to become recumbent—the anesthetic agent, etorphine, is approximately 6,000 times more potent than morphine.

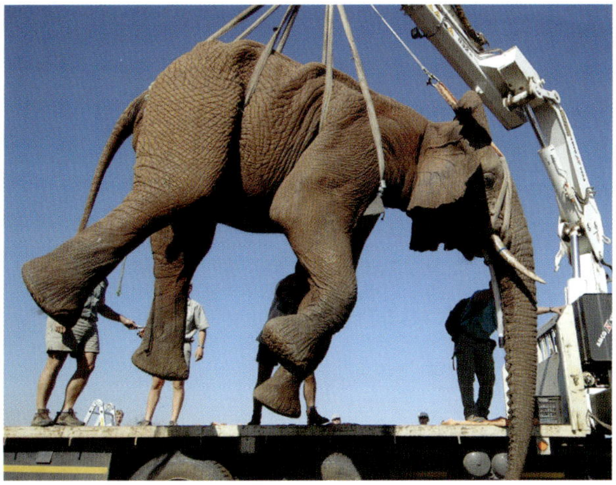

Figure 8.71 An anesthetized bull elephant is moved into a modified standing position for surgery using a sling.

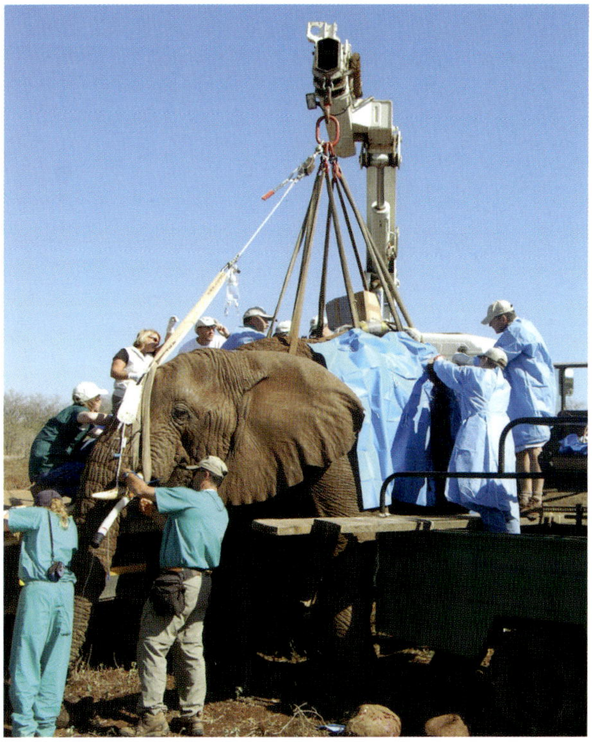

Fig. 8.72 Surgical scene. The elephant's weight is supported using a sling and a modified standing capture truck with a crane. The elephant's testes and vas deferens are intra-abdominal and located near the kidney on each side. The surgeons perform the vasectomy with specially designed elephant-sized laparoscopic equipment.

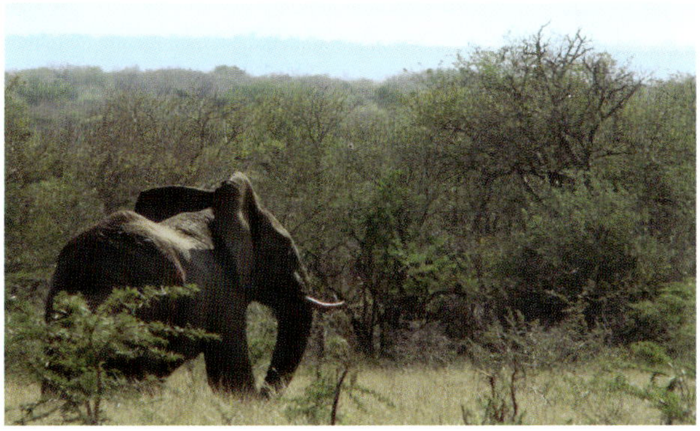

Figure 8.73 The sterilized bull elephant released back into the wild after laparoscopic vasectomy and recovery from anesthesia.

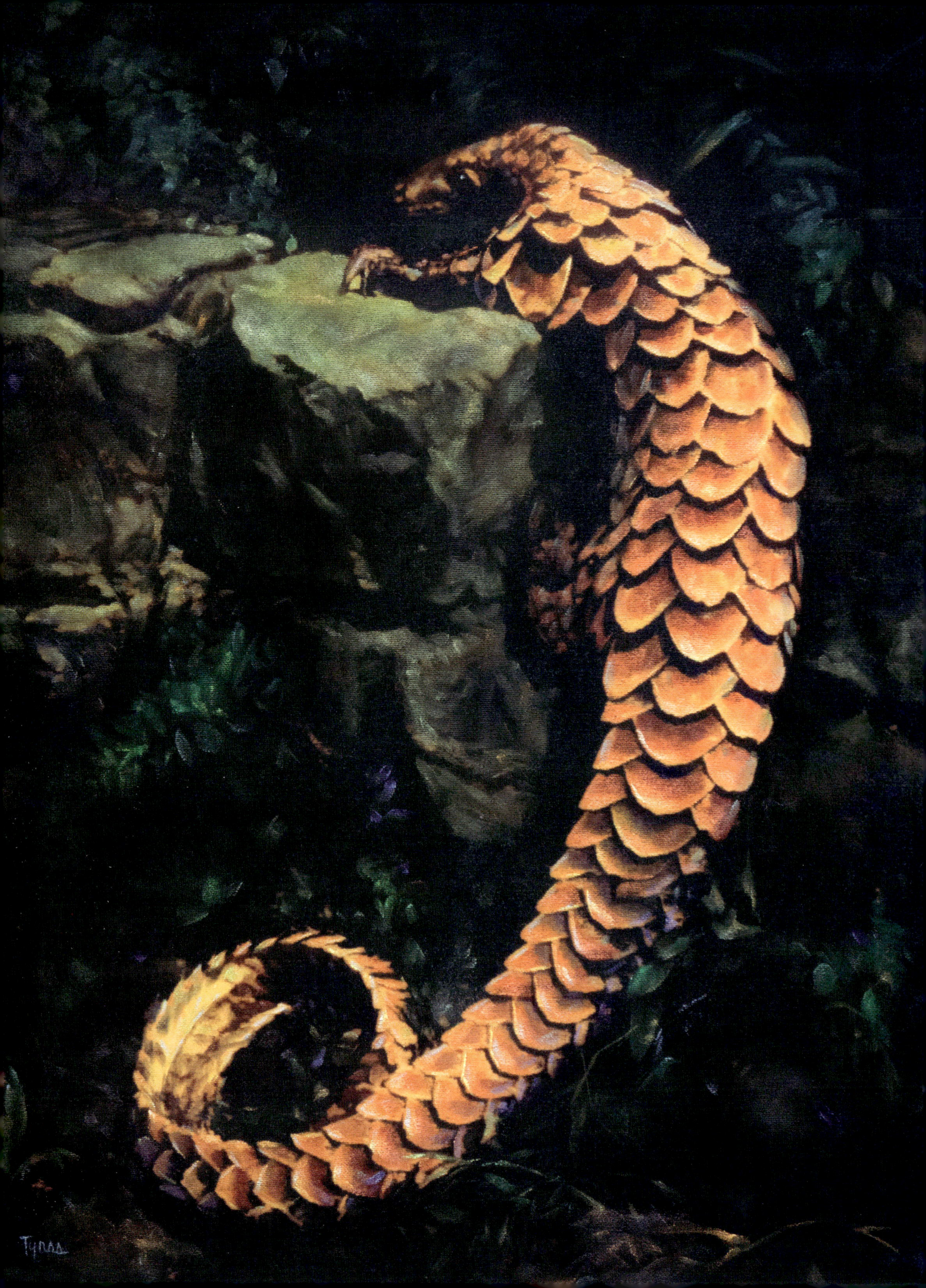

CHAPTER 9

Asian Animals

"He who rides a tiger is afraid to dismount."

—CHINESE PROVERB

Just as Jack London's stories of the South Pacific had inspired a wanderlust in Tyree, so too did London's stories of Asia. Since first reading these tales of faraway places, Tyree had been enchanted and fascinated by them—the cultures, peoples, and animals. In 1939, years before Tyree began his travels of the Pacific, he stowed away on a freighter bound for Japan, hoping to explore Asia for himself. Unsurprisingly, he was caught and spent the remainder of the voyage working on the ship. While he never returned to Asia, he was able to interact with and learn about its peoples and cultures during his time in the Pacific, when many of his friends and models were of Asian descent.

When Tyree began his animal paintings, he looked to Asia for inspiration, depicting a variety of animals from the continent. This included some of Asia's unique animals, like the Przewalski's horse, the red panda, and the pangolin. Tyree concluded his Asian subjects with the world's largest feline, the famous tiger. These endangered creatures are perfect examples of the types of animals Tyree sought to bring to the public's attention.

Green Pit Viper

The **green pit viper** (*Trimeresurus macrops*), also called the large-eyed pit viper, is a member of the *Viperidae* family and one of four green-colored pit vipers found throughout Asia (figure 9.2). This venomous snake is native to Southeast Asia, particularly Cambodia, Laos, Thailand, and Vietnam (figure 9.3). As its name suggests, it is bright green with a light stripe down its side, a dark tail, and notably large eyes, especially in the adults. This arboreal pit viper tends to make its home in trees near water sources. The IUCN does not classify it as threatened, and as such, there are no current conservation efforts to protect it.

Figure 9.4 shows Tyree's oil-on-velvet painting of a green pit viper coiled around a thin tree

Figure 9.1 *Orange Pangolin*, Ralph Burke Tyree. 1974, oil on velvet, 24″ x 18″, author's collection.

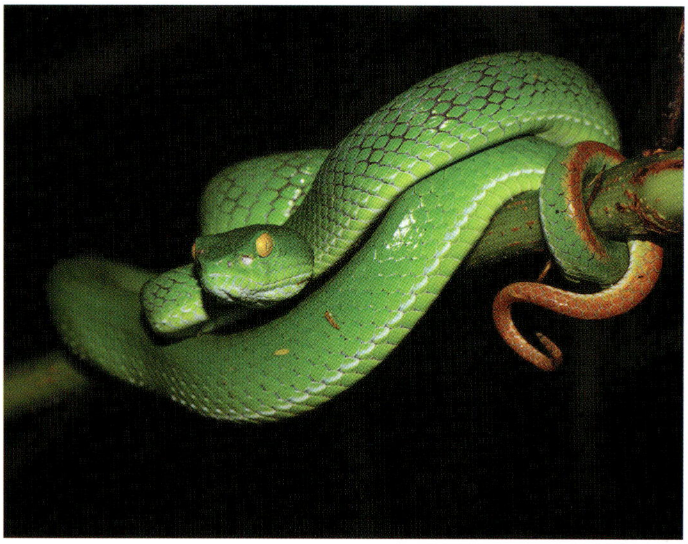

9.2 Green pit viper.Thailand, 2015, Rushenb.

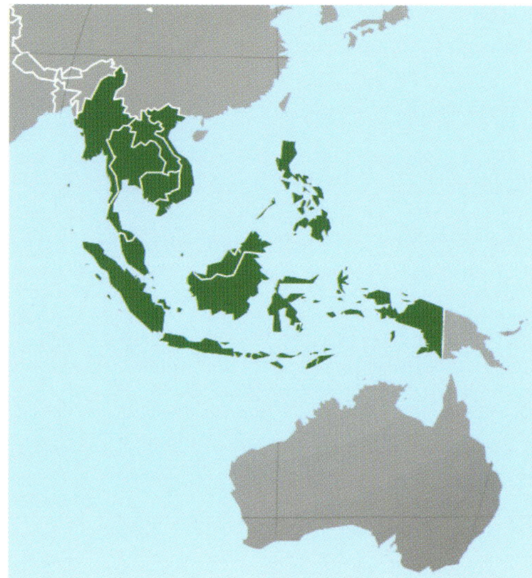

9.3 Greenpit Viper is found in Southeast Asia.

branch. The viper's striking color stands out against the dark, unfocused black and green background. He paints the snake's smooth scales with remarkable realism, illustrating the wide range of naturalistic textures Tyree could create on velvet, thus demonstrating his mastery of his materials and technique. Not only does he expertly depict the soft textures of fur but also the sleek textures of scales.

Mandarin Duck

The **Mandarin duck** (*Aix galericulata*) is an East Asian perching duck from the *Anatidae* family of water birds (figure 9.5). This wary bird prefers wooded areas near lakes or streams where it can build nests in the trees and safely retreat to the canopies. Although native to China, Japan, and Siberia (figure 9.6), the mandarin duck has escaped captivity and established populations in Europe and North

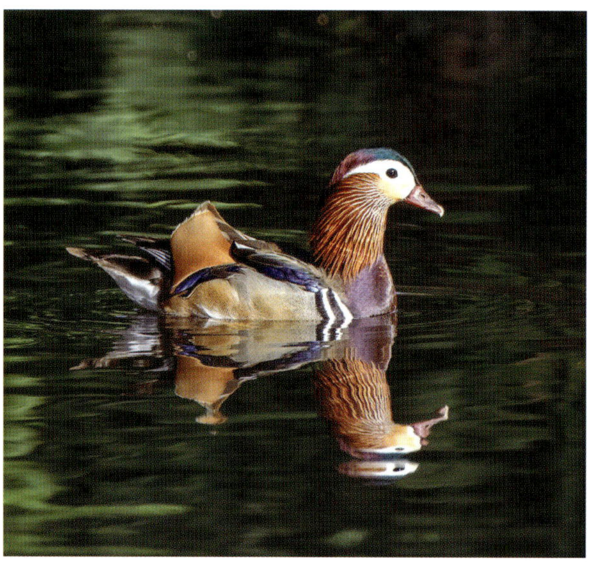

Figure 9.5 Mandarin duck on Alden Pond. Jim Bremner, Poway, California, 2021, courtesy of the photographer, www.jfbphotos.com.

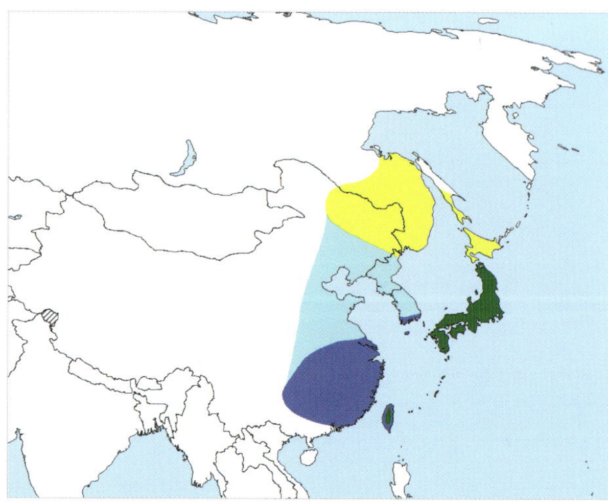

Figure 9.6 Mandarin duck distribution. MPF, 2014.
Yellow=breeding summer visitor
Green=breeding resident
Light blue=migration passage
Dark blue=winter visitor

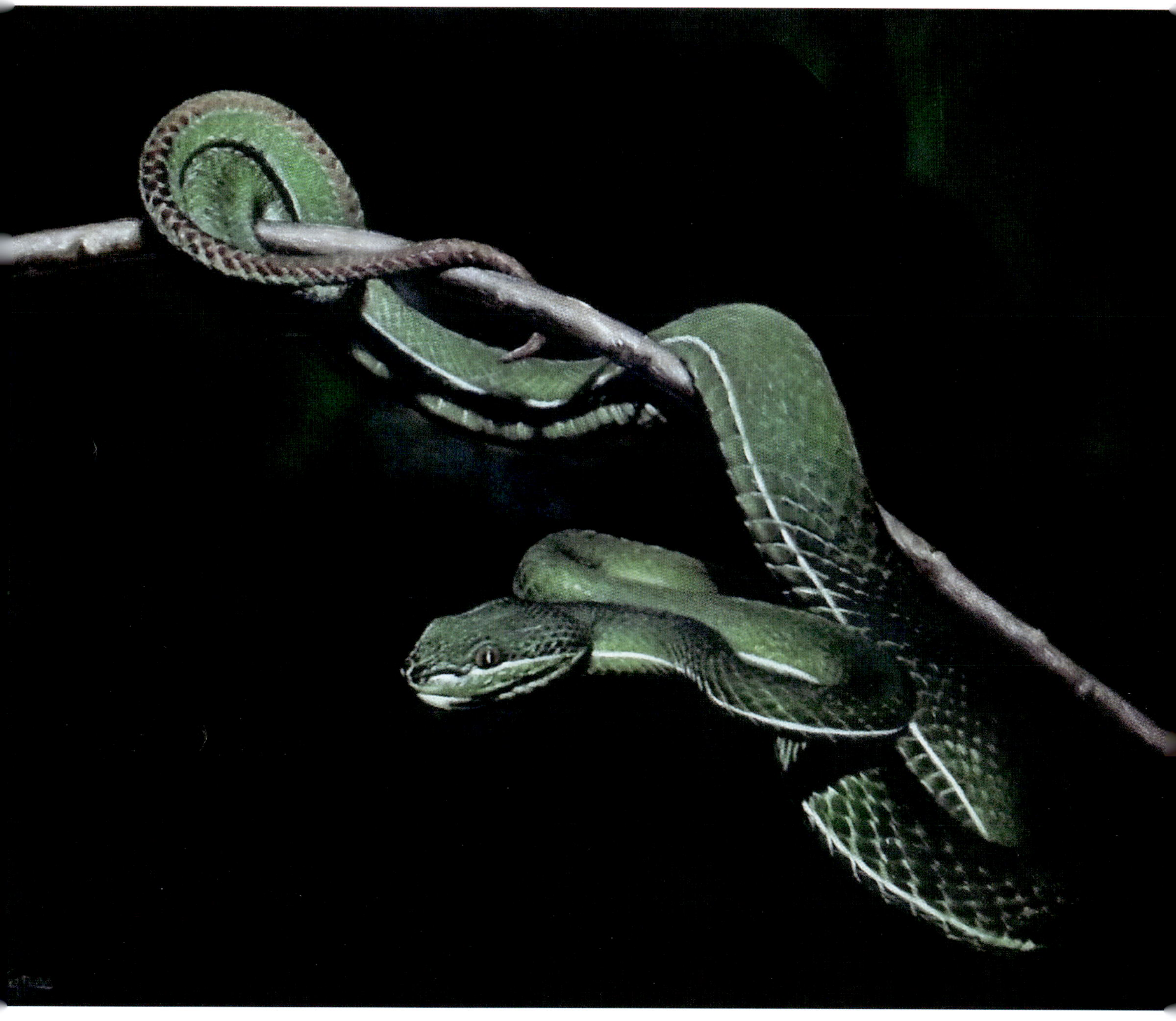

Figure 9.4 *Green Snake, Pit Viper*, Ralph Burke Tyree. 1975, oil on velvet, 20″ x 24″, Tyree family.

America. The Mandarin is closely related to its North American cousin, the wood duck, which it easily intermingled and interbred with when it first arrived in North America (figure 9.7).

Migrating in small flocks of roughly ten birds, it is a fantastic, nimble flyer that can land effortlessly on a tree branch like a sparrow.

With a red bill and dramatic markings of white, black, red, orange, violet, blue, and tan, the male mandarin duck is one of the prettiest birds. This plumage is especially stunning in comparison to his female counterpart's brown and white feathers (figure 9.8). This pair has come to symbolize lovebirds and themes of wedded bliss, fidelity, and peace across Chinese, Korean, and Japanese cultures due to the belief that they mate for life. While they do not seem to mate for life, they are monogamous, forming bonds that last over multiple breeding seasons.

While Mandarin duck populations are decreasing, the IUCN does not consider it endangered. Logging has been and continues to be detrimental to this bird's native habitats. It also continues to be hunted, not as a food source, but for its beauty or even by accident due to its similar appearance to other ducks that are hunted for food.

Tyree liked painting the male Mandarin duck and did so on multiple occasions. In Tyree's painting of a male mandarin duck (figure 9.9), he utilizes the velvet to accentuate the bird's vivid colors, soft feathers, and rippling reflection. The lone duck is seen in profile, swimming, with his unique feathers brightly illuminated and mirrored on the inky water's surface. The mandarin duck and its stunning plumage are the only focus in these dark surroundings.

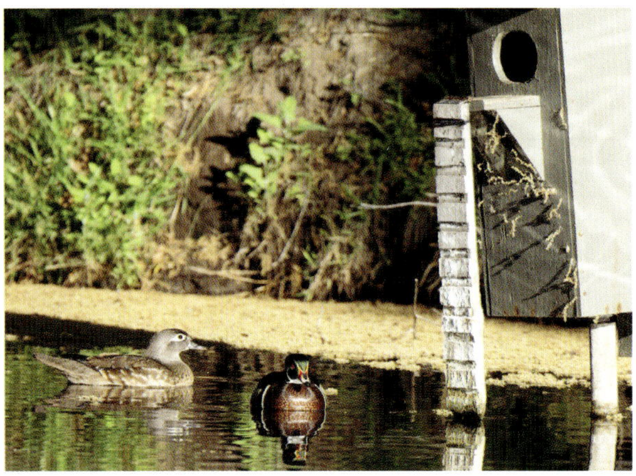

Figure 9.7 A pair of wood ducks (Mandarin duck's North American cousin) swimming below their nest box at Alden Pond. Mandarins use the same type of nest boxes. Jim Bremner, Poway, California, 2021, courtesy of photographer, www.jfbphotos.com.

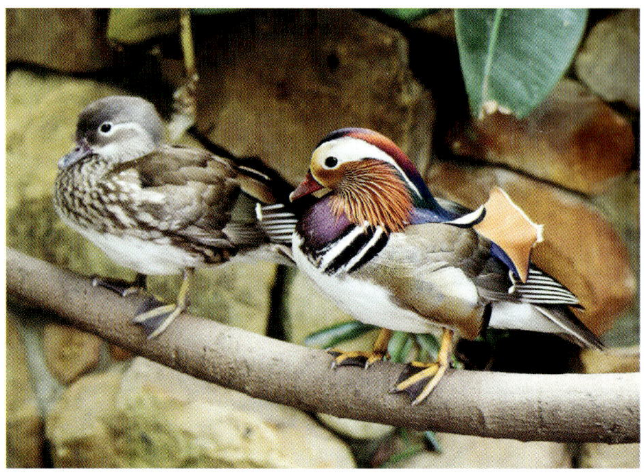

Figure 9.8 A mandarin duck couple on a branch, female (*left*) and male. Note the long white tear on the mandarin hen compared to the wood duck hen seen in figure 9.7. Ron Porter, 2013.

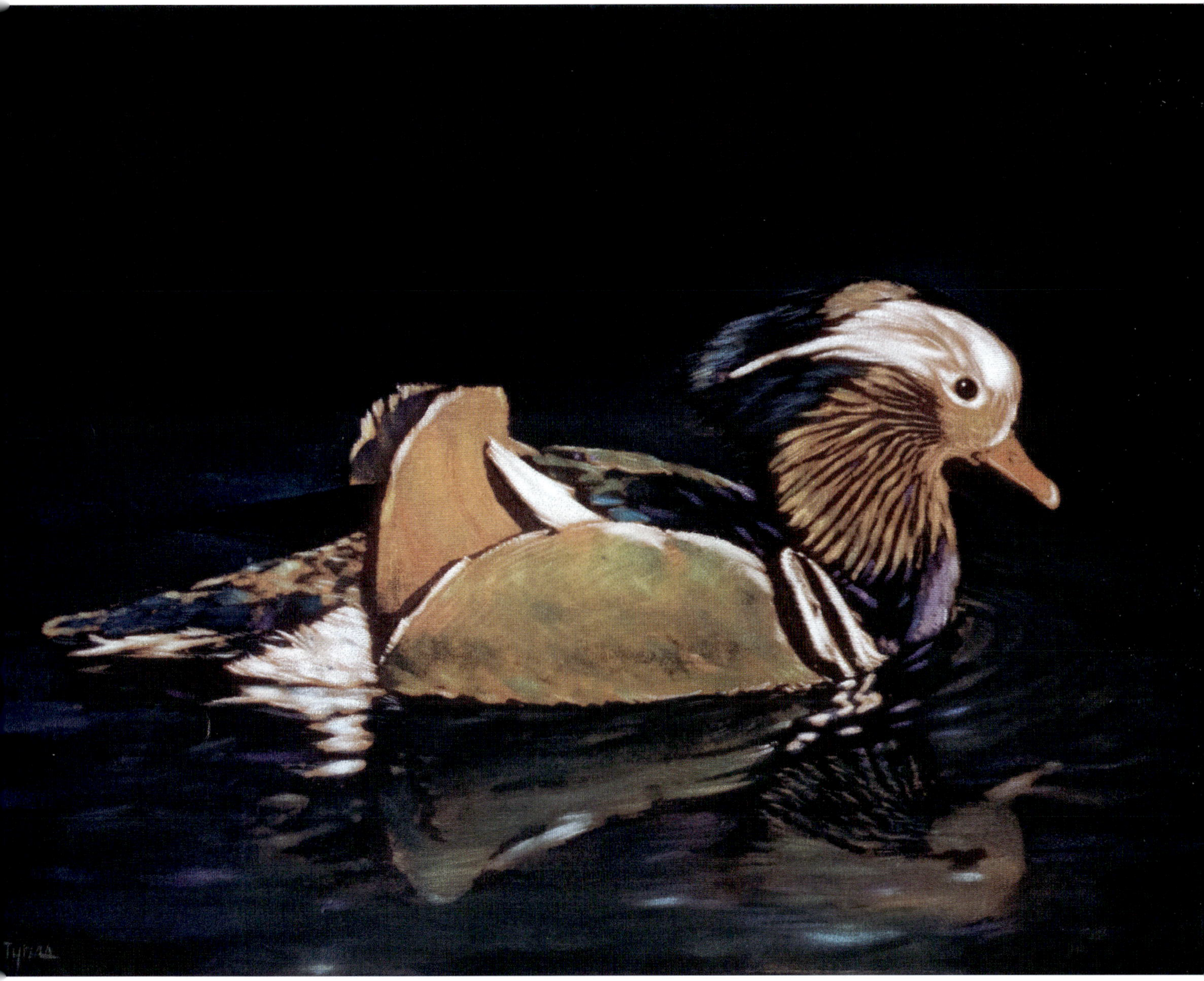

Figure 9.9 *Male Mandarin Duck*, Ralph Burke Tyree. 1976, oil on velvet, 16" x 20", author's collection.

Bay Owl

Bay owls (*Phodilus*) are a genus in the *Tytonidae* family of barn owls, but at only about ten inches long, they are smaller than other barn owls. This genus is recognizable for its V-shaped faces and feather tufts that resemble ears (figure 9.10). The three species of bay owls inhabit the forests and grasslands of South and Southeast Asia and parts of Central Africa. Preferring to eat small rodents, birds, snakes, and insects such as beetles and spiders, they hunt from perches low in the trees. All three species face the threat of habitat loss as humans destroy their forests. This has particularly impacted the Congo bay owl from Africa, leading the IUCN to classify it as endangered. The IUCN does not consider the other two species, the Oriental and Sri Lankan bay owls, at risk, although they face similar habitat loss.

In Tyree's painting of a bay owl, the bird swivels its head to peer straight at the viewer with its huge, shiny eyes (figure 9.11). In the deep darkness of the nighttime jungle, the owl holds all the focus as its light plumage nearly glows. Reminiscent of the young Thomson's gazelle gazing at the viewer, it almost seems as if the viewer has walked up on and disturbed the owl.

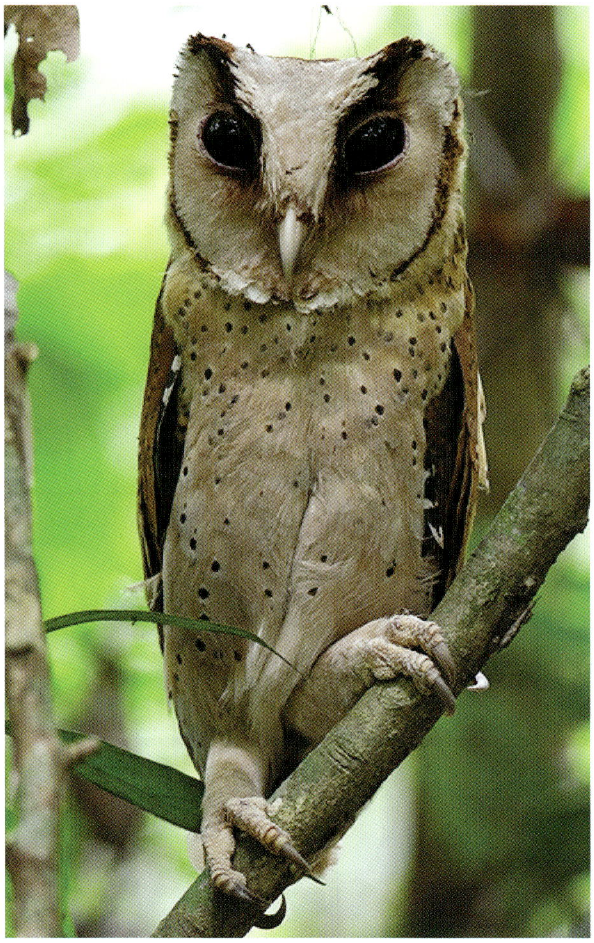

Figure 9.10 Sri Lankan bay owl. Dr.abhilash.ak, Kerala, India, 2015.

Figure 9.11 *Bay Owl*, Ralph Burke Tyree. 1976, oil on velvet, 16" x 12", author's collection.

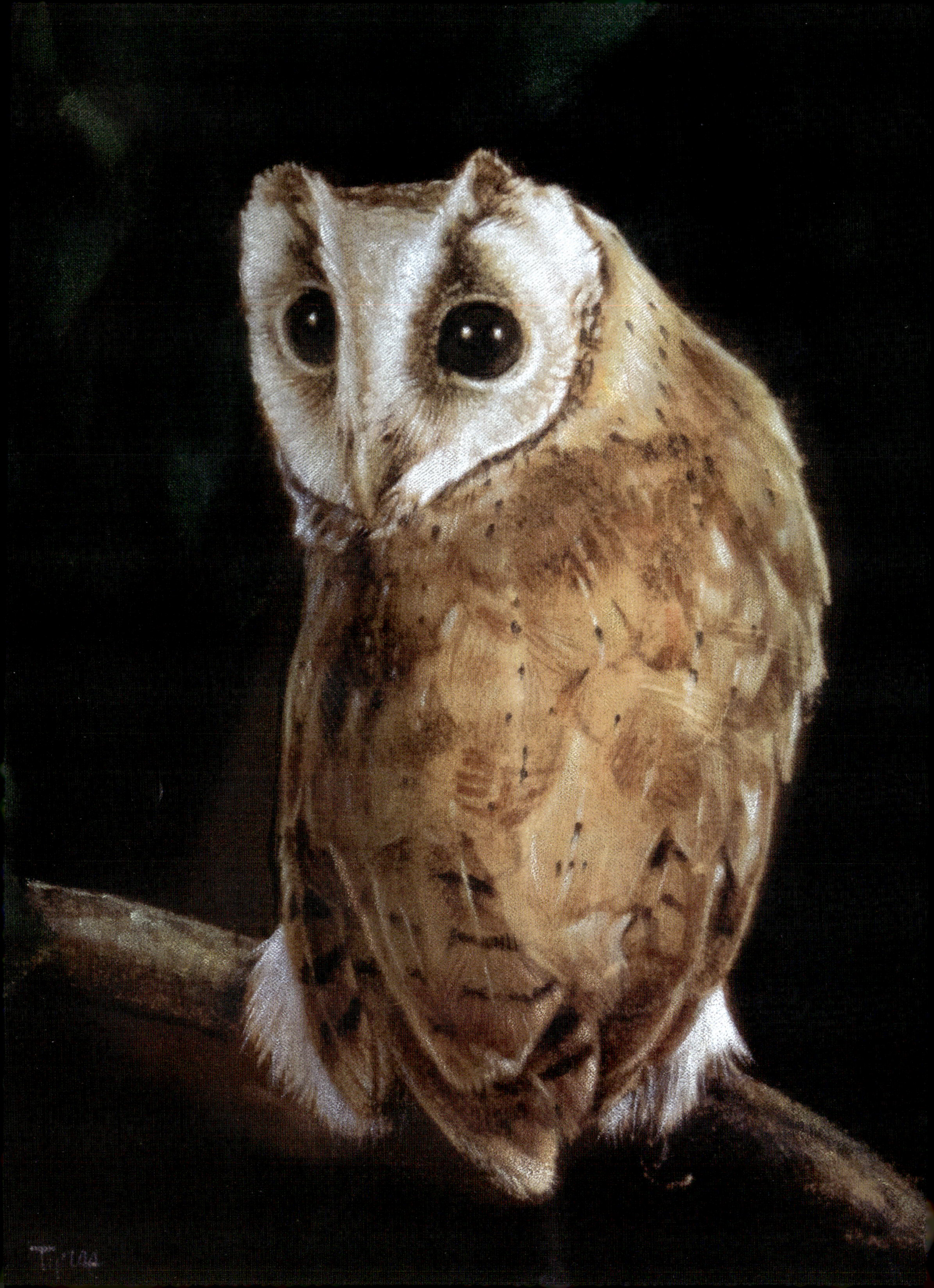

Pangolin

Pangolins (*Pholidota*) are an order of eight species of mammals, sometimes referred to as scaly anteaters (figure 9.12). Found in Asia and Africa (figure 9.13), they range in size from 12 to 39 inches and have large, protective keratin scales unique to the order (figure 9.14). These primarily nocturnal mammals eat insects, using their long tongues to imbibe ants and termites similar to giant anteaters. Pangolins split from mammals 80 million years ago, when dinosaurs still roamed the earth, making them distinctive from an evolutionary standpoint.

Not only are pangolins considered a delicacy in places like China and Vietnam, but their scales have long been used in traditional medicines Due to this, these creatures are now the most trafficked mammals on the planet (figures 9.15–9.17).

Despite an international ban on their trade and a crackdown on smuggling routes, hunting continues, putting certain species at risk of overhunting. Deforestation and associated habitat loss have only worsened this orders' circumstances. As such, the IUCN classifies all eight species of pangolins as threatened, with three considered critically endangered. In 1989, Taiwan passed the Wildlife Conservation Act, introducing Wildlife Rehabilitation Centers that work with the police to prevent poaching and trafficking. With this, Taiwan has become a leader in pangolin conservation and has the highest population density of these animals worldwide. In the early days of the COVID-19 pandemic, some authorities blamed the virus on pangolins in the wet markets of Wuhan, China. While this theory has been disproven, this accusation likely led to the mass slaughter of these gentle and endangered animals.

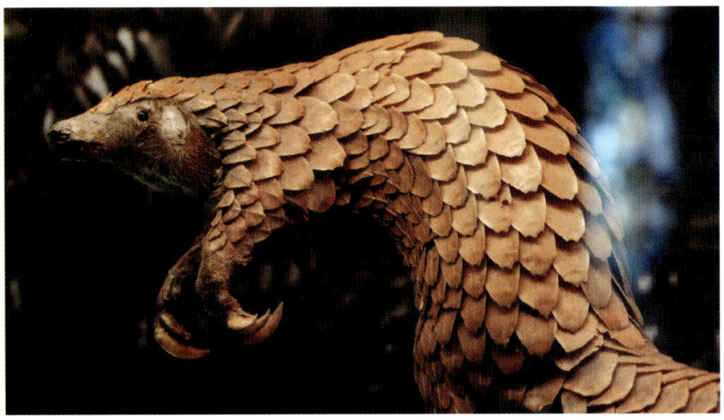

Figure 9.12 Pangolin. Louis Mornaud on Unsplash.

Tyree's painting of a pangolin (figure 9.1) seems to depict an African tree pangolin based on its long, curling tail and orange color. This arboreal genus of pangolins is relatively small and uses its long tail to balance while climbing trees, as shown in Tyree's painting of a pangolin climbing up a boulder. Its bright color stands out against the dark backdrop of rocks and leaves, drawing the viewer's eye to its unique scales. From there, the eye travels down the length of its tail before looping back around to the details of its face and environment. Through this, Tyree focuses all the attention on the scales that make these animals so sought out.

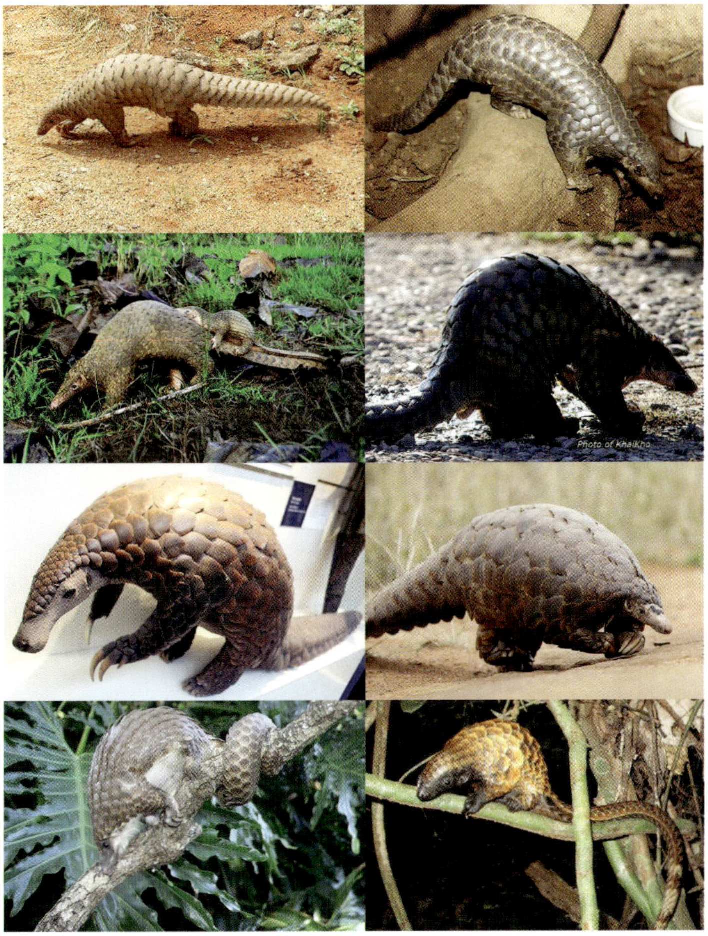

Figure 9.14 The eight species of pangolins, four from Asia and four from Africa. The Explaner, 2021.

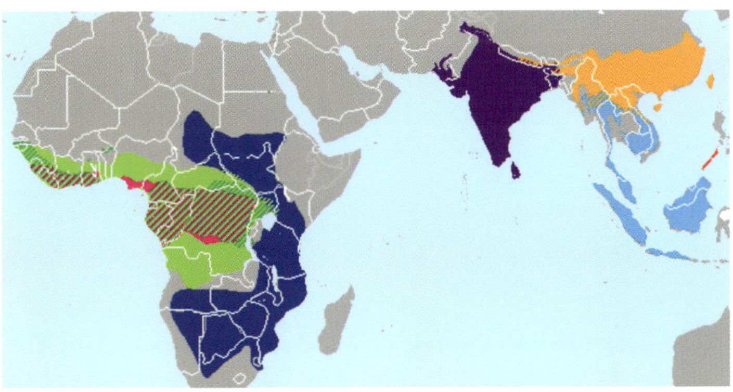

Figure 9.13 Pangolin species distribution. Craig Pemberton, 2011.

Violet=Manis crassicaudata
Orange=Manis pentadactyla
Cyan=Manis javanica
Red=Manis culionensis
Yellow green=Phataginus tricuspis
Magenta=Phataginus tetradactyla
Green=Smutsia gigantea
Blue=Smutsia temminckii

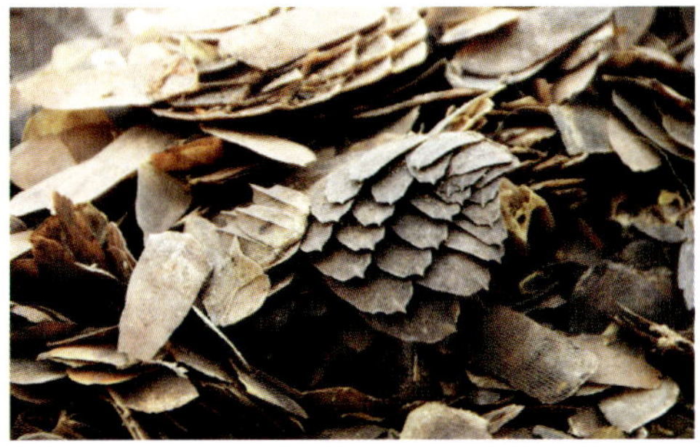

Figure 9.15 Pangolin scales confiscated from a wet market. Kenneth Cameron, 2017, courtesy US Fish and Wildlife Service Headquarters.

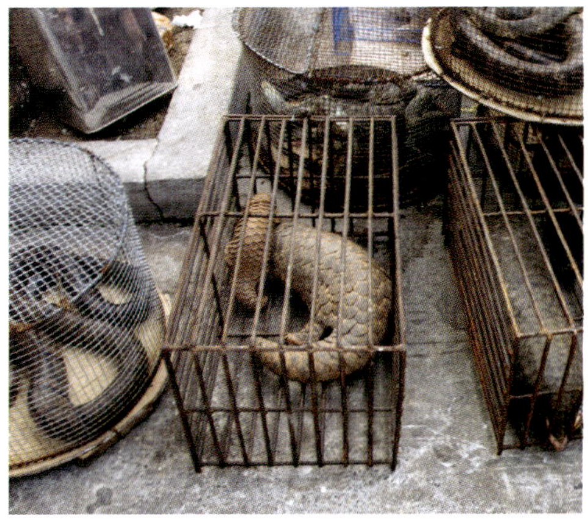

Figure 9.16 Two pangolins in a cage in a Myanmar wet market. Dan Bennett, 2008.

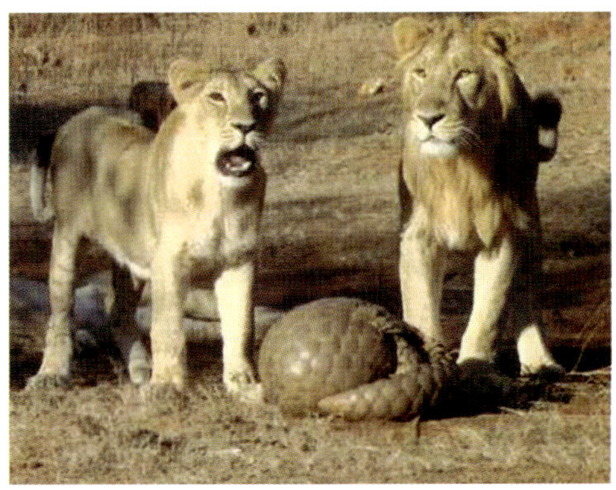

Figure 9.17 Pangolin rolls up into a ball in a defensive move to avoid attack by these young Asiatic lions. However, a human trafficker can easily pick up this docile animal and take it to the market. Sandip kumar, Gujarat, India.

Red Panda

The **red panda** (*Ailurus fulgens*) is the only living species in the *Ailuridae* family (figure 9.18). This small carnivore from southwestern China and the eastern Himalayas (figure 9.19) is known by many names, including the lesser panda, the red bearcat, the red cat-bear, and the firefox. While aptly named for its copper-colored fur, the red panda is only distantly related to the giant panda. Instead, it is more closely linked to fellow members of the *Musteloidea* superfamily, like the raccoon, the weasel, and the skunk. This colorful animal is roughly the size of a house cat and has a long, shaggy tail. While it enjoys eggs, birds, and insects, it primarily eats bamboo. This arboreal species is shy, solitary, and nocturnal, sticking to forests with dense bamboo cover. All of these factors make the red panda challenging for humans to observe.

The IUCN has considered the red panda endangered since 2015, as less than 7,000 remain in the wild. Despite protection, the species is threatened by poaching for its unique fur and habitat loss due to deforestation.

"Today, they are endangered as farmland is pushing into their once-remote range and eliminating their forest habitats, while their beautiful pelts are widely traded as decorations."

Figure 9.20 shows Tyree's oil painting on velvet of a red panda, in which he puts all the focus on the animal's coat and tail. The panda sits in the fork of a tree, gazing at the viewer with its tail held up in front of its body. The green foliage of the forest creates a striking contrast with the panda's red-orange fur. As the light bounces off the animal's coat, it creates a beautiful orangey glow. Through this, Tyree brings attention to the very parts of the red panda hunters target it for, its vibrant fur and bushy tail.

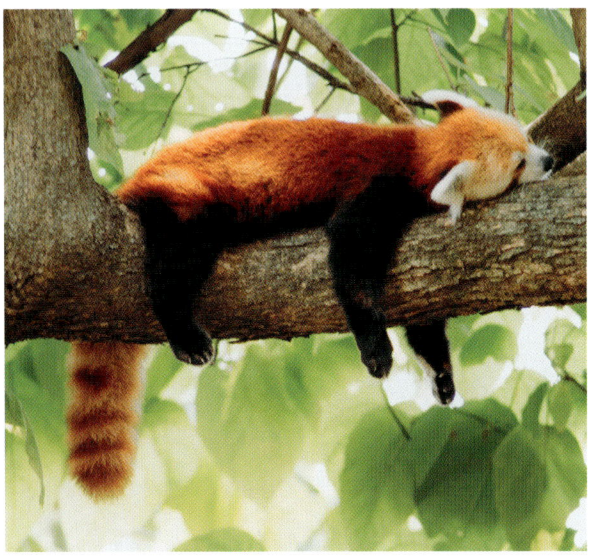

Figure 9.18 Red panda resting in a tree. Kalen Kemp on Unsplash.

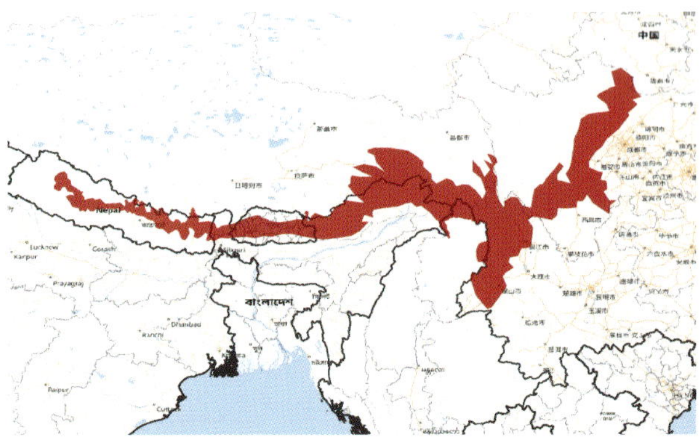

Figure 9.19 Red panda distribution. BhagyaMani, 2022.

Figure 9.20 *Red Panda*, Ralph Burke Tyree. 1973, oil on velvet, 20" x 16", author's collection.

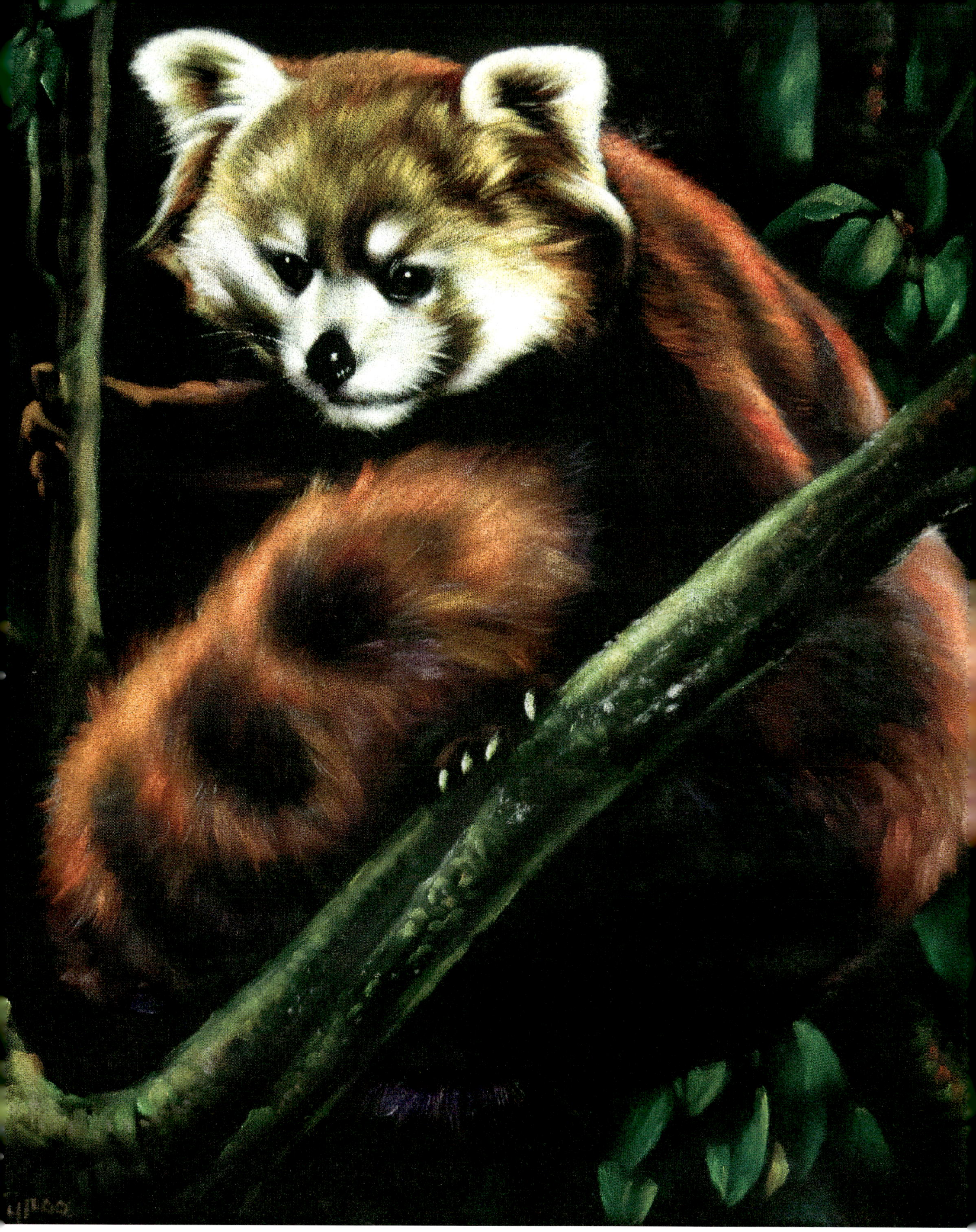

Loris

Lorises (*Lorinae*) are a subfamily of small primates from the *Strepsirrhini* order that are native to Southeast Asia (figures 9.21 and 9.22). All ten species have soft grey-brown fur, wet noses, round heads, large, outlined eyes, and small ears. Like their cousins, the lemur and bushbaby, they are nocturnal primates of the high rainforest and prefer to eat fruits and insects.

Most loris species are threatened, with six considered endangered or critically endangered. Habitat loss associated with deforestation is a significant threat to these forest dwellers. Still, their primary danger comes from trade and poaching, despite protection by law. Thought to have mystical and curative properties, parts of the loris are sold for use in traditional medicine. They are also popular in the exotic pet trade and are sometimes even sold for their meat.

Tyree created two paintings of lorises, one on board (figure 9.23) and the other on velvet (figure 9.24). The compositions are almost identical, with both lorises grasping onto a branch while stalking a butterfly against an abstract background. Their curious, child-like smiles humanize the tiny primate. Aside from the medium used, the only variations between the two paintings are the colors of the loris and the butterfly. As such, these paintings serve as a perfect example of the unique textures a skilled artist can achieve when painting on velvet, as compared to board.

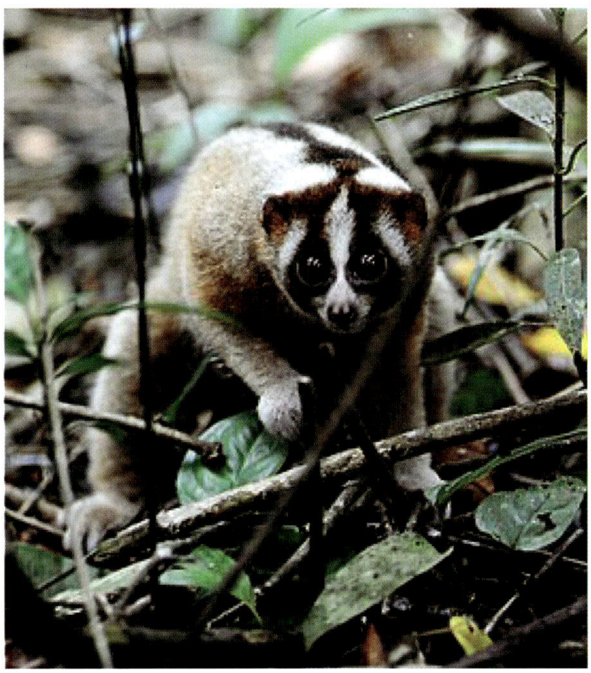

Figure 9.21 Javan slow loris. Jefri Tarigan, West Java, Indonesia, 2016.

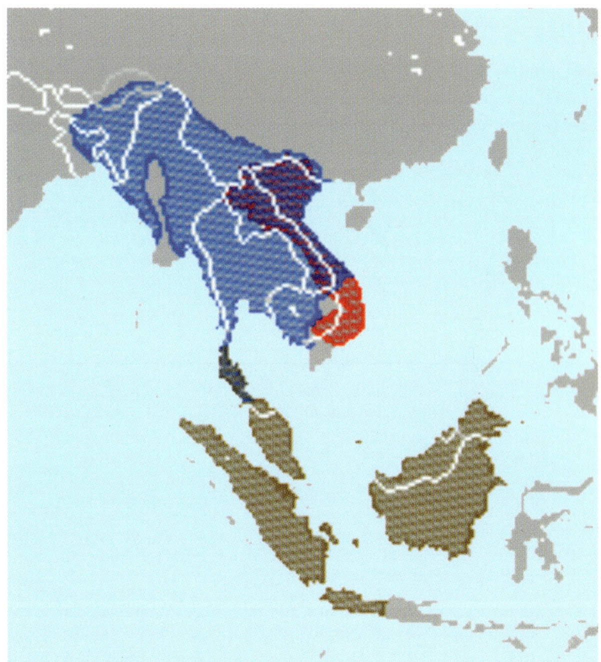

Figure 9.22 Slow loris distribution. Chermundy, 2011.
Red=pygmy slow loris
Blue=Bengal slow loris
Purple=sunda slow loris

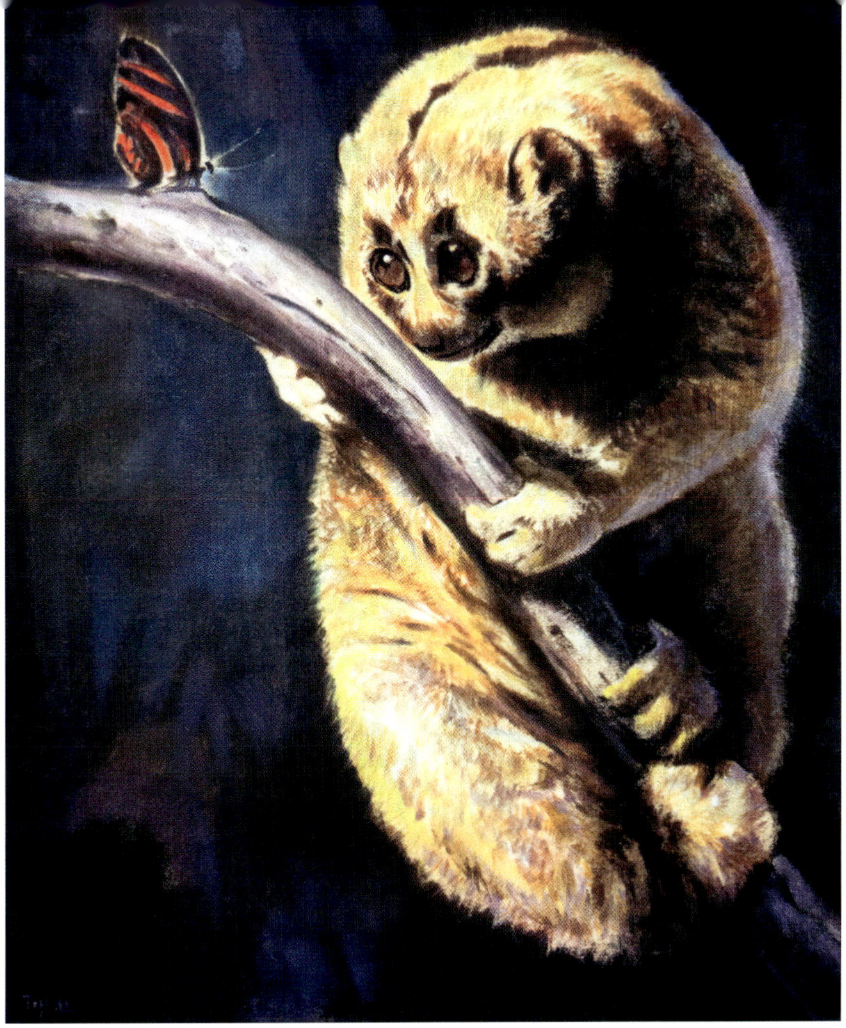

Figure 9.23
Loris Stalking a Butterfly, Ralph Burke Tyree. 1972, oil on board, 20" x 16", private collection.

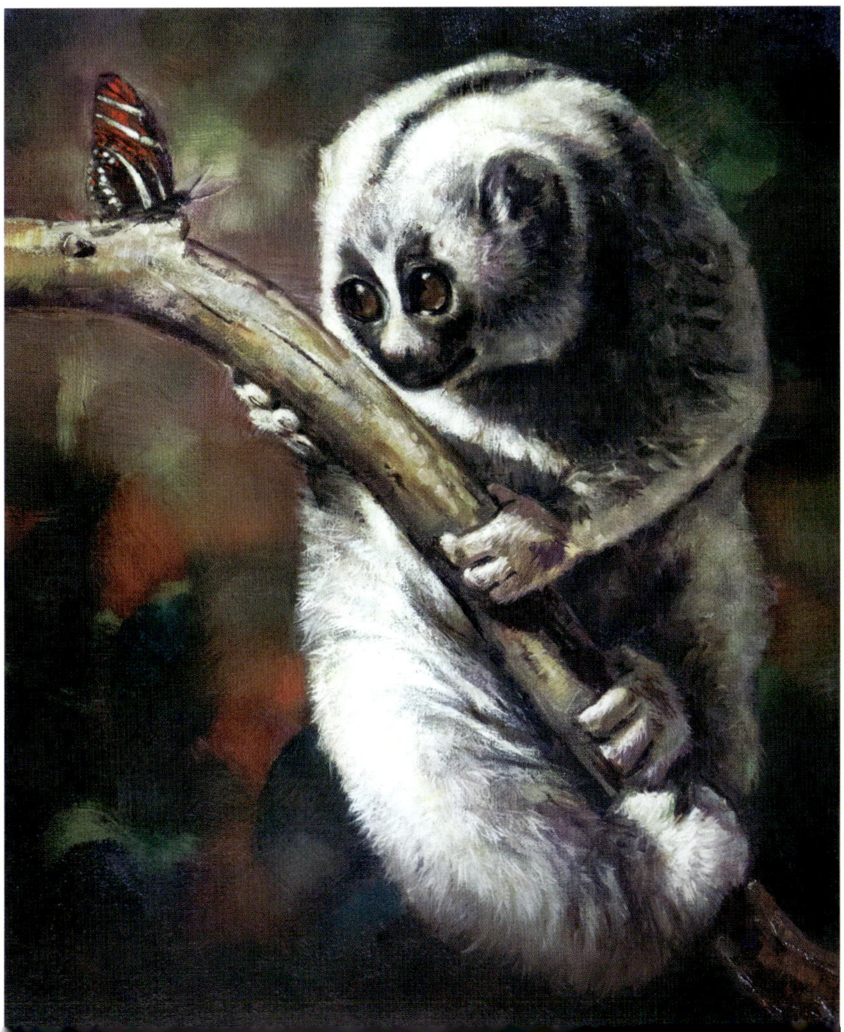

Figure 9.24
Loris Stalking a Butterfly, Ralph Burke Tyree. 1974, oil on velvet, 20" x 16", private collection.

Przewalski's Horse

The **Przewalski's horse** (*Equus ferus przewalskii*), also known as the Mongolian wild horse, the takhi, and the Dzungarian horse, is a rare member of the *Equidae* family (figure 9.25). The species is named for Russian explorer Nikolai Przewalski, who first described it scientifically. Although this wild horse once roamed Europe and Asia, today, it is only found in the steppes of Central Asia (figure 9.26). The ancestral lineage of the Przewalski's horse and the domestic horse split from a common ancestor about 500,000 years ago, leading to notable differences between the two. At no more than five feet tall, Przewalski's horse is stockier and smaller than a standard domesticated horse. Genetically, the Przewalski's horse has 33 chromosome pairs, compared to the domestic horse's 32.

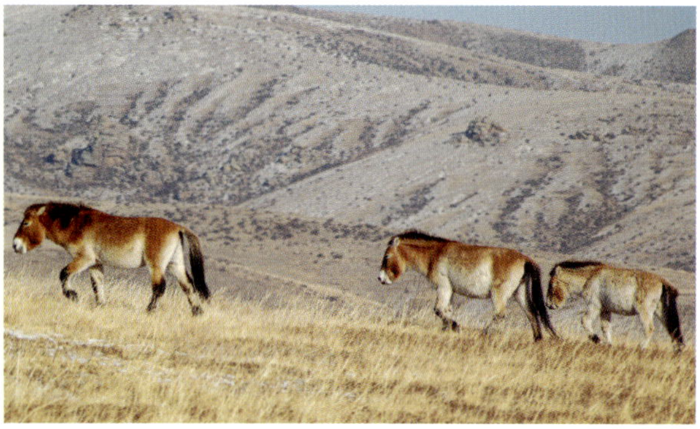

Figure 9.25 Przewalski's horse. Tengis Galamez on Unsplash.

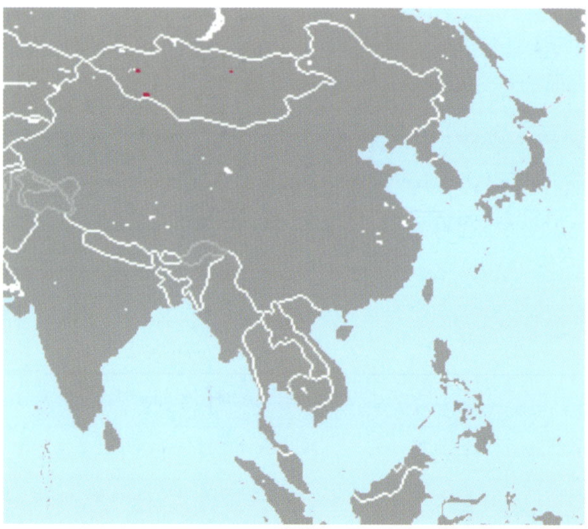

Figure 9.26 Przewalski's horse distribution. Chermundy, 2010.

The Przewalski's horse is a conservation success story akin to the fantastic achievements of the California condor breeding program. When Nikolai identified it in 1881, this horse's populations were already significantly reduced. By this point, it had disappeared from Europe and Asia and was only found in the Central Asian steppes. By the mid-nineteenth century, Przewalski's horse was nearly extinct in the wild, and captive populations were not faring much better, as interbreeding reduced fertility rates. In 1975, with only about a dozen left between two zoos, conservationists introduced a wild mare into the captive population, increasing genetic diversity. Over the next eight years, numbers expanded to 130 horses across 32 zoos, and despite the wild population's simultaneous extinction, survival seemed possible. Reintroduction to Mongolia and the Gobi Desert began in the 1990s, and by 2011, with an estimated 400 horses in wild herds, the IUCN officially reclassified the Przewalski's horse as endangered. Thanks to global awareness, its numbers have slowly continued to improve, allowing its further reintroduction to China, Ukraine, and Kazakhstan.

In 2020, the birth of a cloned Przewalski's horse (figure 9.27) brought an even more exciting development for the future of this species and other endangered animals. He was named Kurt for Dr. Kurt Benirschke, the San Diego Zoo Safari Park geneticist who developed the idea of cryopreserving genetic material and created the frozen zoo that facilitated the horse's birth. The San Diego Zoo Global

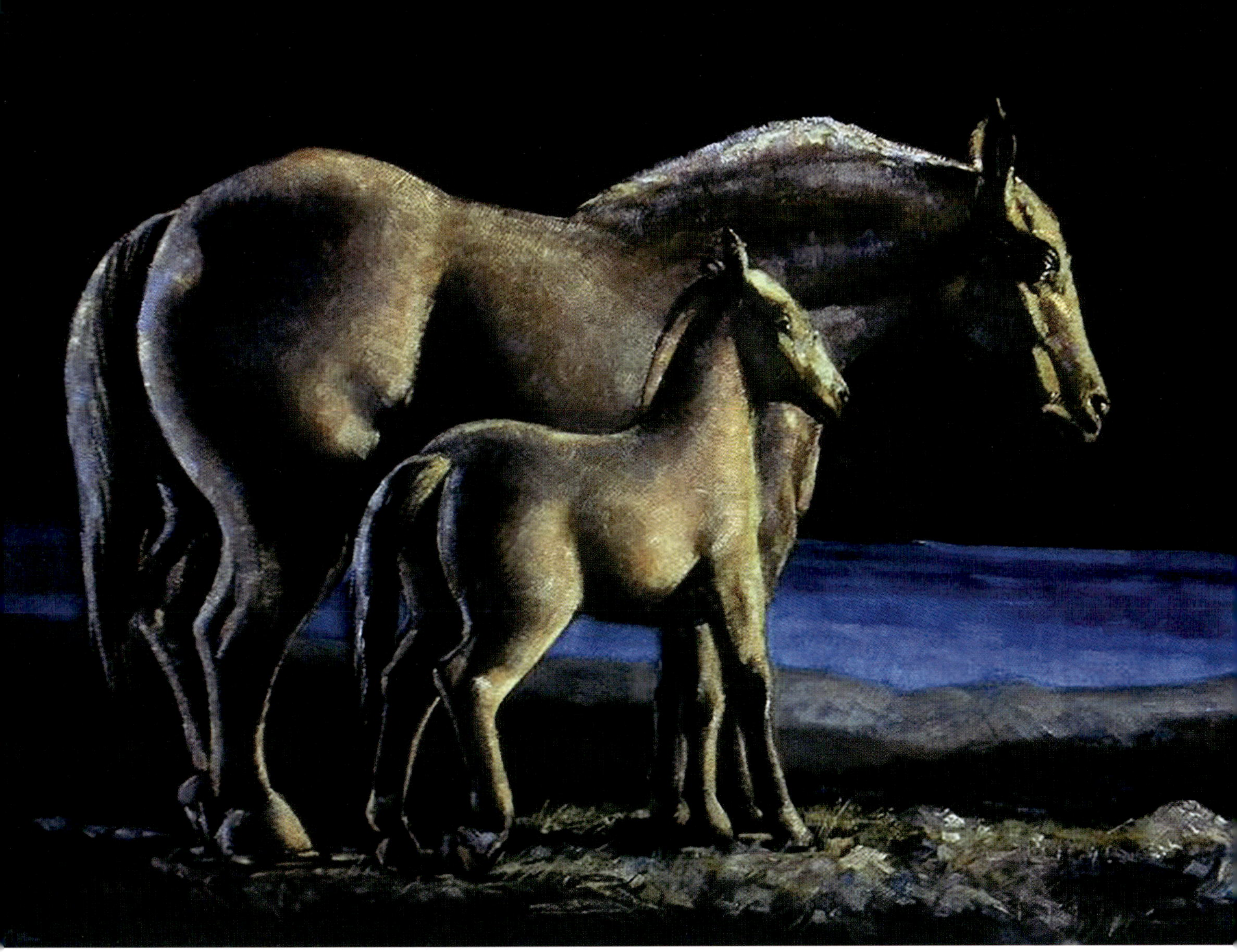

Figure 9.28 *Przewalski's Horse and Foal*, Ralph Burke Tyree. 1975, oil on velvet, 16" x 20", private collection.

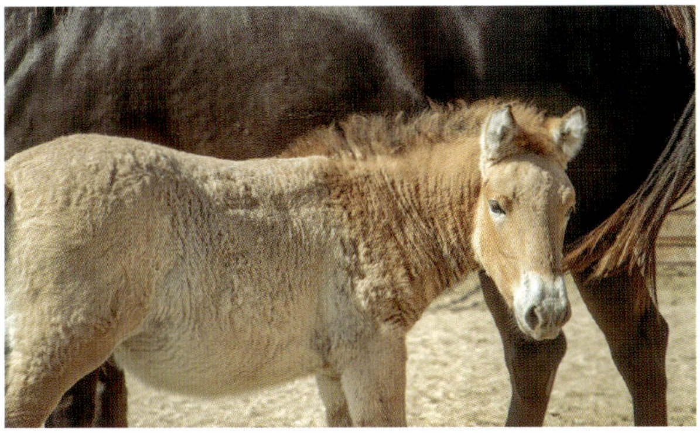

Figure 9.27 Kurt, the first Przewalski's horse clone, with his surrogate mother, Wolf. The cloning effort was a joint project with San Diego Zoo Wildlife Alliance, ViaGen, and Revive & Restore. 2020, courtesy of San Diego Zoo Wildlife Alliance.

Frozen Zoo held the DNA used for the colt for 40 years since being cryopreserved in 1980. This achievement provides hope for continued cloning success, which could help improve the genetic diversity of numerous endangered species, allowing them to remain viable and vibrant.

In 1975, soon after the Przewalski's horse was declared extinct in the wild, Tyree painted a pair of these horses (figure 9.28), recognizing and drawing attention to this very endangered species. Painting on velvet, he utilizes chiaroscuro beautifully in this work. The adult and foal are brightly lit from the front, with bold shadows cast across them, which help the two stand out in the deep darkness of night. The light dances across the barren landscape, illuminating the cobalt body of water behind the pair.

Tiger

"When a man wants to murder a tiger, he calls it sport.
When a tiger wants to murder him, he calls it ferocity."

—GEORGE BERNARD SHAW

One of the most famous cats in the world is the massive tiger (*Panthera tigris*) of the *Felidae* family. With characteristic orange, black, and white stripes unique to each cat, this muscular feline is the largest living cat species (figures 9.29–9.34). The male tiger can exceed 600 pounds, making it significantly larger than the next biggest cat, the African lion, which rarely exceeds 500 pounds. Historically, the tiger ranged from Turkey to the coast of the Sea of Japan, but today it is found primarily prowling forests in India and continental Southeast Asia, Sumatra, and Far Eastern Russia, with India's Bengal tiger being the most renowned (figure 9.35). This solitary hunter is an apex predator that mainly eats ungulates, like deer and bison, but also other medium-sized prey, like monkeys and wild boars. Sometimes it ventures out of preserves and kills domestic livestock, which triggers retaliation from farmers and much grief for the tiger. Modern tigers likely descend from a common ancestor that emerged in Central Asia about 100,000 years ago. This cat has been a popular part of many different iconographies throughout history, leading to its inclusion in everything from ancient mythologies and legends to modern logos and media, with recent examples including, *The Jungle Book*, the *Life of Pi*, *Winnie-the-Pooh*'s Tigger and the Frosted Flakes mascot, Tony the Tiger. The tiger is even the national animal of Bangladesh, India, Malaysia, and South Korea. US universities are not particularly creative when selecting their mascots, and 70 schools have selected a big cat, with the tiger being the most popular. Following the tiger's 19 schools are the lion at 16, the cougar at 12, and the panther and wildcat tied at 11.

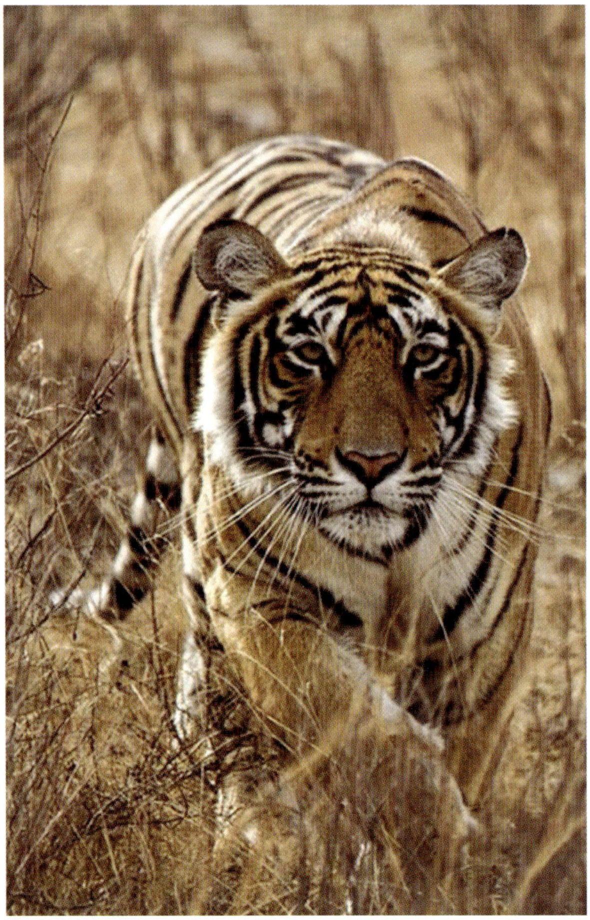

Figure 9.29 Camouflaged Bengal tiger. India, 2012, private collection.

Poaching and habitat loss have had devastating effects on the tiger population. Less than 4,000 mature tigers are estimated to remain, leading the IUCN Red List to categorize the species as endangered. Not only is this beautiful cat targeted for its skins and use in traditional medicine, but it is also losing extensive amounts of habitat to human encroachment. Over the last 100 years, the tiger has lost 93% of

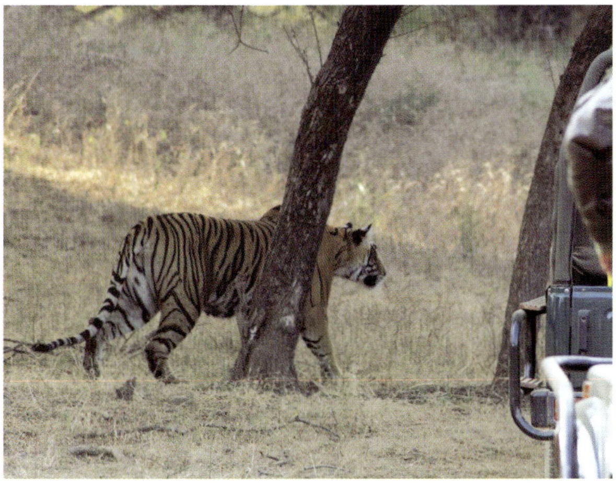

Figure 9.30 A chilly early morning tour of Ranthambore Tiger Country, India. 2012, private collection.

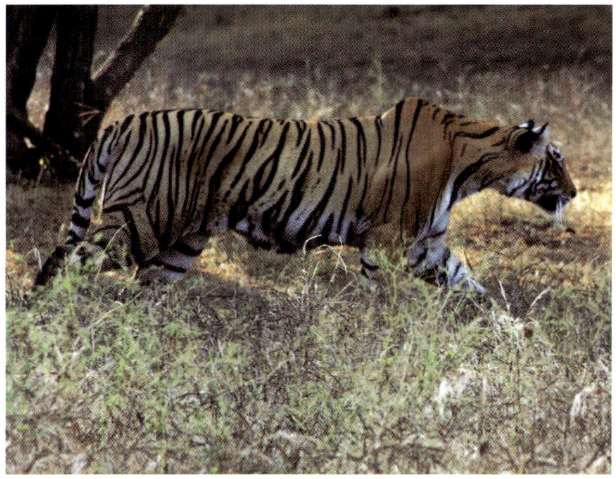

Figure 9.31 Stalking Bengal tiger. Ranthambore Tiger Country, India, 2012, private collection.

its historical range, fragmenting the population. This encroachment also brings the cats into increased conflict with humans. Since the 1970s, the tiger has been a significant focus of conservation efforts, with many global and local campaigns aimed at improving its numbers and strict measures protecting its population. Governments have established reservations, blocked off habitats that can host tigers, called Tiger Conservation Units, and some have even created dedicated anti-poaching teams like those in African reserves. A hopeful sign that conservation efforts are working—in 2016, global tiger numbers rose for the first time in a century.

When Tyree painted a Bengal tiger (figure 9.36) in 1974, the species was on the brink of extinction and a primary focus of conservationists. The Indian government recently began a concerted effort to protect Bengal tigers, passing the Wildlife Protection Act in 1972 and launching Project India, which

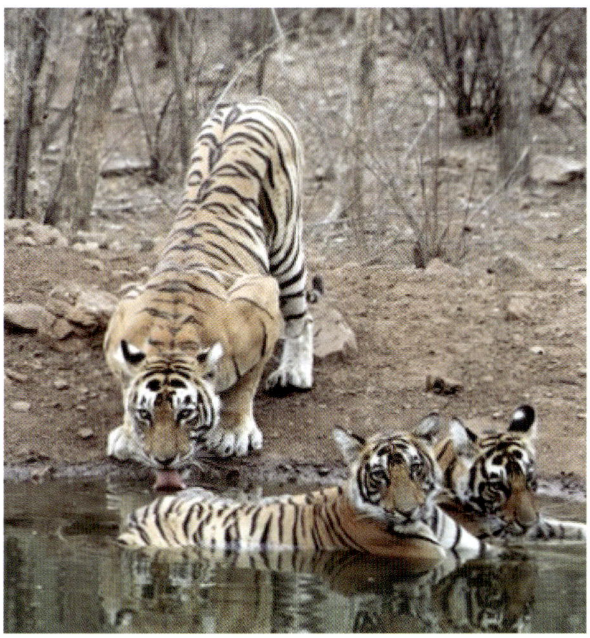

Figure 9.32 Tigers love water. Ranthambore Tiger Country, India, 2012, private collection.

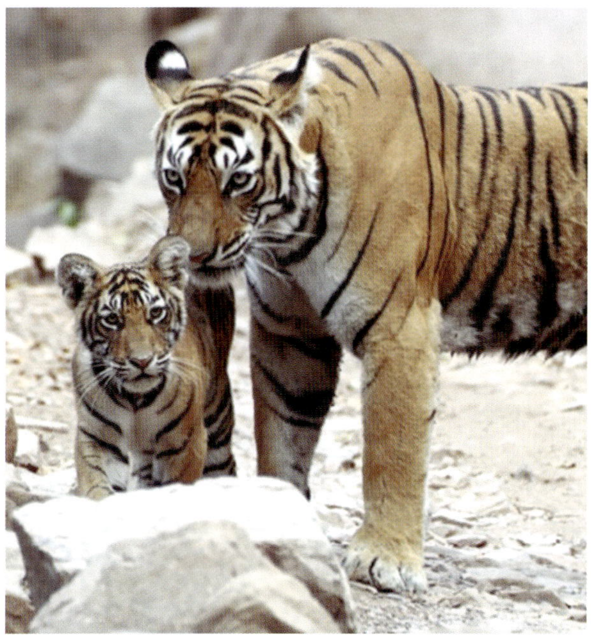

Figure 9.33 Tigress and her cub. India, 2012, private collection.

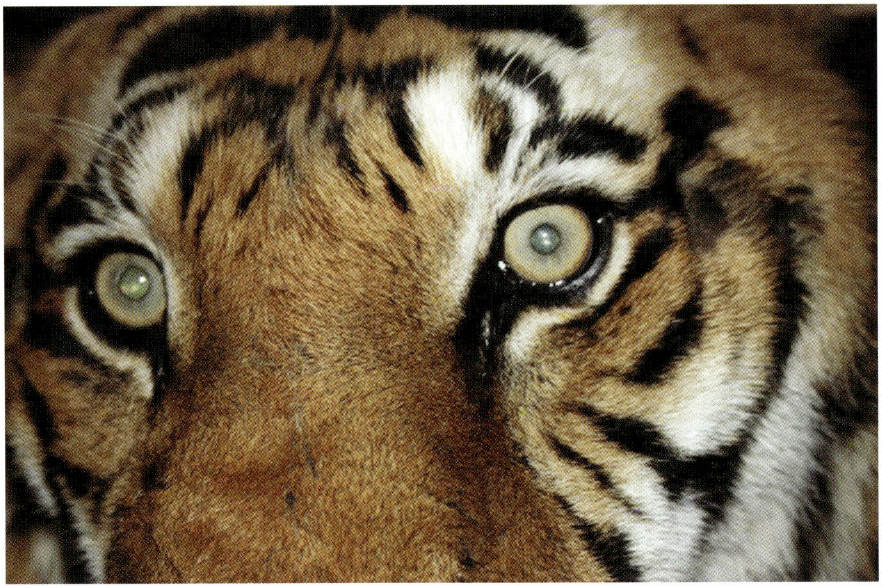

Figure 9.34 The eye of the tiger. 2012, India, private collection.

established numerous tiger reserves in 1973. Once again, Tyree's art reflects awareness and interest in conservation efforts.

As the tiger's real-world circumstances forced humans to confront our part in the near extinction of these beasts, Tyree created a symbolic depiction of this moment. His painting shows the magnificent cat amongst tropical foliage. The tiger peers straight at the viewer from behind fern leaves with the lower half of its face obscured, illustrating the cat's elusive and increasingly rare nature. The dark green leaves create a distinct barrier between the tiger and the viewer as the tiger seems to fade into the orange-tan background. Still, the cat's direct stare creates a connection between the two, as if to confront the viewer with our role in the tiger's disappearance and our responsibility to ensure it has a future.

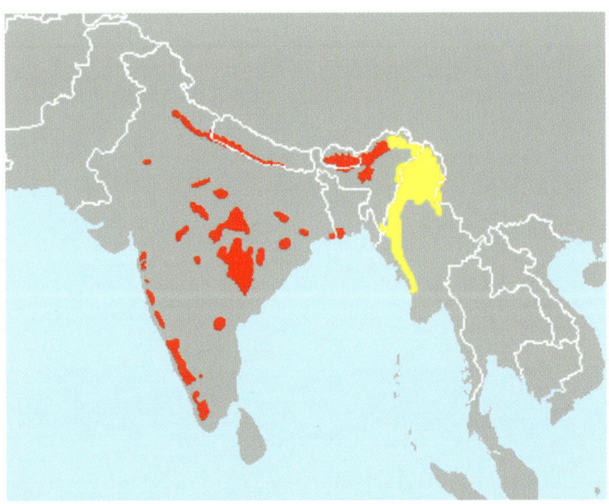

Figure 9.35 Bengal tiger distribution. The Nature Box, 2022.

Yellow=Myanmar tiger population now considered to belong to *Panthera t. corbetti*

Red=distribution of *Panthera tigris tigris*

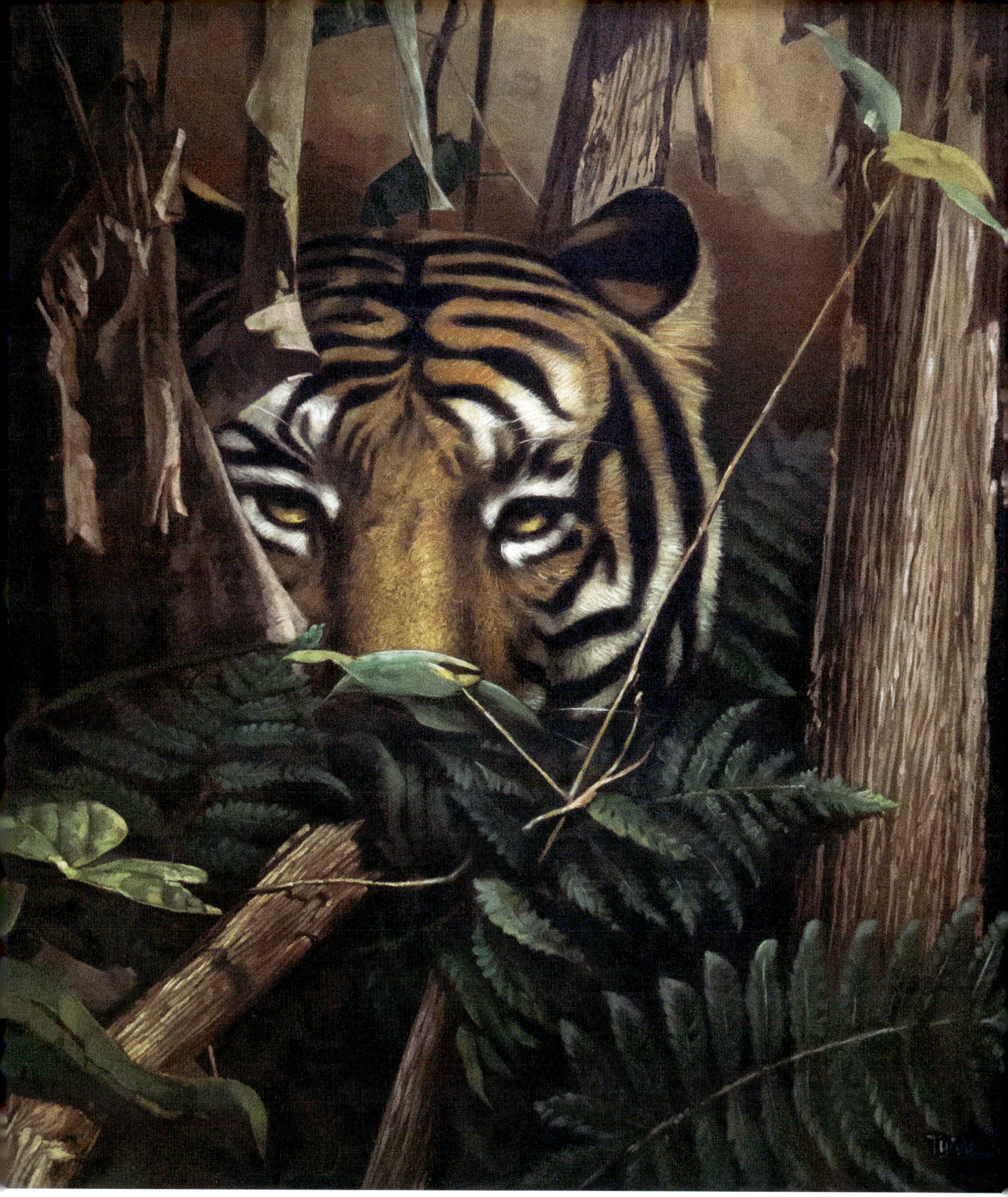

Figure 9.36 *Tiger*, Ralph Burke Tyree. 1976, oil on board, 24" x 20", author's collection.

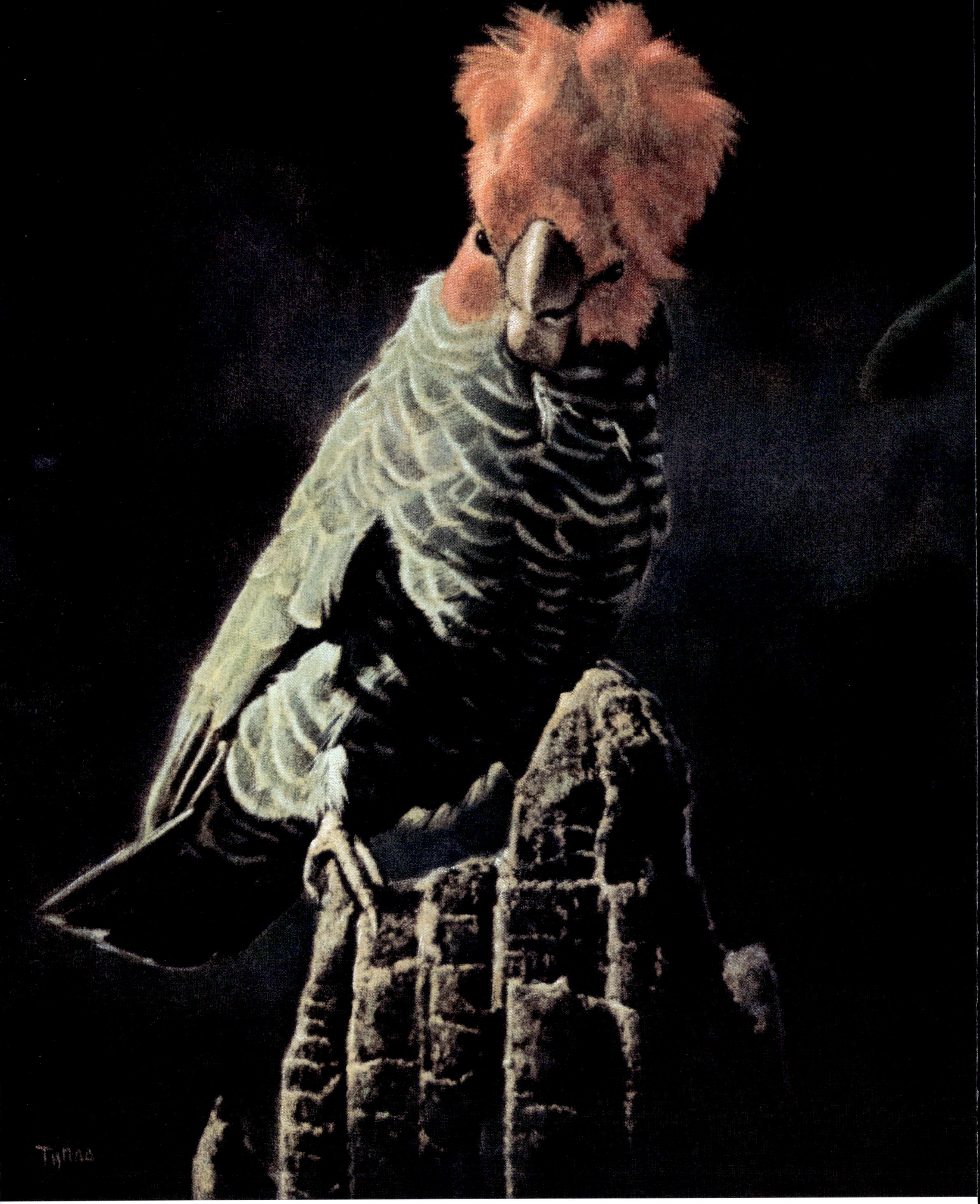

CHAPTER 10

Land Down Under

"We are all visitors to the time, this place.
We are just passing through. Our purpose here is to observe,
to learn, to grow, to love . . . and then we return home."

—AUSTRALIAN ABORIGINAL PROVERB

Another land that fascinated Ralph Burke Tyree was the faraway island continent of Australia, home to unusual critters with unique sounds and intense colors. Thanks to Australia's remote nature, its native species experienced isolated evolution over thousands of years. Because of this, it is now home to many animals found nowhere else in the world. This includes a staggering nearly 90% of its land mammals, birds, and reptiles, as well as almost 70% of the world's marsupials. It is no wonder that Tyree looked to this land, famous for its poisonous snakes, cuddly koalas, and talkative birds with names such as kookaburra, cuckoo, and gang-gang cockatoo when his subject matter pivoted to animals. The splendid oil paintings in this chapter exemplify his wonderment at this incredible place.

Laughing Kookaburra

"Kookaburra sits in the old gum tree;
Merry merry king of the bush is he.
Laugh, Kookaburra, laugh, Kookaburra,
Gay, your life must be!"

—MARION SINCLAIR, 1932

The **laughing kookaburra** (*Dacelo novaeguineae*) is a bird native to eastern Australia (figure 10.3). At up to 19 inches long, it is the largest bird in the *Alcedinidae* family of kingfishers. Because this kookaburra is mostly shades of beige and brown, it is most recognizable for the blue patches on its wings

Figure 10.1 *Gang-Gang Cockatoo*, Ralph Burke Tyree. 1977, oil on velvet, 20" x 16", private collection.

(figure 10.2), brown mask around its eyes, and laughing call, from which it gets its name. Though kookaburras belong to the kingfisher family, they are not heavily associated with water and usually don't eat fish. Instead, they prefer to eat mice, insects, and small reptiles. Still, kookaburras are known to steal goldfish or koi fish from backyard ponds. They've developed a particular reputation for killing snakes, which led to their introduction to New Zealand, Tasmania, and Western Australia.

The laughing kookaburra has thrived across its range and is commonly seen in all types of environments, from arid suburbia to open woodlands. As such, the IUCN does not consider it at risk. This common Australian sight is the national bird of the island continent and the subject of Marion Sinclair's well-known 1932 nursery rhyme, *Kookaburra*. Its distinctive laugh also proved helpful to Hollywood, where it has been used for decades as a jungle sound effect in movies like *Tarzan* and *Raiders of the Lost Ark*.

Tyree painted this iconic Australian bird on velvet, perched on a branch with its distinctive blue patch shown with incredible vibrancy (figure 10.4). This unique work is a departure from his usual oil-on-velvet works in several ways. Not only does he paint on brown velvet, but he utilizes bold, expressive brushstrokes for the background. This represents a direct contrast to his typical precise, highly realistic style. Still, the bird itself is detailed, and its feathers appear soft to the touch

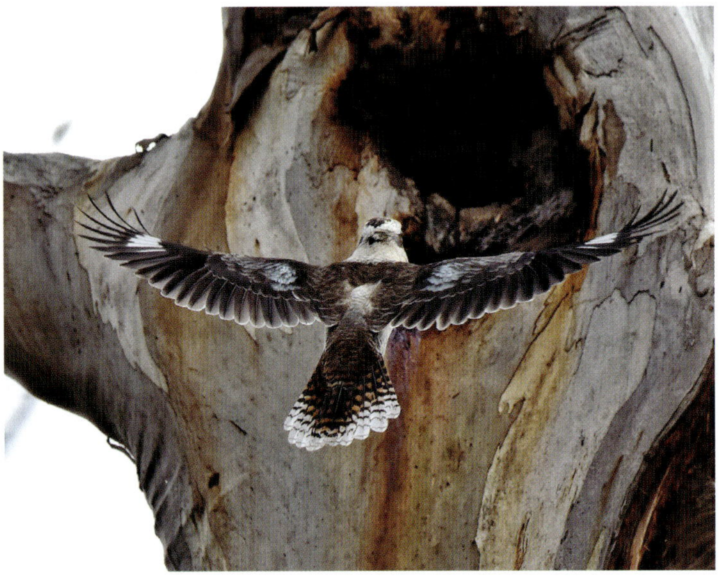

Figure 10.2 Laughing kookaburra approaching its nesting hollow. Geoffrey Moore on Unsplash.

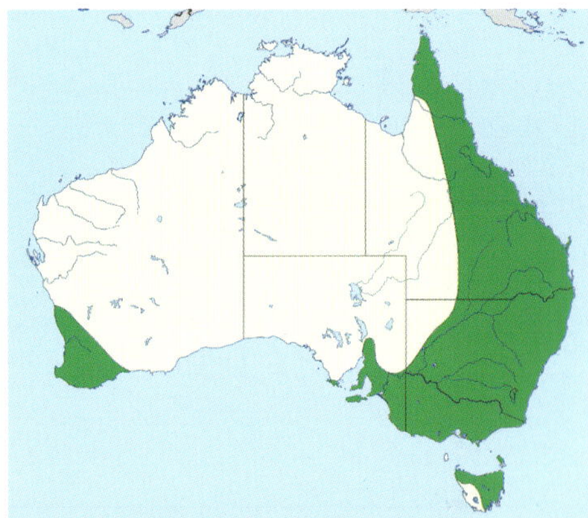

Figure 10.3 Laughing kookaburra distribution. Aa77zz, 2016.

Figure 10.4 *Laughing Kookaburra*, Ralph Burke Tyree. 1975, oil on velvet, 18" x 14", author's collection.

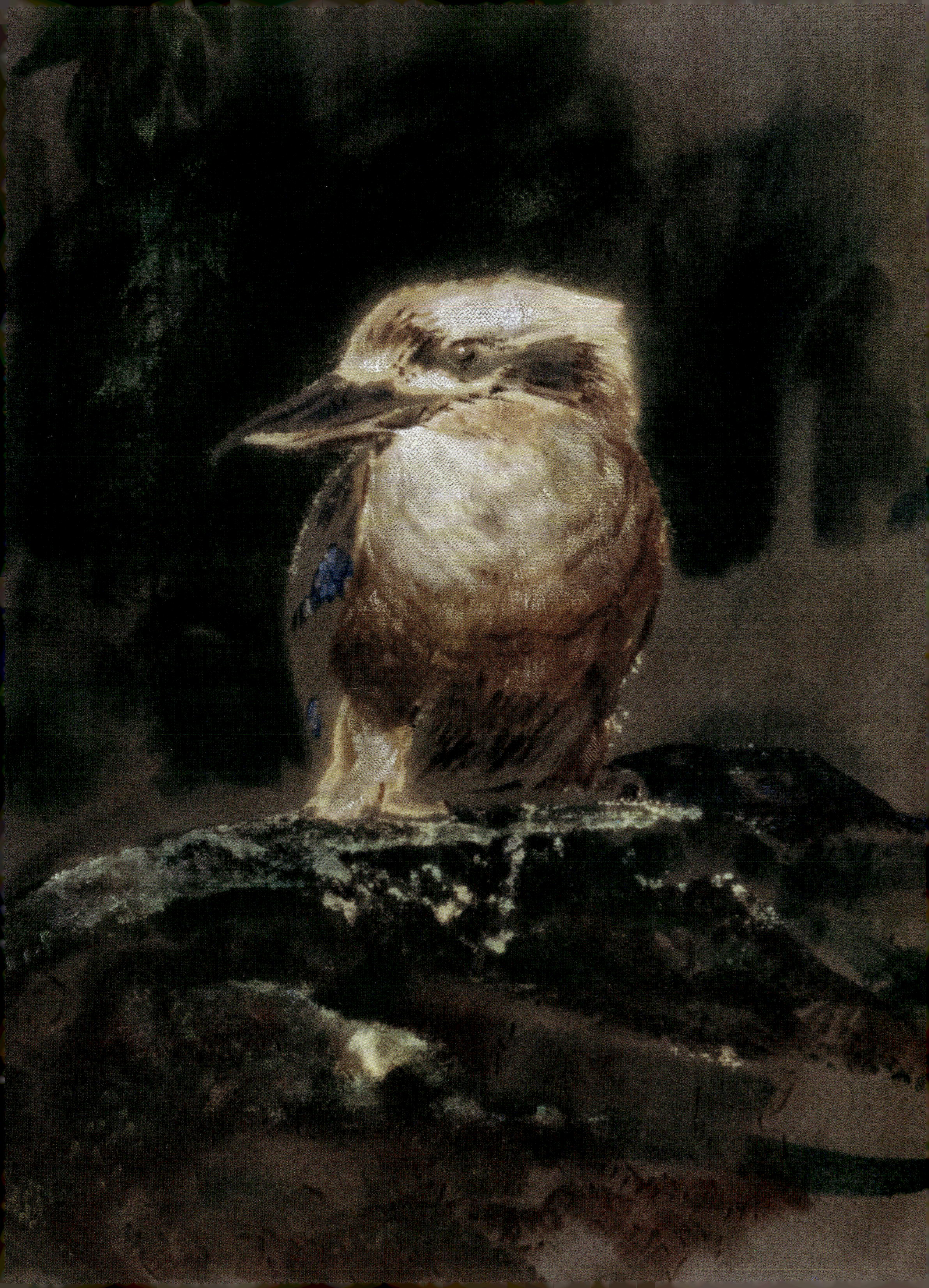

Cuckoo

"O Cuckoo! Shall I call thee Bird, Or but a wandering Voice?"

—WILLIAM WORDSWORTH

Cuckoos (*Cuculidae*) are a family of 150 bird species across 33 genera, including common cuckoos, roadrunners, and malkohas. These birds are found on every continent except Antarctica—wherever they have access to an abundance of insects, especially caterpillars. Most cuckoos prefer tropical rainforests, but they are also known to frequent forests, woodlands, and even deserts.

Cuckoos have long played a significant role in human culture and symbology across many regions. Greek mythology tells of the god Zeus seducing his wife, the goddess Hera, by transforming into a cuckoo. William Shakespeare later used these birds as a symbol for springtime and cuckoldry, alluding to the cuckoos' reputation for laying eggs in other birds' nests, thus, implying that the unfaithful wife has successfully concealed her affair and deceived the husband. Playing into similar themes, these birds symbolize unrequited love in Japan, and in Hinduism, they are associated with Kamadeva, the god of love and desire. Cuckoo clocks, famously from the Black Forest region of Germany, mimic the common cuckoo's famous call.

Tyree's painting of a cuckoo (figure 10.6) seems to depict the Australian pallid cuckoo. While this bird is associated with Australia, it sometimes migrates to the island of Papua New Guinea during winter. This species is 11 to 13 inches long and has distinctive markings that differentiate it from other cuckoos, like its dark bill, dark eyes, gold eye-rings, and olive feet (figure 10.5). Tyree illustrates most of these features in his painting, most notably, the golden eye-ring. The bird perches on a bare tree stump in front of an obscure dark green background with its back facing the viewer and its head turned in profile. With this pose, Tyree puts the bird's beautiful plumage and facial markings on full display, highlighting its beauty.

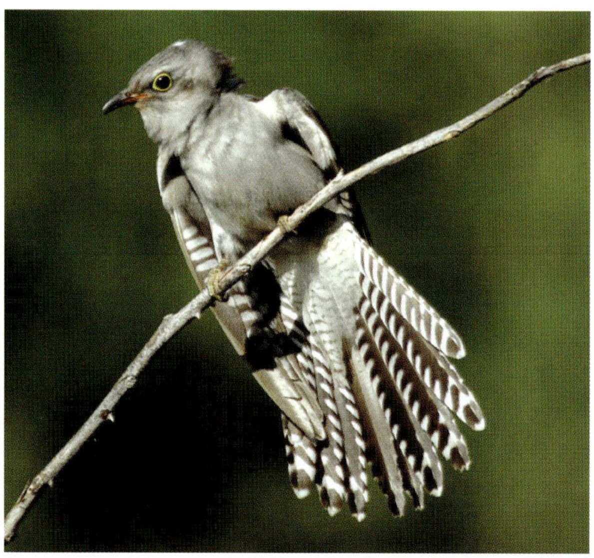

Figure 10.5 Pallid cuckoo. Aviceda, Queensland, Australia, 2007.

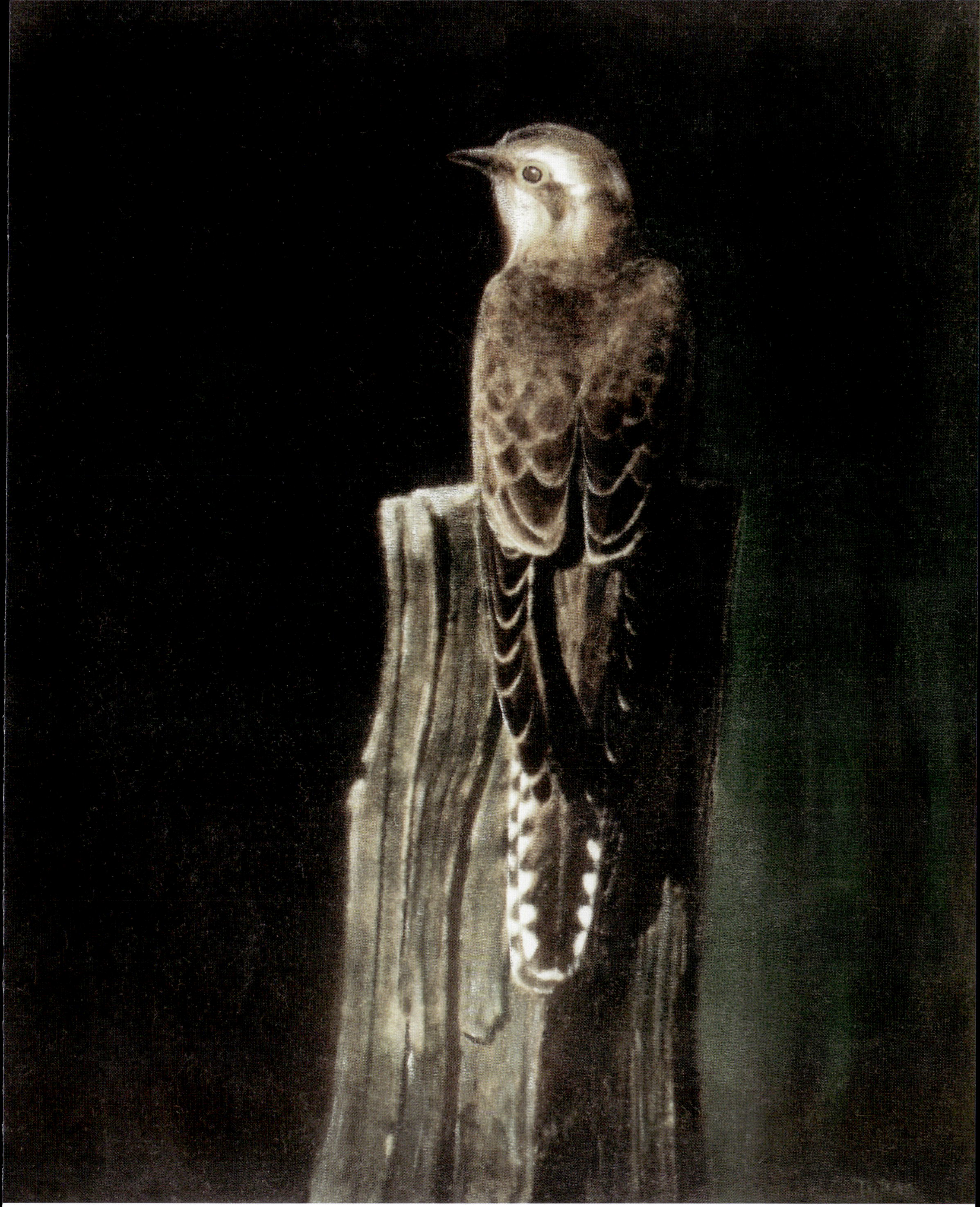

Figure 10.6 *Cuckoo*, Ralph Burke Tyree. 1976, oil on velvet, courtesy of Jon Toal.

COCKATOO

The **cockatoo** (*Cacatuidae*) is a family of 21 parrot species native to the Australian continent and its neighboring island nations of the Philippines, Papua New Guinea, Indonesia, and the Solomon Islands. These birds are usually less colorful than other parrots and are instead primarily black, white, and grey. Cockatoos are identifiable by their prominent crests and curved bills. Throughout his career, Tyree featured three cockatoos in his paintings—the white sulphur-crested cockatoo, the red-crested gang-gang cockatoo, and the pink Major Mitchell's cockatoo.

Sulphur-Crested Cockatoo

The large, white **sulphur-crested cockatoo** (*Cacatua galerita*) is recognizable for its yellow crest and yellow-tinged underwings and tail (figures 10.7 and 10.8). It possesses a loud, jarring call, an adaptation that allows its voice to carry through the forests in which it dwells. Although this cockatoo is native to wooded habitats of western Australia and New Guinea, it is abundant beyond this region due to its introduction to places like Perth, Singapore, New Zealand, and Hawaii (figure 10.9).

I once owned a sulphur-crested cockatoo named Crocodile Dundee, a noisy nuisance who often attempted to bite the hand that fed him. Eventually, he was given away to a parrot-loving friend. Even though this pretty cockatoo makes a demanding pet and its importation to the US is prohibited, it remains popular in aviculture.

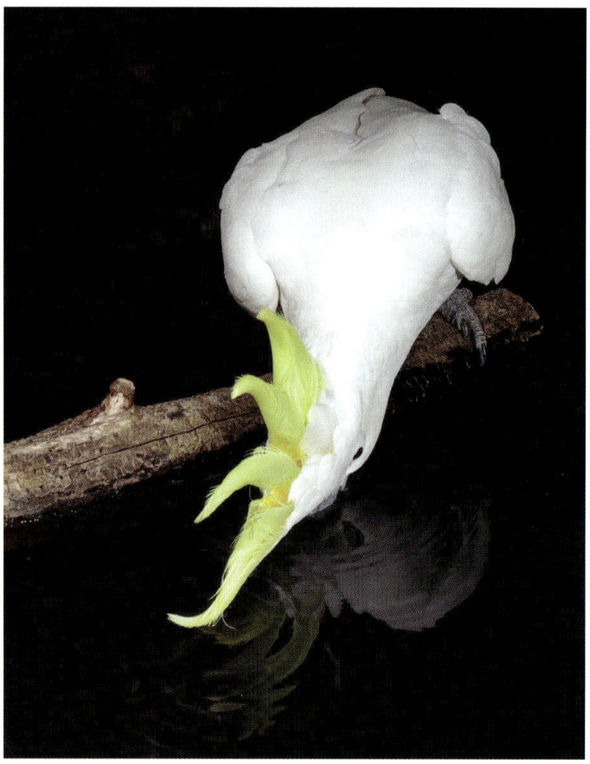

Figure 10.7 Sulphur-crested cockatoo.
Chris Charles on Unsplash.

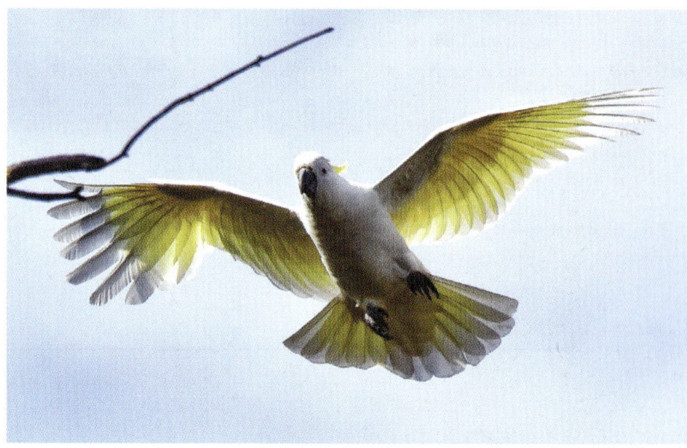

Figure 10.8 Sulphur-crested cockatoo mid-flight.
Geoffrey Moore on Unsplash.

Figure 10.11 *Sulphur-Crested Cockatoos & Jungle Wahine*,
Ralph Burke Tyree. 1953, oil on board, 22" x 17", private collection.

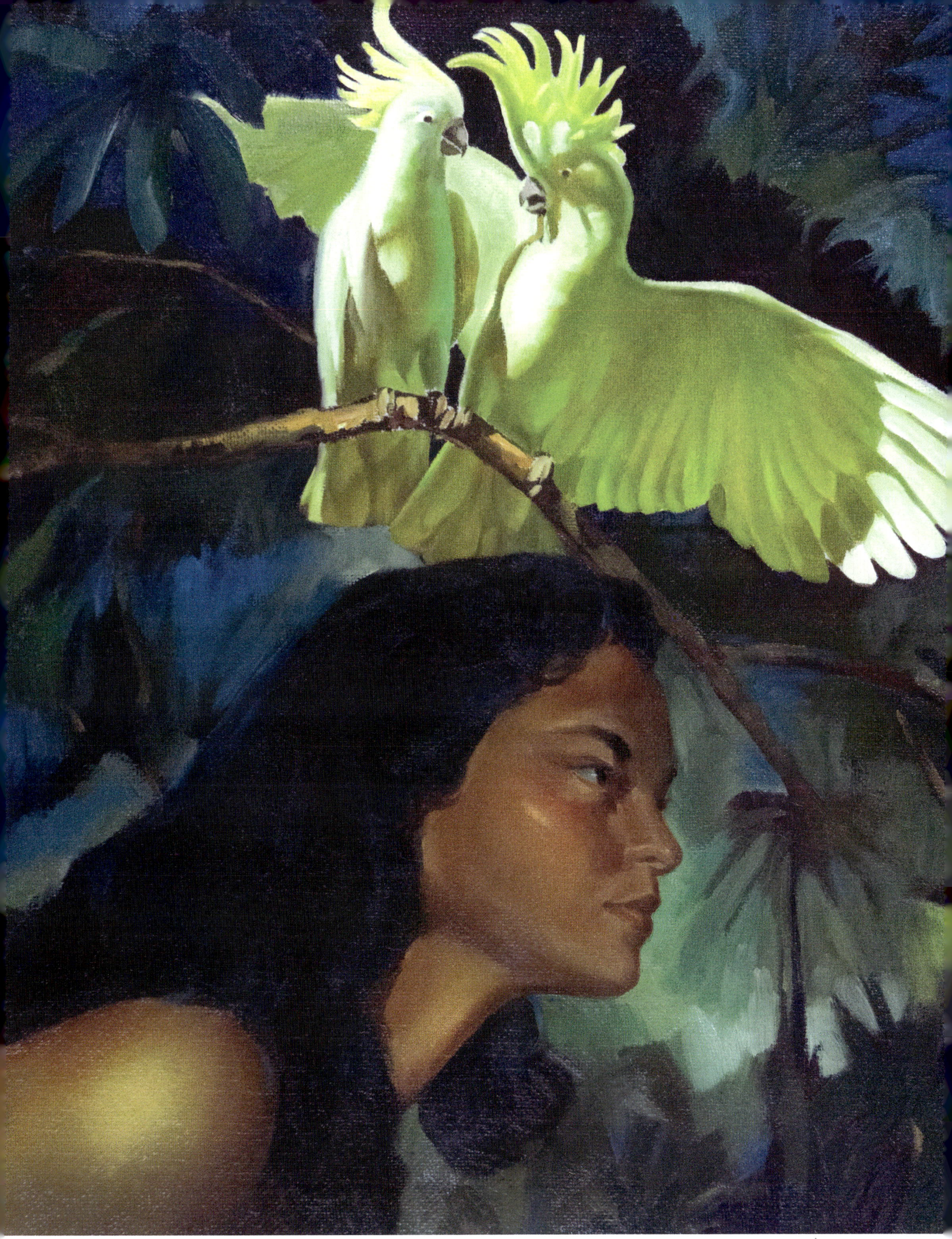

The curious and intelligent sulphur-crested cockatoo has adapted so well to urban Australia that it is often considered a pest because of its tendency to damage wooden structures (figure 10.10), get into garbage bins, and eat crops. With this cockatoo's plentiful population, the IUCN does not consider it endangered.

Over a decade before Tyree's primary subject matter shifted to animals, he painted a duo of sulphur-crested cockatoos perched above a wahine in a lush, tropical jungle (figure 10.11). This is a perfect example of how he began incorporating animals into his portraits before fully turning his focus. In this case, Tyree draws the viewer's eye to the cockatoos, allowing them to overtake the model and nearly become the primary subject. The cockatoos are brightly lit from above, with dramatic shadows cast across their bodies. Thanks to the sun's rays, the mostly white birds appear entirely yellow and stand out against the dark foliage. This is in stark contrast to the woman who stands in their shadow and seems to blend into the jungle background. Both birds' crests are shown on full display as one spreads its wings, giving the piece an element of dynamic movement and further drawing the eye toward the cockatoos. Tyree's skilled use of color, light, and composition ensures the viewer's initial focus is the pair of striking birds.

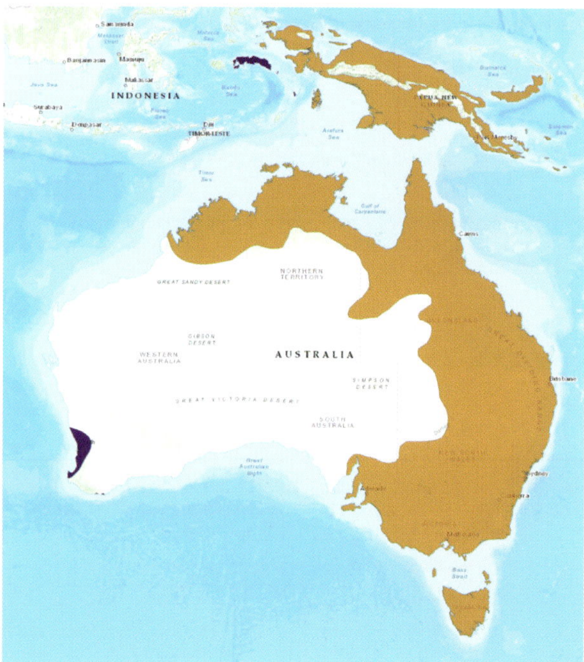

Figure 10.9 Sulphur-crested cockatoo distribution. Izvora, 2013.

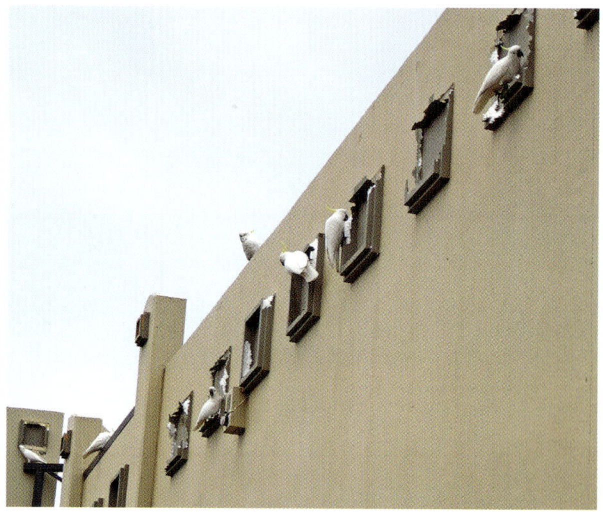
Figure 10.10 Sulphur-crested cockatoo causing damage to buildings in Australia. Bidgee, 2010.

Gang-Gang Cockatoo

The **gang-gang cockatoo** (*Callocephalon fimbriatum*) gets its name from an Aboriginal language local to its native domain in southwestern Australia (figure 10.12). This social bird favors this region's cool, wet coastal woodlands and forests. Both genders are primarily grey, but the male gang-gang has a bold red head and crest (figure 10.13). Although this bird is not especially chatty, it has a distinctive call that sounds like a creaky gate or a cork pulled from a wine bottle.

The gang-gang cockatoo population has been declining for decades, largely due to the destruction of older, hollow trees. The 2019–2020 Australian bushfire season was particularly catastrophic for this cockatoo, which experienced high mortality rates and extensive habitat loss. These fires are estimated to have impacted up to 36% of the gang-gang cockatoo's range, pushing the gang-gang cockatoo into the IUCN Red List's vulnerable category. The bird's current protection status makes the knowing destruction of its habitat a criminal offense.

As seen in figure 10.1, Tyree plays with color in his painting of the gang-gang cockatoo. He brings out the green and yellow tones of the bird's dark feathers and pulls similar shades into the beak. By also painting the bird against an atmospheric teal background, he creates a beautiful contrast with the bright red head, allowing it to stand out. The feathers appear fluffy and soft, especially compared to the rough bark of the tree the bird rests upon.

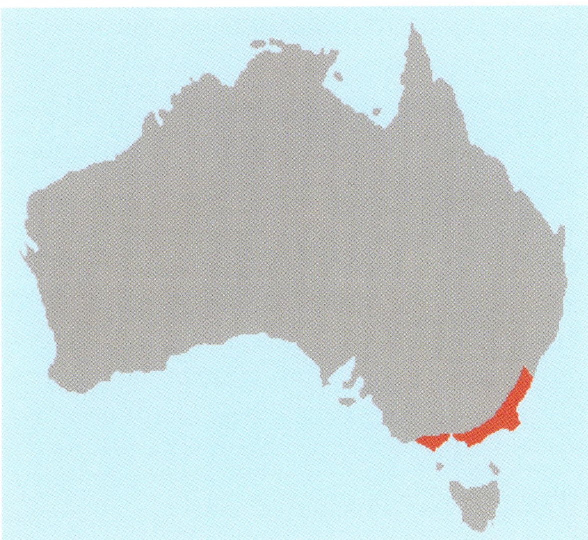

Figure 10.12 Gang-gang cockatoo distribution. Kiwifruitboi, 2006.

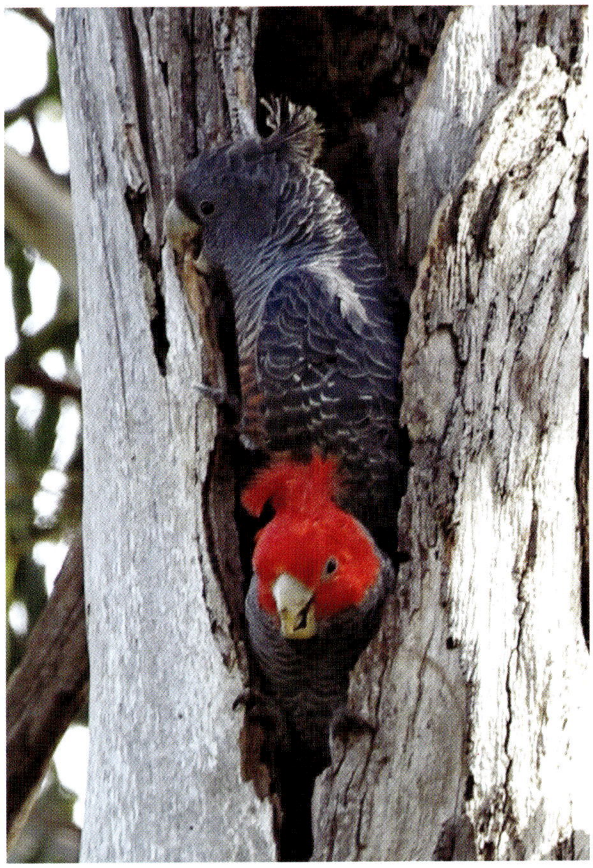

Figure 10.13 Nesting gang-gang cockatoo pair. Geoffrey Moore on Unsplash.

Major Mitchell's Cockatoo

The **Major Mitchell's cockatoo** (*Lophochroa leadbeateri*) is found widely across Australia (figure 10.15) and is commonly referred to as the Leadbeater's cockatoo or the pink cockatoo. This bird is named for Major Thomas Mitchell, who described this striking cockatoo in his book on expeditions. Primarily white and pink, this cockatoo is most notable for its stunning pink, red, and yellow crest (figure 10.14). It inhabits semi-arid, arid, and subtropical inland areas and requires spacious woodlands for nesting.

While the IUCN does not consider Major Mitchell's cockatoo threatened, its populations are declining due to human encroachment on its habitats. The territory of Victoria has categorized this bird as vulnerable and prepared an action statement to conserve the species.

The bird's expressive crest is the focal point in Tyree's oil painting on board of a Major Mitchell's cockatoo (figure 10.16). Its crest is fully raised and fanned out, showing off the vibrant red and yellow at the base of these feathers. The rest of the bird is painted in shades of pale salmon and white, allowing it to contrast with the dark background. Tyree depicts both the bird and the branch it stands upon with incredible detail.

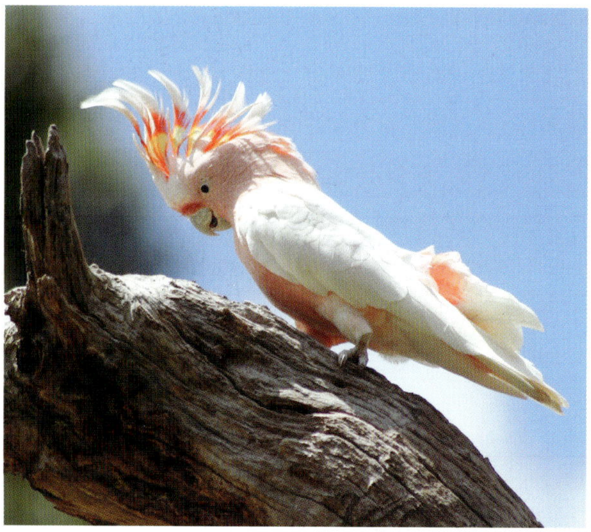

Figure 10.14 Major Mitchell's cockatoo. Craig Manners on Unsplash.

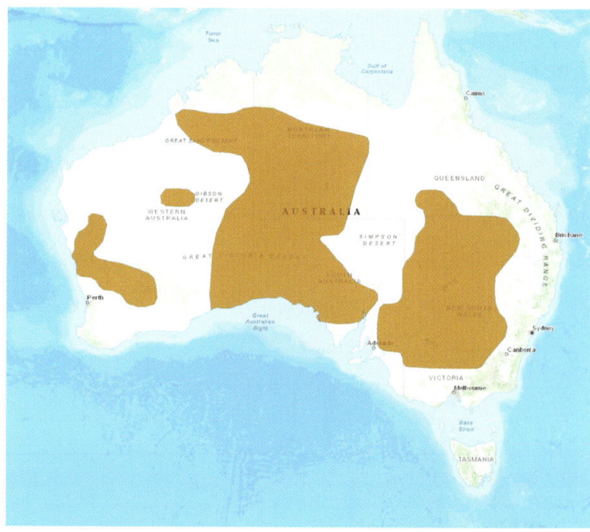

Figure 10.15 Major Mitchell's cockatoo distribution. Kiwifruitboi, 2006.

Figure 10.16 *Major Mitchell's or Pink Cockatoo*, Ralph Burke Tyree. 1977, oil on board, 20″ x 16″, author's collection.

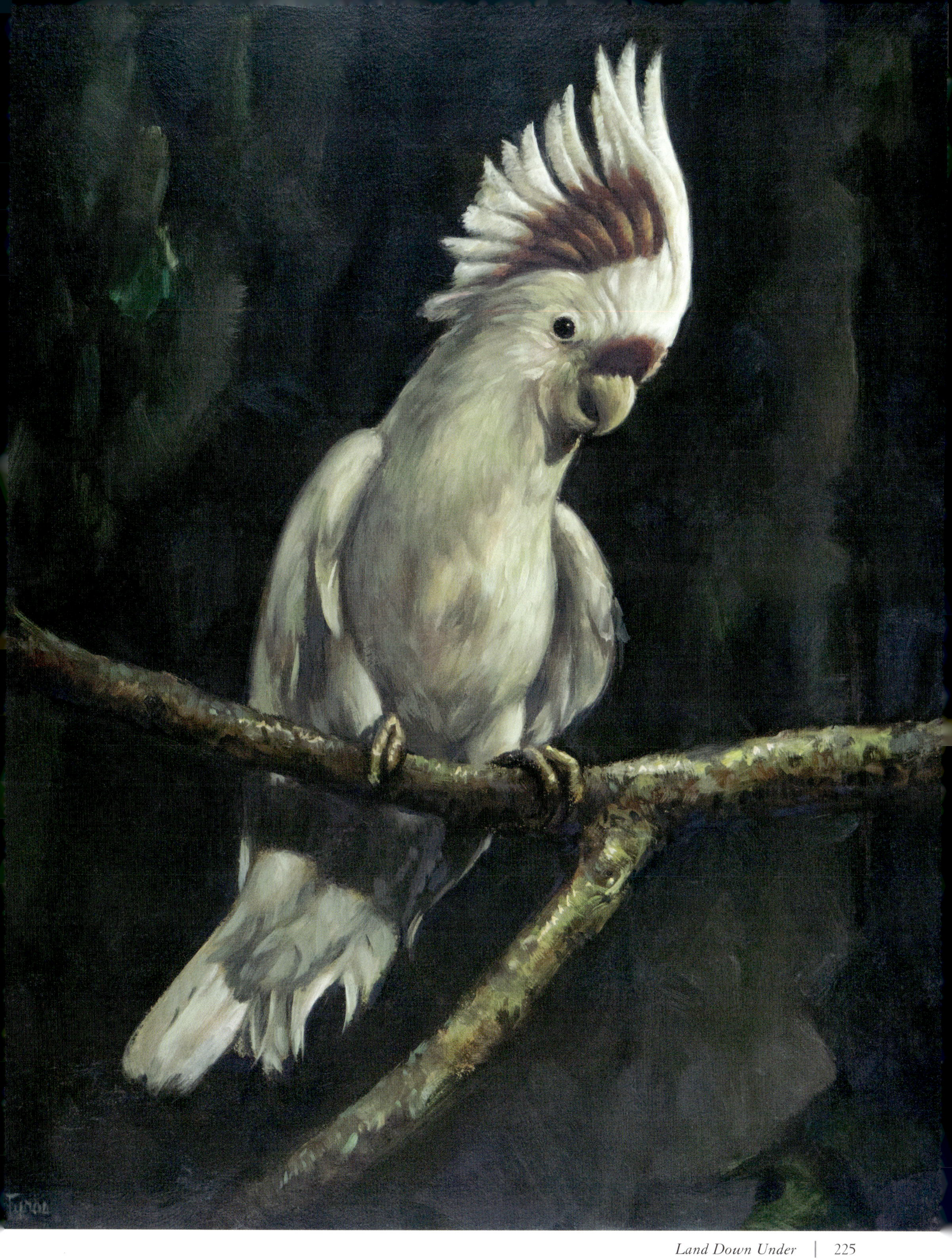

Dingo

The **dingo** (*Canis familiaris*) is a wild species from the *Canidae* family of dogs that is found across Australia (figure 10.17). This ancient dog was likely brought to Australia at least 4,000 years ago by an unknown group of humans. Today, it inhabits various environments all over the island continent, except for the southeast and Tasmania (figure 10.18). This medium-sized canine is typically a warm tan color but can also range from cream to dark brown to black. It is an apex predator with a hard, lean body built for speed, endurance, and agility. The dingo eats mammals, birds, rodents, fish (figure 10.19) and, impressively, even giant red kangaroos and livestock, much to the chagrin of farmers. When the British arrived, the dingo lived alongside Indigenous Australians, not as pets but as allies. While people sometimes train these dogs and keep them as pets, this practice is widely criticized as they are ultimately wild and unpredictable. The dingo plays a vital role in the Dreamtime narratives of Indigenous Australians, often in stories of the world's creation. Due to the difficult relationship that developed with European settlers, a view of the dog as a sneaky thief and coward was perpetuated and continues today.

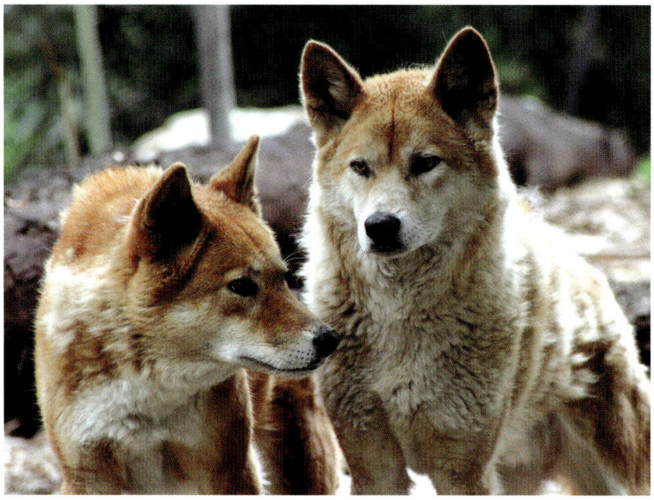

Figure 10.17 Dingo pair. Craig Manners on Unsplash.

European settlement in Australia reduced the dingo's range as settlers introduced agriculture and livestock. With this, the dog began targeting sheep and cattle, and farmers fought back by enacting laws to control the dingo. As a native Australian animal, it is protected within all national parks, but beyond park borders, protection of the dingo is at the discretion of each territory. Some territories have extended this protection, while others categorize the dingo as a pest and require landowners to deal with any that trespass, whatever that may entail. In 2018, the IUCN Red List removed the dingo altogether, believing it to be a descendant of the domestic dog and, thus, a feral species. Before this, the IUCN had listed it as vulnerable for years due to a decline of true dingoes after decades of crossbreeding with domestic dogs. Because of this crossbreeding, some continue to argue that the pure dingo is endangered despite its seeming abundance across the country. Fraser Island is home to about 200 dingoes that have at least largely avoided crossbreeding with dogs and, as such, represent a unique and essential population. This World Heritage Site is where I first encountered a dingo in the 1990s.

In figure 10.20, Tyree depicts a lone dingo pup standing in a rocky environment using oil paints on canvas. The characteristically ginger-tan dog has white markings on its nose, feet, and tail. The orange, beige, and red rocks surrounding the canine help identify the environment as Australia and, thus, the animal as a dingo. All attention is on the dingo, which Tyree paints with precise detail, a direct contrast to the impressionistic background.

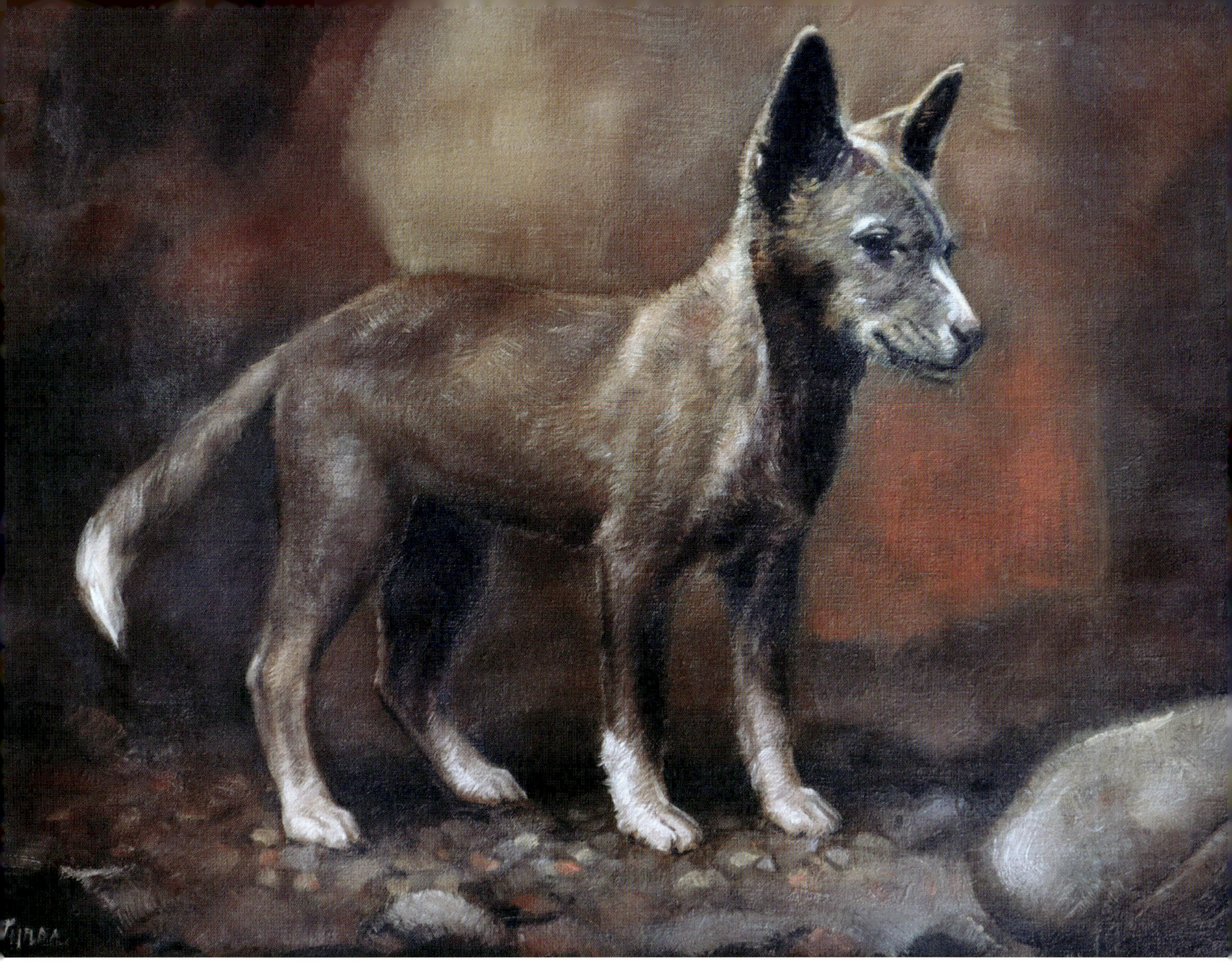

Figure 10.20 *Dingo Pup*, Ralph Burke Tyree. 1973, oil on canvas, 16" x 20", private collection.

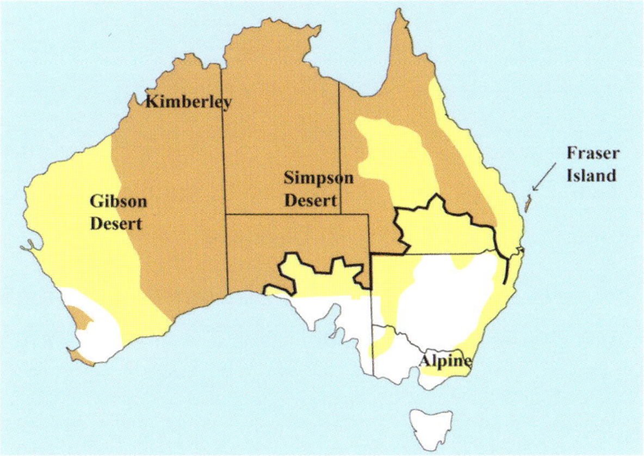

Figure 10.18 Dingo distribution. Note Fraser Island. William Harris, 2020.

Orange=present

Yellow=generally present but often high levels of lethal control

White=absent or no data

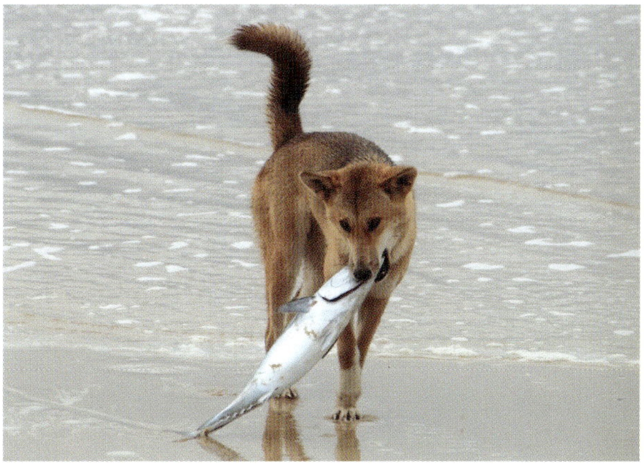

Figure 10.19 Dingo finds a dead fish on Fraser Island. Marc Tarlock, 2010.

Koala

The **koala** (*Phascolarctos cinereus*) from the *Phascolarctidae* family is the national symbol of Australia's unique wildlife. This marsupial has long held important status in Australia, featuring prominently in the Dreamtime stories of multiple groups of Indigenous Australians. This adorable critter is often mistaken for a bear with its round head, fluffy ears, and oval-shaped nose (figure 10.21). Although it is more closely related to the wombat than the bear, it does make a cute, cuddly stuffed toy for a child. Using long, sharp claws, the koala climbs about the woodlands of coastal eastern Australia (figure 10.22), munching on its favorite food, eucalyptus leaves. This diet provides the koala with very little energy, leading it to sleep up to 20 hours daily. Whereas the males communicate with loud bellows to intimidate rivals and attract mates, the females use soft bellows, snarls, and howls. Like other marsupials, the female koala gives birth to a partially developed joey, which then spends six months in her pouch finishing development.

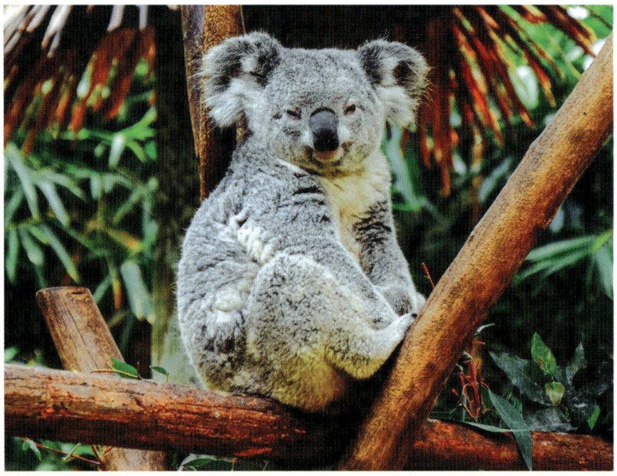

Figure 10.21 Koala. Ellicia on Unsplash.

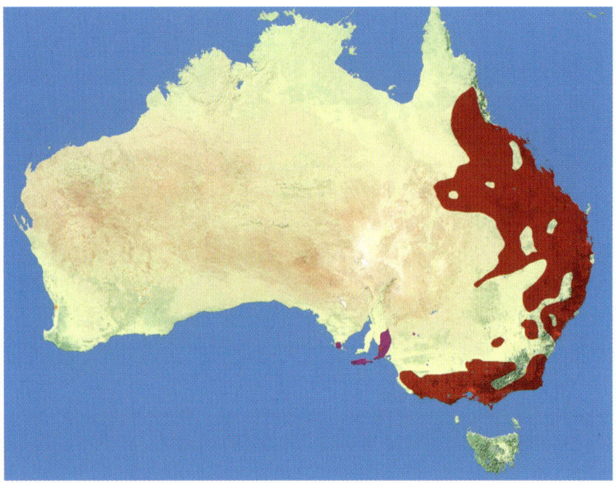

Figure 10.22 Koala distribution. Nrg800, 2011.

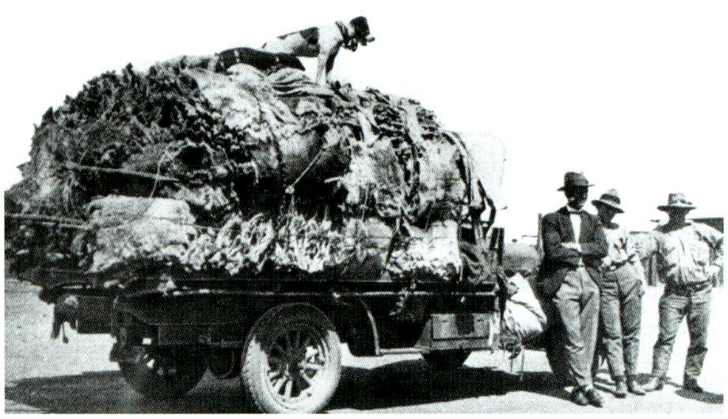

Figure 10.23 A truckload of 3,600 koala skins in Queensland. Gordon Cumming Pullar, 1927, courtesy of the John Oxley Library and State Library of Queensland.

After the heavy hunting of the koala by early twentieth-century European settlers for its fur (figure 10.23), the first conservation efforts for this marsupial were initiated around 1930 with the creation of sanctuaries in Brisbane and Sydney. Unfortunately, since then, the koala's forests have been extensively cleared for urbanization, agriculture, and logging. In 2014, when the IUCN last assessed the koala, they categorized the species as vulnerable. Since then, the koala population has experienced a dramatic decline, with a 30% drop between 2018 and 2021.

The devastating bushfires of 2019–2020 impacted over 60,000 koalas, killed at least 30,000, and destroyed more than 10% of their habitat. While these mega-fires played a major role in the population drop, they were by no means the only contributor. Drought, heat waves, and continued

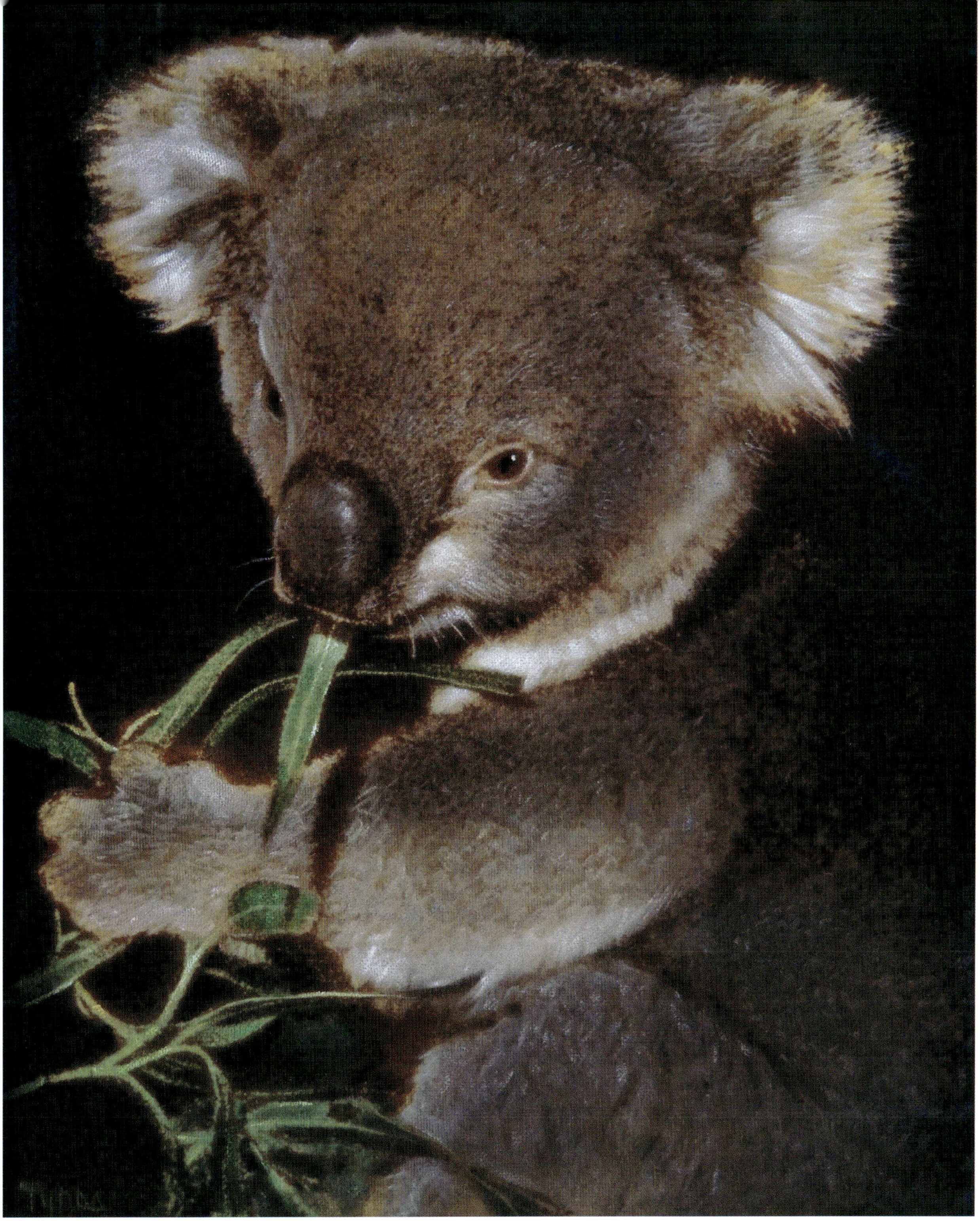

Figure 10.24 *Koala Eating a Eucalyptus Leaf*, Ralph Burke Tyree. 1976, oil on velvet, 16" x 12", private collection.

land clearing have also been significant drivers. As a result, the Australian government has re-categorized the koala as endangered and allocated tens of millions of dollars to conservation and recovery.

In Tyree's koala painting, seen in figure 10.24, the animal fills the entire frame as it snacks on eucalyptus leaves against a blue background. The long French velvet pile makes the perfect surface for a skilled painter like Tyree to accentuate the soft fur on this adorable marsupial. This, along with the zoomed-in point of view, makes the exquisitely painted coat the focal point of the work, highlighting both Tyree's mastery of the materials and the part of the animal that long made it a target of hunters.

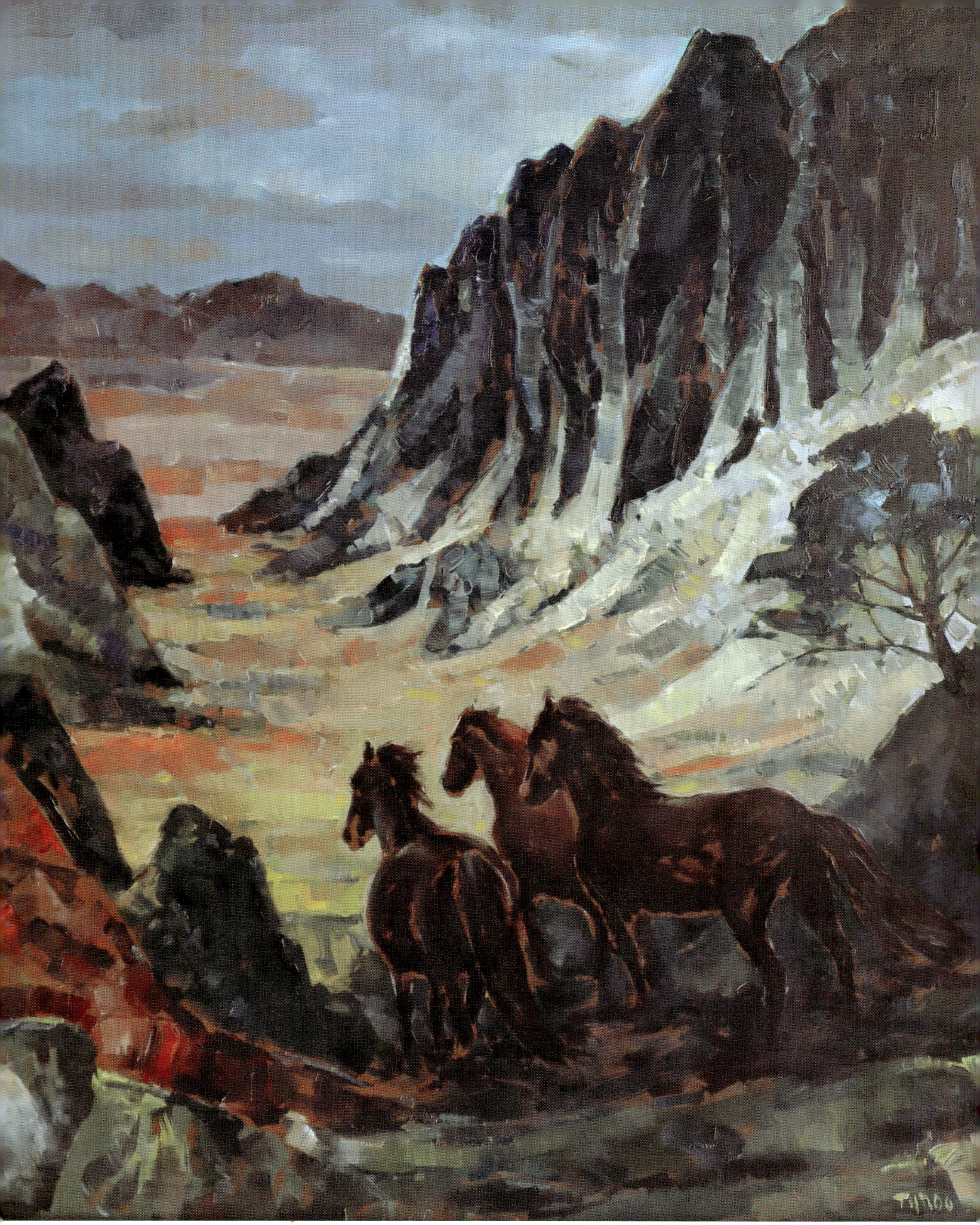

Figure 11.1 *Canyon Horses*, Ralph Burke Tyree. Circa 1970, oil on board, 24″ x 20″, private collection.

CHAPTER 11

Dabbling With Domestics

"If I paint a wild horse, you might not see the horse . . .
but surely you will see the wildness!"

—PABLO PICASSO

Finally, we must explore Tyree's depictions of a handful of domesticated animals. Because he created some of these artworks before his subject matter switched to animals, they illustrate how animals first made their way into his compositions. In these early paintings, the animals are often accessories to the primary human subject or are essentially part of the landscape. The later works are more representative of the portrait-like style he came to prefer.

Siamese Cat

The **Siamese cat** is a particularly recognizable breed of domestic cat (*Felis catus*) from the large *Felidae* family. It is notable for its blue, almond-shaped eyes, triangular head, large ears, distinct grey markings, and long, muscular body (figure 11.2). The Siamese is one of several breeds of cat native to Thailand and is believed to have descended from the sacred temple cats of the King of Siam. In 1878, US President Rutherford Hayes famously received the first documented Siamese in the United States, a cat named Siam sent by the American Consul in Bangkok. After this cat's exportation to Europe and North America in the late nineteenth century, it quickly became one of the most popular breeds. This affectionate, social, spirited cat is known for outgoing, dog-like behavior, even sometimes playing fetch. It usually bonds strongly to a single person, becoming neurotic if left alone. The Siamese's distinctive meow is reminiscent of a human baby's cry, and it is known to

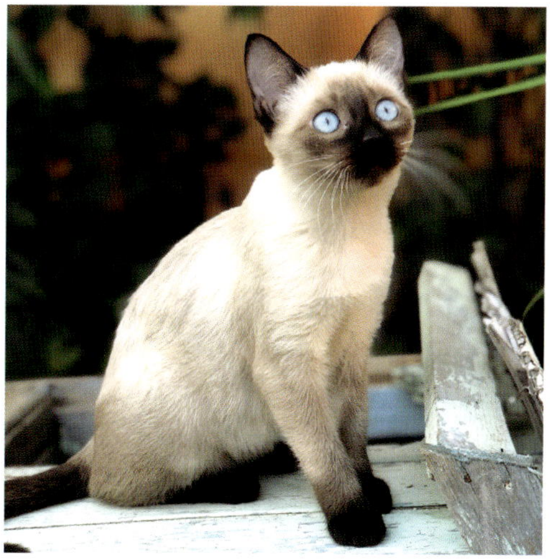

Figure 11.2 Siamese cat. Vitor Gabriel Carrilho on Unsplash.

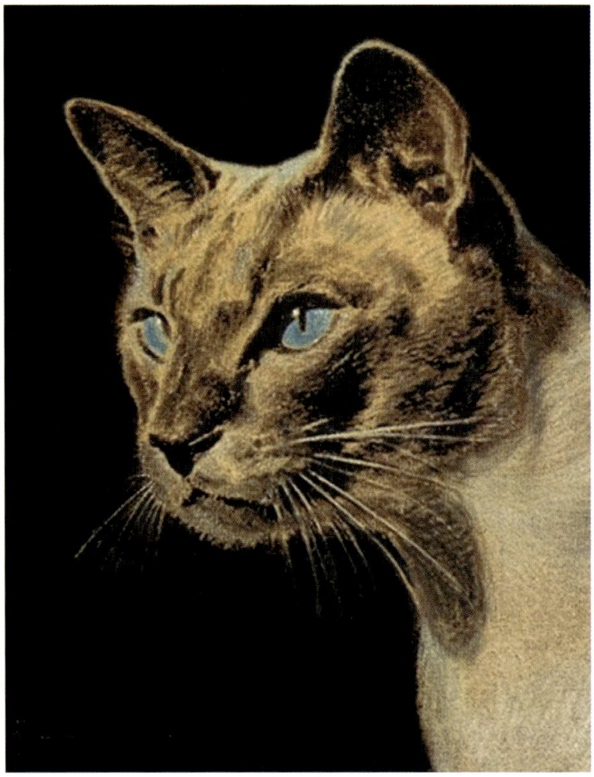

Figure 11.3 *Siamese Cat*, Ralph Burke Tyree. 1977, oil on velvet, 12" x 9", courtesy of Greg Bergmann, soapstonewerks.com.

use this meow persistently when seeking attention. Unsurprisingly, this loving cat continues to be a popular pet.

In Tyree's painting of a Siamese cat, its brilliant bright blue eyes are the focus of attention (figure 11.3). Painting with oils on velvet, he uses the cat's neutral coloring and the deep black velvet backdrop to let its eyes shine literally and figuratively. Tyree paints the Siamese in a regal three-quarter profile, befitting the once-holy cat. The composition and glowing eyes are reminiscent of Tyree's treatment of another cat with legendary origins, the caracal.

Goats

The **domestic goat** (*Capra hircus*) is a species in the broader *Bovidae* family comprised of over 300 distinct breeds. This species was first domesticated about 10,000 years ago throughout the Middle East and Central Asia, making it one of the earliest to be domesticated by humans. As such, goat herding has long provided dairy products, meat, fur, and more, thus cementing the species' vital role on the planet's farms. Although usually horned and often bearded (figure 11.4), goat breeds vary significantly in weight and size; the largest reach over 300 pounds, and the smallest only up to 45 pounds. With a famous reputation for eating almost anything, this browsing eater will give anything resembling plant matter a taste to decide if it's worth eating. However, it is just looking for leaves, hay, and shrubbery and will not actually consume inedible material. Intelligent, curious, and agile, the goat is a skilled climber known to escape enclosures. Beyond the goat's usefulness, it also forms close bonds with its human caretakers and has become a popular pet in the modern world.

Legend credits the goat with discovering coffee and its energizing effects. While no one knows the true origins of coffee, one story says that during the ninth century, a shepherd in Ethiopia observed his goats jumping about after eating red berries from a specific small tree. With the help of a local monastery, they created a new drink from these berries, and when the shepherd tasted it, his energy spiked, allowing him to stay awake through the night. Soon after, the coffee industry was born, and this new drink spread around the world (figure 11.5).

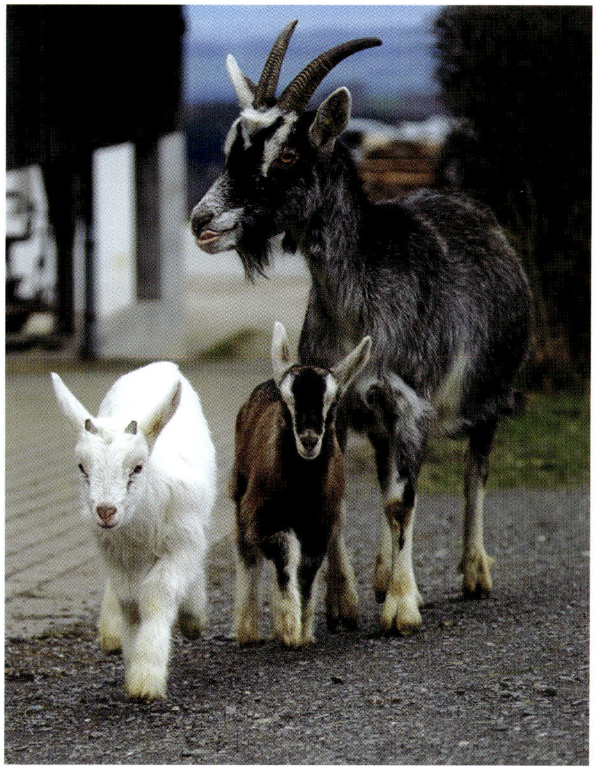

Figure 11.4 Goat and her kids. Dorothea Oldani on Unsplash.

As a domesticated animal, the IUCN does not include the goat on the Red List. Still, with an estimated global population of about one billion, the species is by no account at risk. The domesticated goat's wild cousins of the *Capra* genus are less populous. Of this genus' nine species, the IUCN considers seven either near threatened, vulnerable, or endangered. Most at risk is the West Caucasian tur from the Caucasus Mountains, which separate Europe and Asia (figure 11.6). The

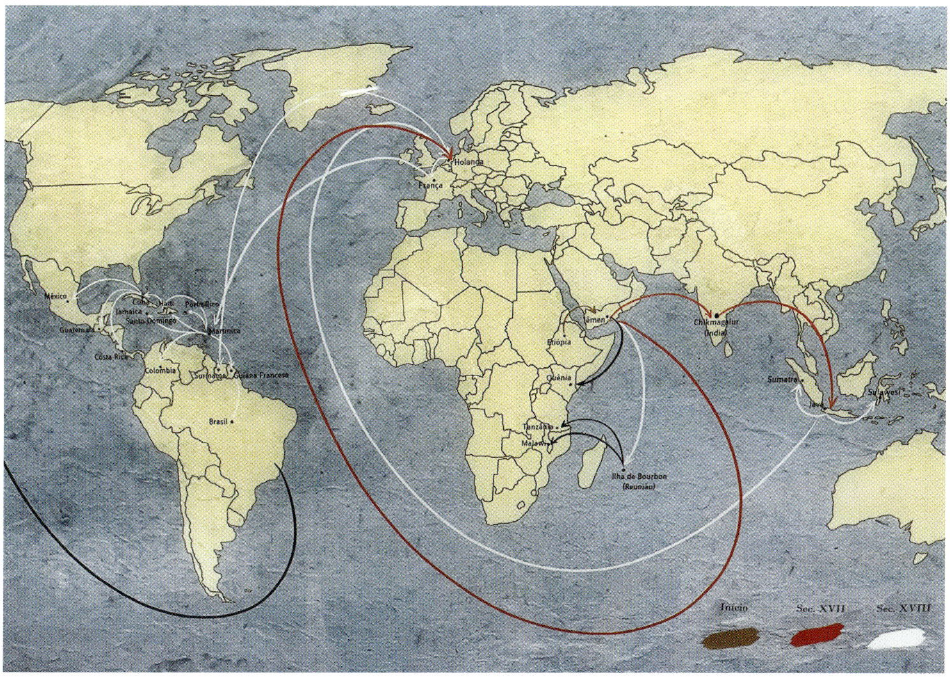

Figure 11.5 Map showing how coffee discovered by goat herders spread from Ethiopia to other continents. From the Coffee Museum, Santos, Brazil, 2022, author's photo.

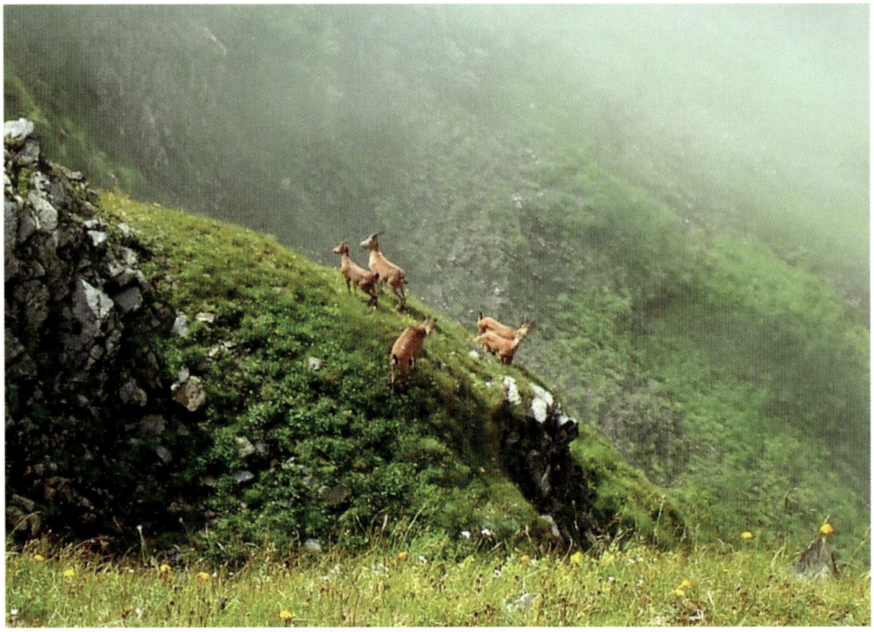

Figure 11.6 West Caucasian turs. Sergei Kazantsev, 2008.

IUCN classifies this species as endangered, as less than 6,000 are estimated to remain. The Walia ibex of the rocky mountains of Ethiopia is extremely rare, with only about 500 left in the wild, leading the IUCN to list it as vulnerable (figures 11.7 and 11.8).

In Tyree's painting of a shepherd boy and his goats (figure 11.9), the goats are a part of the boy's iconography, identifying him as a shepherd. The boy, modeled after one of Tyree's sons, sits atop a rocky cliff of pink-red rock, his hair blowing in the wind as he looks out over the cloudy vista beyond the painting's frame. The adept climbers are seen scaling the rock face in the lower half of the composition. One goat has climbed up just below the boy and looks up at him, indicating their relationship and making clear that these are his goats. Tyree painted this work using oils on board long before making animals his primary subject matter. This painting illustrates Tyree's admiration for animals and the integral role he understood them to play in human lives.

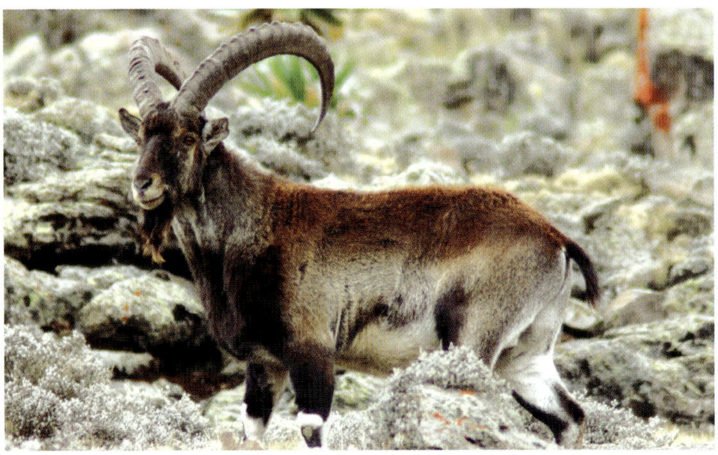

Figure 11.7 Walia ibex. Leonard A. Floyd, Ethiopia.

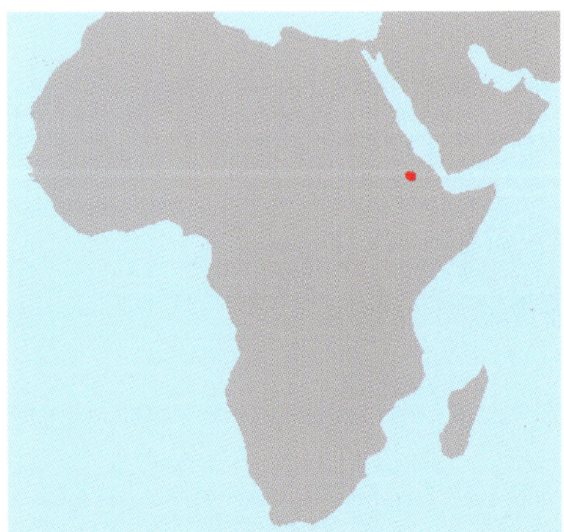

Figure 11.8 Walia ibex distribution. Uporabnik:XJamRastafire, 2019.

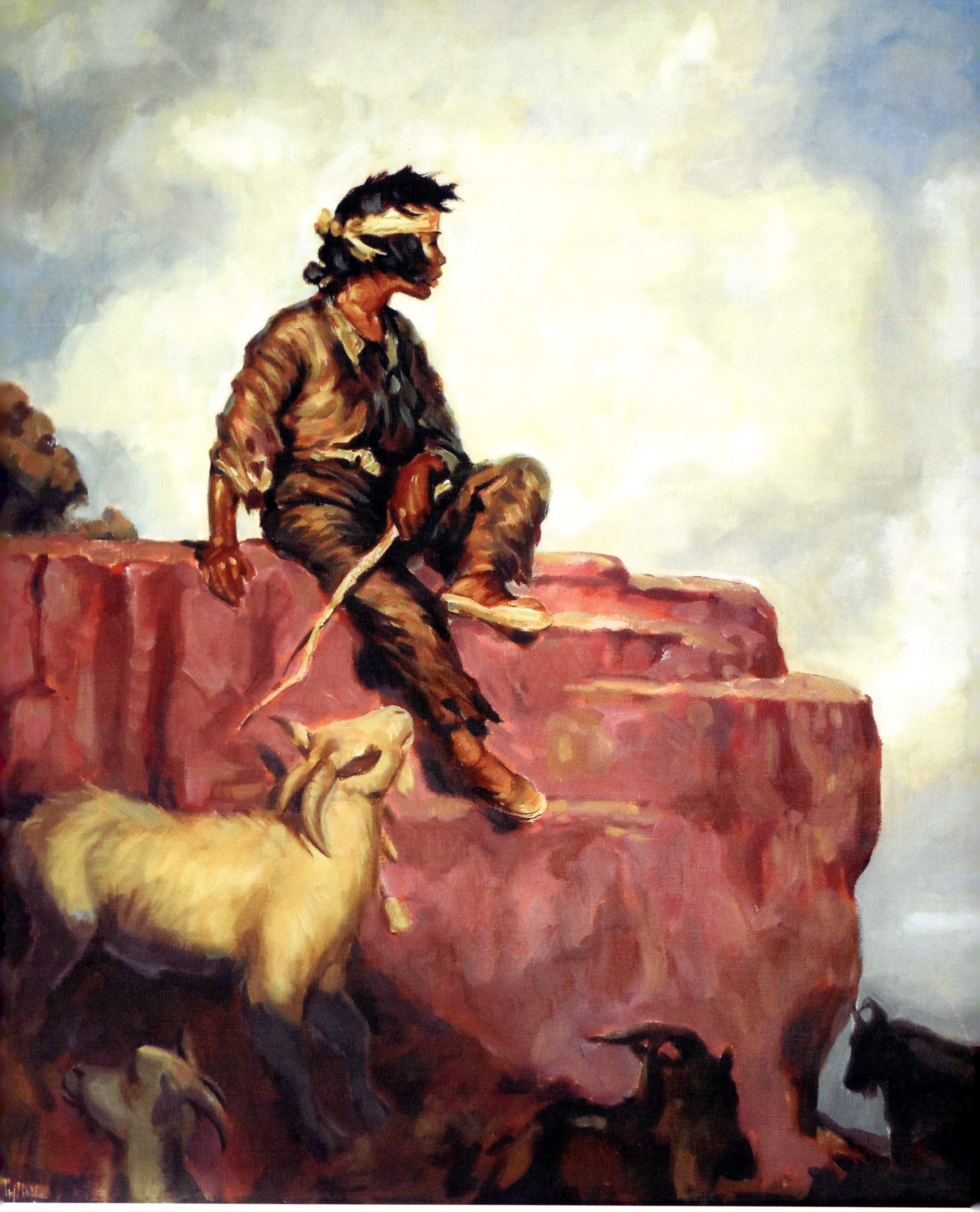

Figure 11.9 *Goats and Shepherd Boy*, Ralph Burke Tyree. 1946, oil on board, Tyree family.

Water Buffalo

The **water buffalo** (*Bubalus bubalis*) is a domestic species from the *Bovidae* family that is found across 77 countries on five continents, although the vast majority live in Asia (figure 11.11). Both subspecies of this massive water-loving animal are often over 2,000 pounds and spend a substantial amount of time immersed in water (figure 11.10). The animal's large hooves and affection for water make it ideal for toiling in the wet soil of rice paddies. The swamp variety particularly enjoys wallowing in mud.

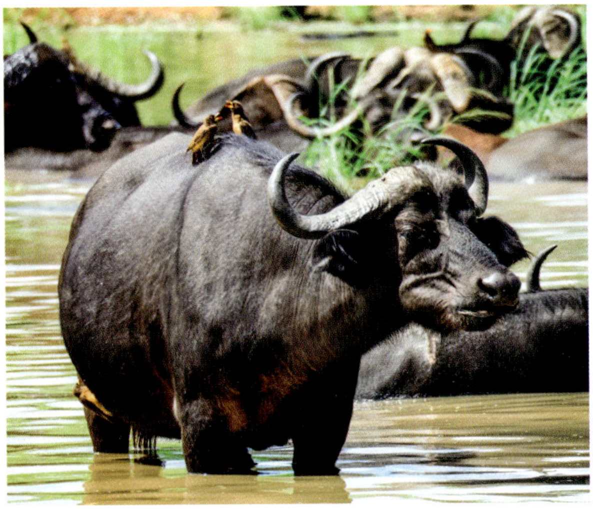

Figure 11.10 Water buffalo. Ante Hamersmit on Unsplash.

Although the river buffalo and the swamp buffalo are both descendants of the wild water buffalo, humans domesticated them independently. The river buffalo was domesticated about 6,300 years ago in India, whereas the swamp buffalo was domesticated in Southeast Asia between 3,000 and 7,000 years ago. The water buffalo has been a massive boon to humanity, with people arguably more dependent on it than any other domesticated animal. It provides dairy products, meat, and hides, all while acting as a beast of burden.

Even without assessment by the IUCN due to its domestic status, with approximately 130 million water buffalo worldwide, this beast is far from vulnerable. However, the wild buffalo is only found in small, protected areas of India, Nepal, Bhutan, and Thailand, and the IUCN has considered it endangered since 1986. Interbreeding with domesticated breeds has been harmful to this population.

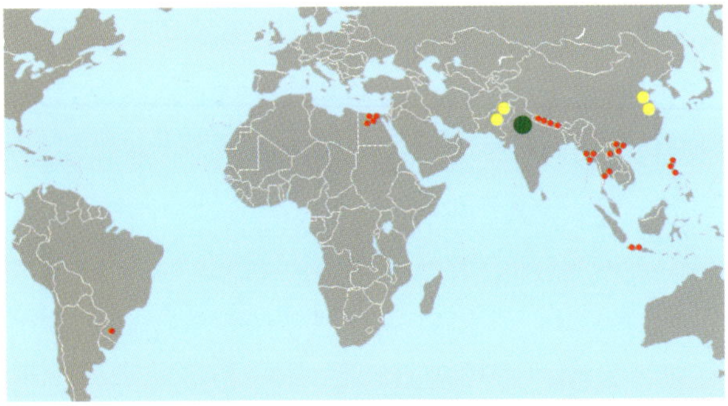

Figure 11.11 Water buffalo distribution. Anwar saadat, 2007.

While stationed in American Samoa in 1943, Tyree completed several pencil drawings featuring water buffalo. He sent one of these to his then-girlfriend Margo (figure 11.12). This exquisite drawing shows a local boy riding a water buffalo along a lush, tropical shoreline with the jungle on one side and the ocean on the other. With its horns wrapped in rope, the water buffalo is clearly domestic. Another drawing shows two buffalo pulling a produce-filled cart (figure 1.9). The local farmer who drives the cart also transports two American soldiers who sit atop the cart. A few palm fronds peek into the corner of the composition, establishing the tropical locale. This work was included in the 1943 book *Marines at War*. Despite the similar medium and subject matter, Tyree uses drastically different styles for the two pieces, hinting at his artistic experimentation during his time in

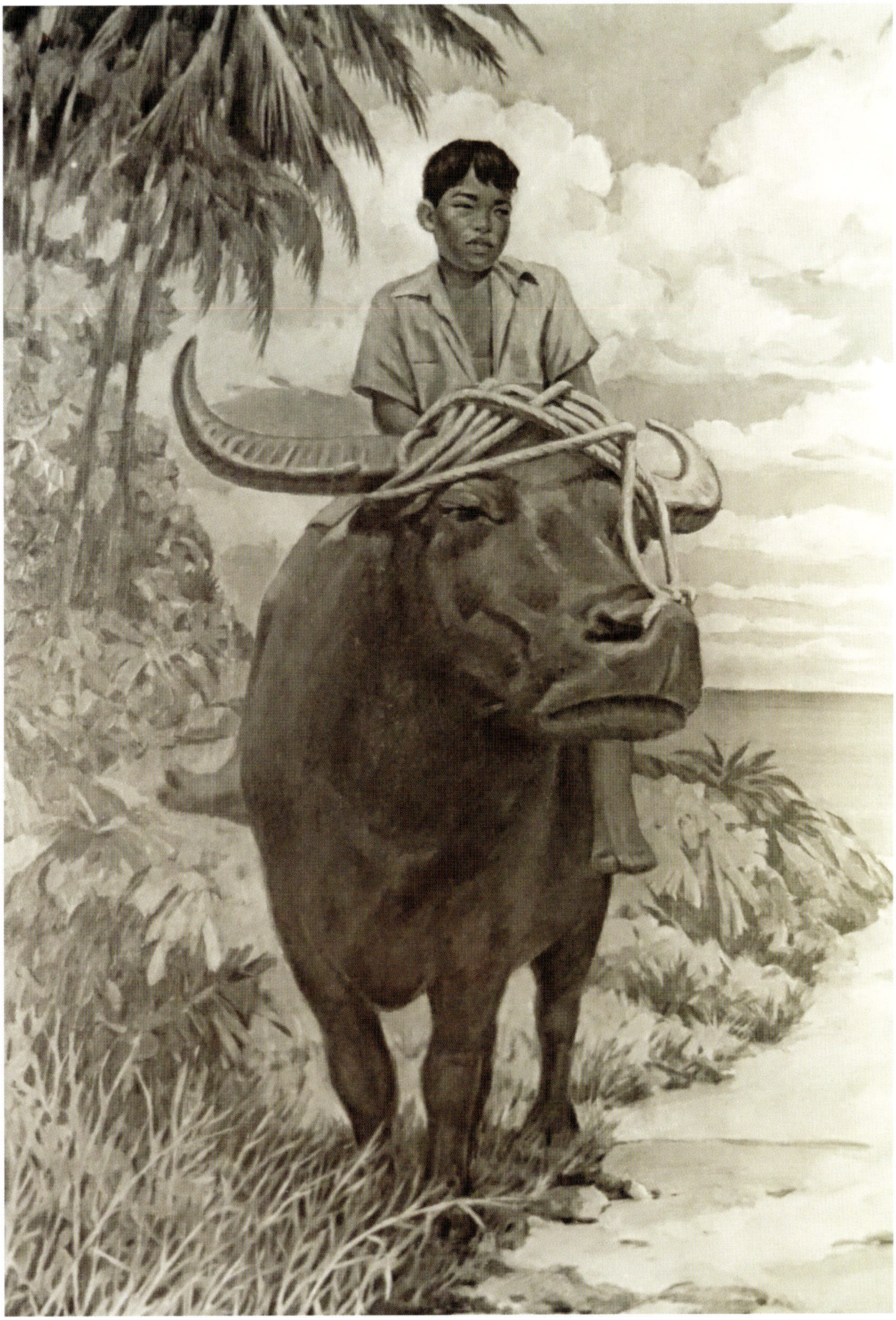

Figure 11.12 *Water Buffalo with Samoan Boy*, Ralph Burke Tyree. 1943, pencil, 12" x 8", Tyree family.

the Marines. The drawing sent to Margo exemplifies the detailed realism he would primarily paint with later in his career. In contrast, the piece featured in *Marines at War* illustrates the more expressive, somewhat abstract style he would sometimes utilize.

The Once-Domestic Horse

The **mustang** (*Equus caballus*) is the feral descendant of Spanish horses that escaped after being brought to the Americas (figure 11.13). However, other horse species, including the pony and the draft horse, have also contributed their genetics to this member of the *Equidae* family. At the end of the seventeenth century, feral herds formed in western states like Texas and Colorado, and by the nineteenth century, millions of these wild horses roamed the Western US. With this, humans began to hunt them for food and sport, and even use them in war. This caused their numbers to fall precipitously and brought the species to the brink of extinction by the mid-twentieth century. After the passage of the Wild and Free-Roaming Horses and Burros Act of 1971, herds came under government protection, and their numbers stabilized.

Today, approximately 90,000 mustangs roam across the Western US, with more than half in Nevada. The population has gotten so large that adoption cannot keep up. The mustang is no longer vulnerable, although, as a feral species, the IUCN has never assessed it. To avoid euthanasia, the Bureau of Land Management has created long-term holding facilities for these feral horses, and over 50,000 are now in captivity. Some have suggested these horses be employed to help prevent wildfires by using them to clear dangerous brush while simultaneously fertilizing the forests.

Thousands of years before the sixteenth-century arrival of the mustang's Spanish ancestors, North America was home to its own native species of horses. These wild horses evolved and prospered across the continent before their extinction at the end of the Last Glacial Period about 25,000 years ago. This unique circumstance has led to conflicting opinions about the mustang's role in the modern American ecosystem. While the species is not native, this feral horse is similar to the prehistoric native North American horse and has taken over its ecological niche. Additionally, the mustang has become part of the cultural iconography associated with the American West and a symbol of freedom. Still, many ranchers argue that these horses' grazing threatens their livestock.

Tyree featured the mustang in his paintings on several occasions. Early in his professional career,

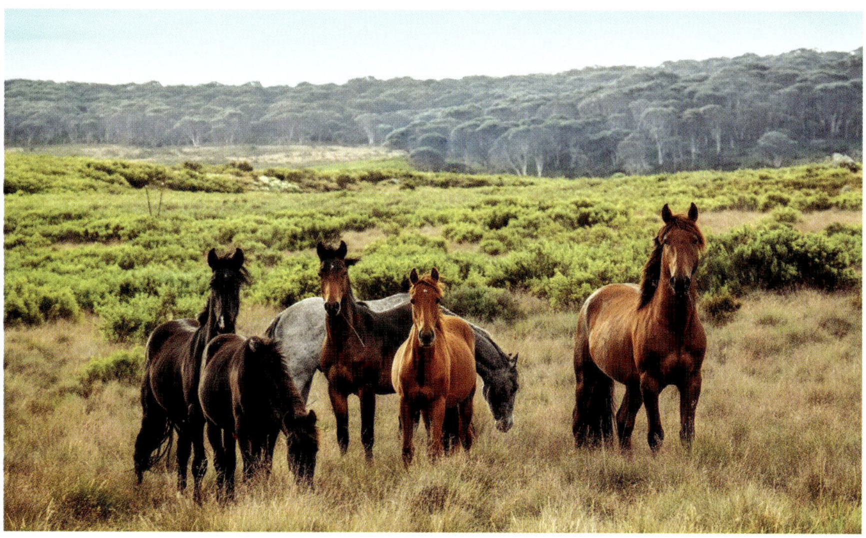

Figure 11.13 Feral horses. Chrisine Mendoza on Unsplash.

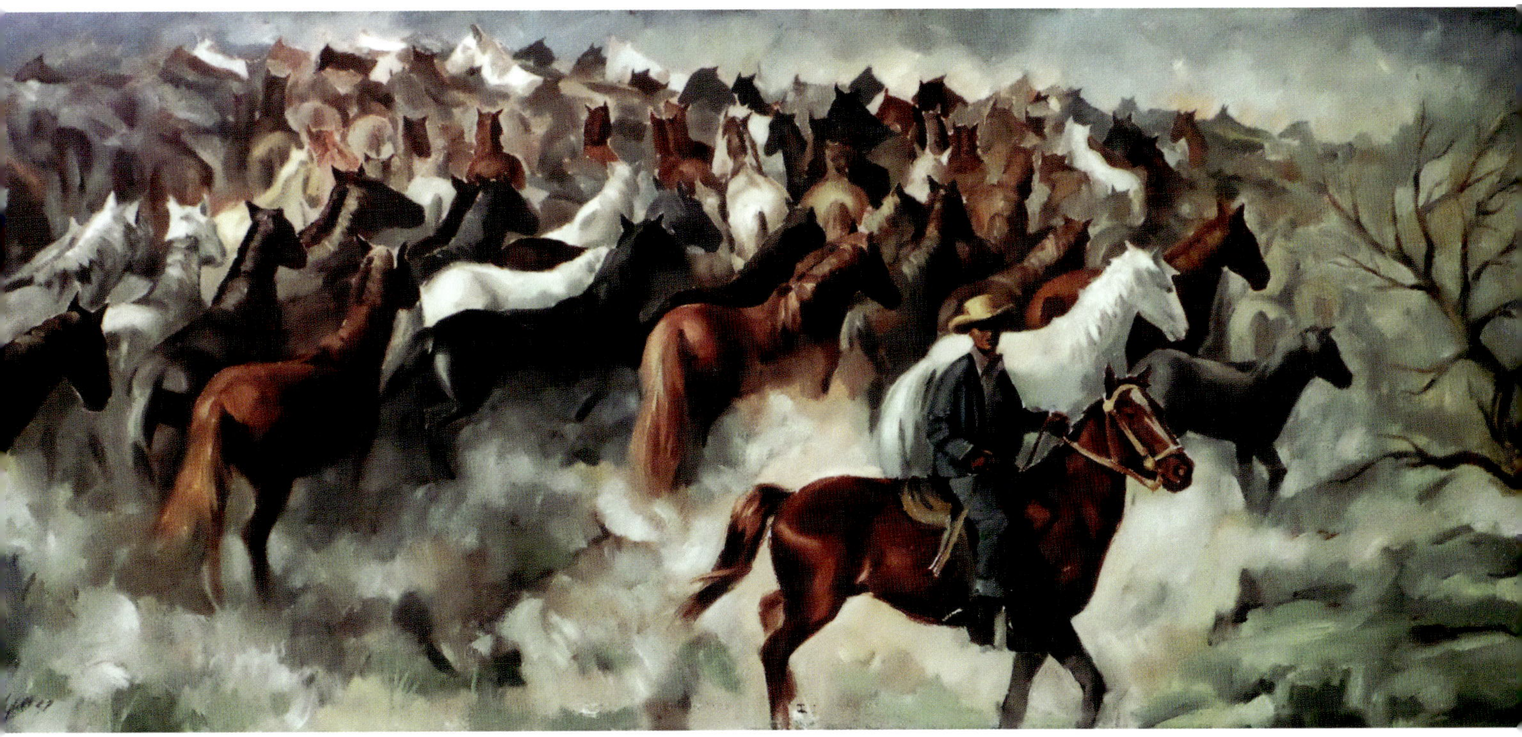

Figure 11.14 *Horses*, Ralph Burke Tyree. 1951, oil on board, 25" x 54", private collection.

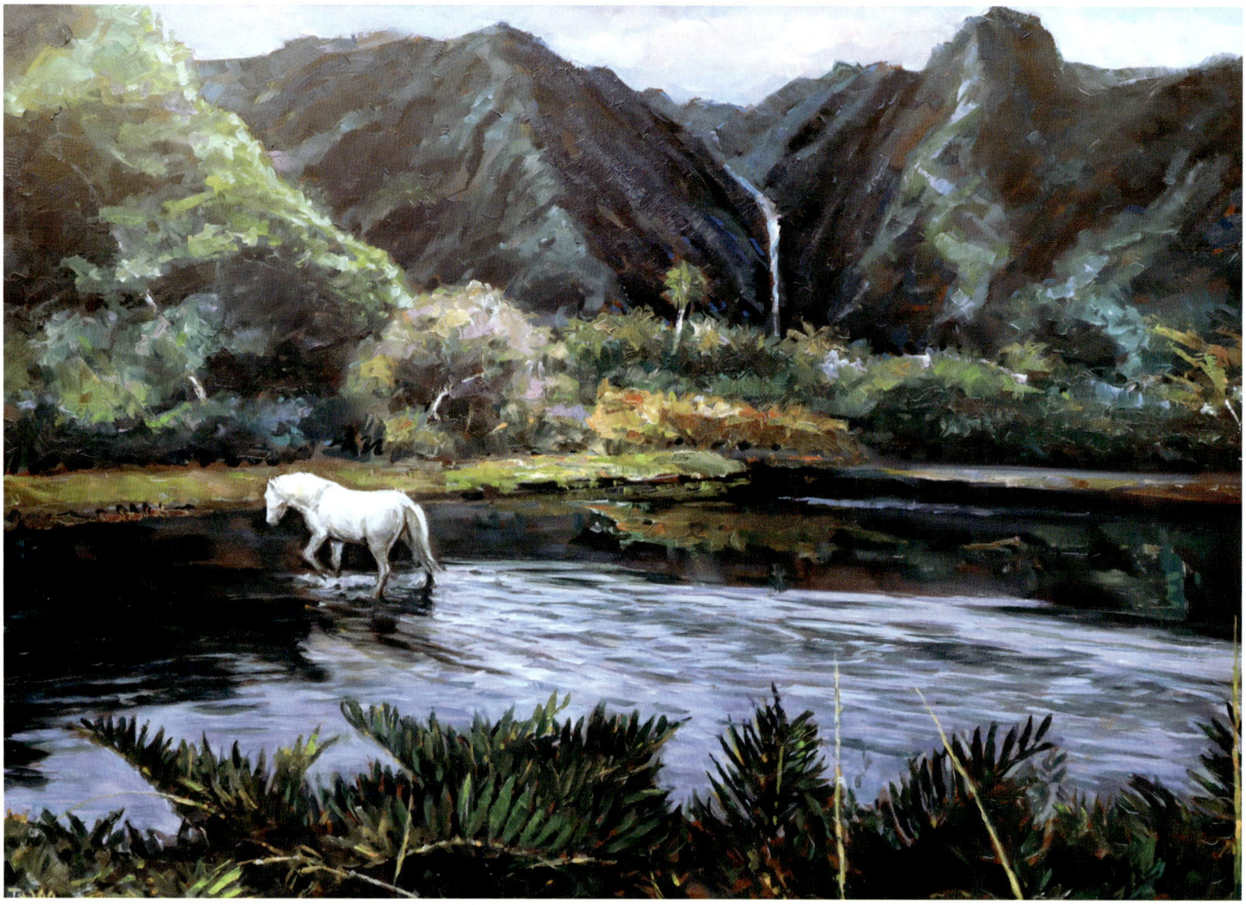

Figure 11.15 *Horse in Water*, Ralph Burke Tyree. Circa 1970, oil on board, 24" x 30", private collection.

he painted a sizable oil-on-board piece depicting a horse stampede as a gift for his father-in-law (figure 11.14). Tyree wanted to move with Margo and their four children to the South Pacific for a few years and hoped the gift would help convince her father to support this unconventional choice. It worked, and the family departed the following year with his blessing. In the painting, a rancher, perhaps Tyree's father-in-law, sits atop a horse with a herd of horses stampeding away from the viewer behind him. The figures of the grey, white, and brown horses create a graphic pattern as the dust they kick up surrounds them and blends into the sky. The rancher pauses in the foreground, turning to face the viewer atop his shining brown horse.

Over a decade later, shortly before turning his focus to animals, Tyree painted several oil-on-board landscapes featuring wild horses. One of these works shows three brown mustangs walking down a ravine away from the viewer (figure 11.1). He colors the terrain with dark, moody greys and warm rusty ambers, evoking the American west. In the other painting, he depicts a tropical landscape of steep, dark green mountains, lush vegetation, and a glittering waterfall rushing toward a large body of water (figure 11.15). A single white mustang walks through the shallow water, creating ripples across the surface. In both works, he paints the feral horses as part of the landscape itself, representing their natural place in these environments. These paintings stand apart from the animal paintings Tyree created later in his career, illustrating the role animals played in his art prior to the 1970s and the transition period when he began featuring animals more often.

The Donkey is Kick-Ass

"Donkeys live a long time. None of you has ever seen a dead donkey."

—GEORGE ORWELL

The **donkey** (*Equus asinus*) is another beast of burden from the *Equidae* family as the horse (figure 11.16). After the domestication of the African wild ass about 7,000 years ago in East Africa, the donkey quickly became integral to ancient Egypt, Mesopotamia, Greece, and Rome. Its unbeatable cargo-carrying abilities made the development of long-distance trade throughout the ancient world possible. The donkey has also been used for meat, hides, and agriculture and was notably associated with the Greek god of wine and fruitfulness, Dionysus. This essential helper was first brought to the Americas on Christopher Columbus' Second Voyage, with Spanish conquistadors soon adding to its numbers. Today, the donkey is used primarily in underdeveloped countries, usually for transport and agriculture. The male donkey is referred to as a jack or jackass, and the female as a jenny. A jack mated with a female horse, or mare, produces a mule, which has the donkey's strength but is less stubborn and infertile.

With at least 40 million donkeys worldwide, this domestic species is not at risk, although, like other domestic species, the IUCN has not assessed it. Despite this large population, The Donkey Sanctuary published a study in 2019 warning that worldwide donkey populations could see a 50% decrease by 2025. This is due to the increased use of its hides in traditional medicine and its meat as a food source in Italy and China.

A feral donkey population is established when domestic donkeys return to the wild. While established groups of feral donkeys are protected in North America and Italy, in other places like Hawaii and Australia, they are seen as destructive pests.

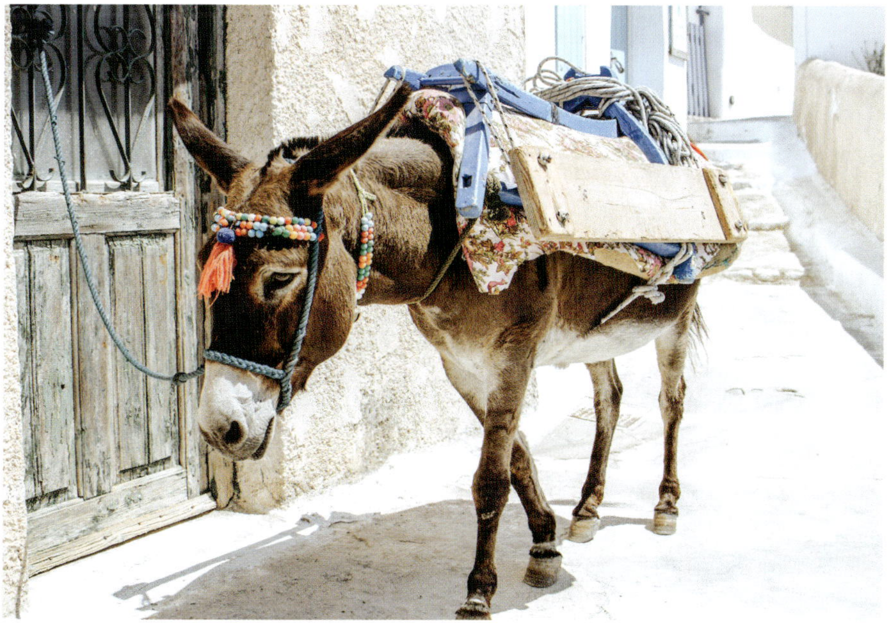

Figure 11.16 Pack burro. Florian Giorgio on Unsplash.

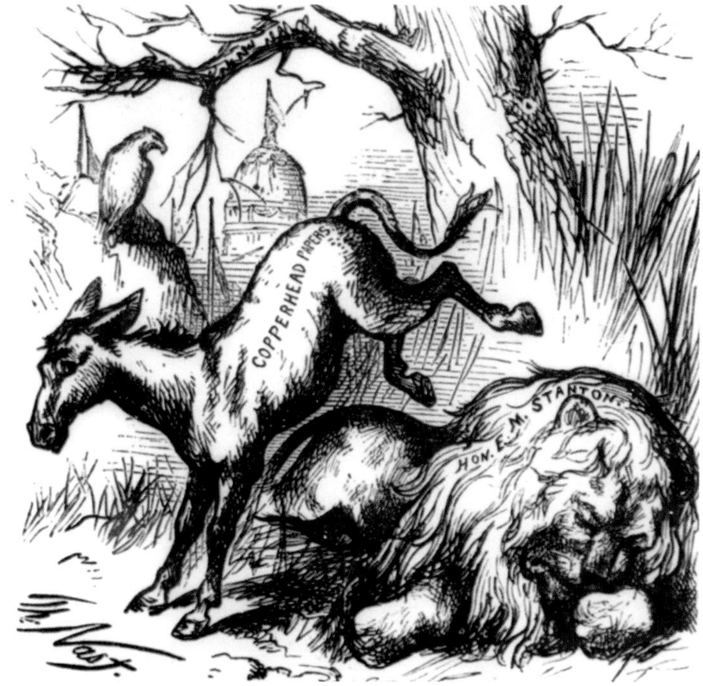

Figure 11.17 *A Live Jackass Kicking a Dead Lion*, Thomas Nast. Courtesy of *Harper's Weekly*, January 15, 1870.

The donkey was first used as a political symbol in the US by Democrat Andrew Jackson during his 1828 Presidential campaign. Jackson's opponents called him a jackass, but Jackson ran with the idea rather than be insulted. He used it on campaign posters and ultimately defeated his Republican opponent, John Quincy Adams. Decades later, in 1870, Thomas Nast used the donkey as the symbol of the Democratic party in his political cartoons (figure 11.17), beginning an association that continues into the present day.

As seen in figure 11.18, Tyree's painting of a burro, a small donkey primarily used as a pack animal, represents a stark deviation from his typical photorealistic style. He uses rough, impressionistic brush strokes to depict the burro standing in profile, saddled up, and tied to a post with buildings in the background. In another change from Tyree's usual artistic choices, the donkey is in a human-built environment instead of in nature. Although, in this long-domesticated animal's case, one could say the human world is its natural environment. In these ways, this oil-on-velvet painting is a unique departure from his other animal works in several exciting ways and suggests Tyree's creative experimentation.

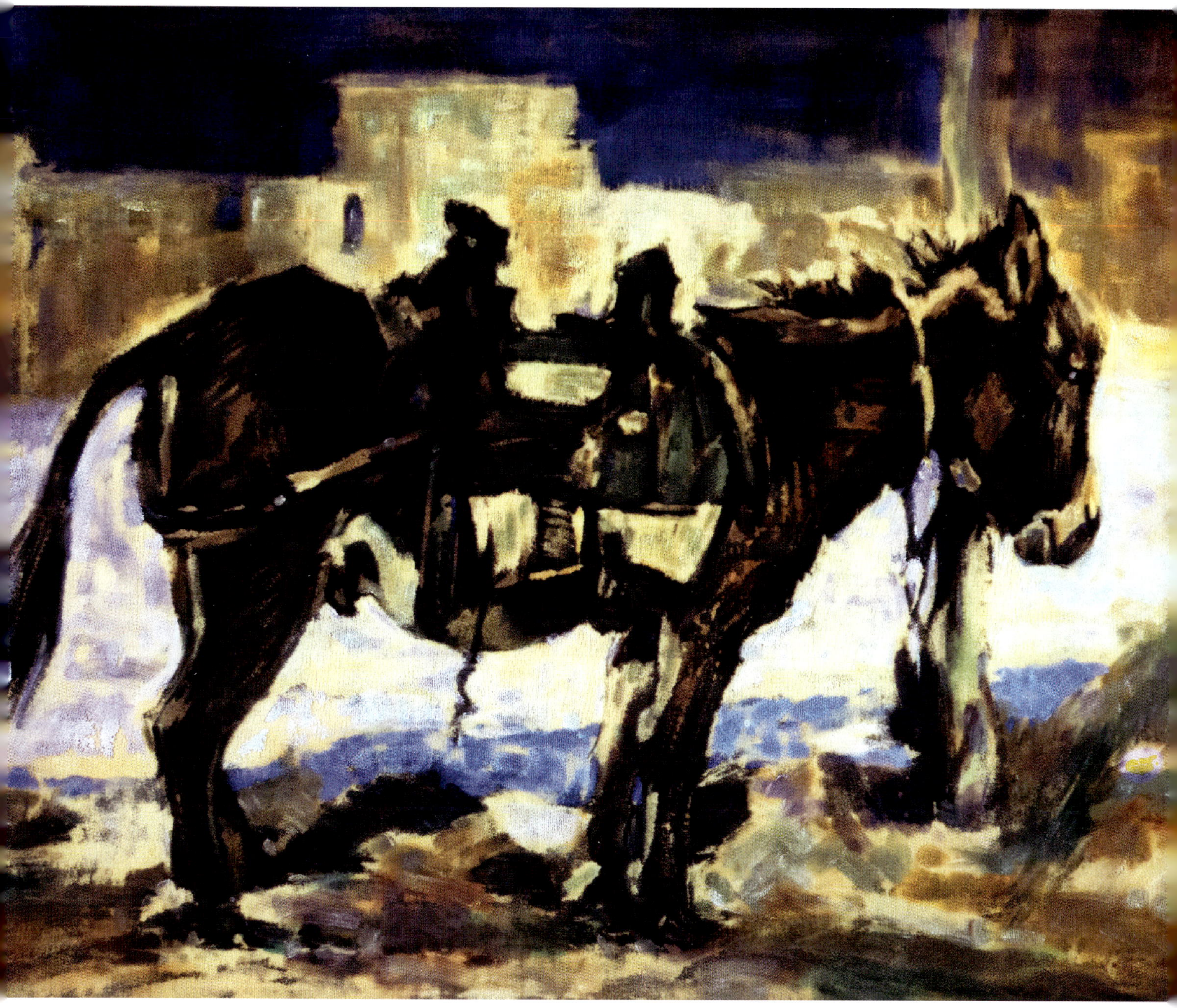

Figure 11.18 *Burro*, Ralph Burke Tyree. 1971, oil on velvet, 20″ x 24″, author's collection.

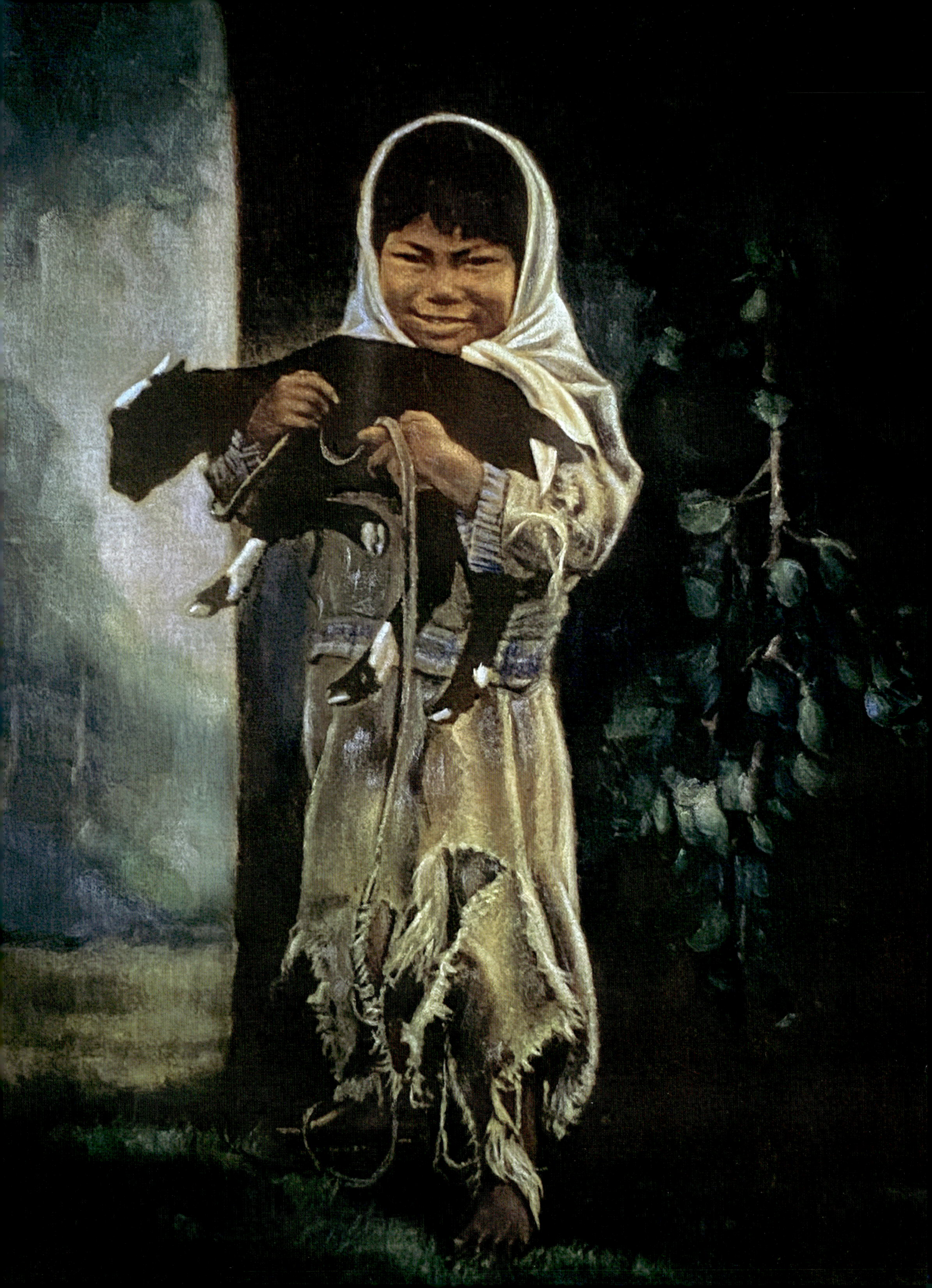

Conclusion

During the last decade of his life, Ralph Burke Tyree primarily focused on painting animals—celebrating creatures big and small, common and endangered, and local and foreign. He skillfully brought all types of fauna to life using his brushes.

While Tyree painted many varieties of animals, he primarily sought to bring public focus to the plight of the world's endangered species. He depicted many vulnerable animals from faraway places, such as the cotton-top tamarin, the dwarf lemur, the red panda, and the pangolin. Still, Tyree also painted those from closer to home, like the sea otter, the grizzly, the bald eagle, and the California condor. Tyree did not overlook the famous and threatened big cats, also featuring the stately lion and the elusive tiger in his artwork. Notably, his final piece was of the highly endangered African elephant, illustrating his ongoing effort to help people recognize and understand the perils of our planet's co-inhabitants.

I'm sure Tyree would be thrilled to know of the conservation successes some of the vulnerable species he painted have experienced since his death. Notably, the populations of the bald eagle and the California condor, both native to his home state, have bounced back thanks to protection and breeding programs. When he painted them, these notable American birds' likely extinction was a major public focus as DDT came under scrutiny and conservation programs began. Additionally, Tyree would be mind-blown to hear of Kurt, the Przewalski's horse cloned from a frozen 40-year-old piece of skin. These exciting advancements in cloning, only imaginable in science fiction during Tyree's lifetime, have renewed hope for the futures of many endangered animals.

Although some of the vulnerable species Tyree painted have impressively rebounded or experienced other positive developments, many remain at risk, with the IUCN classifying them as threatened. This book seeks to continue Tyree's cause, reminding readers of the precious beauty in the beasts that Ralph Burke Tyree loved and painted.

Figure 12.1 *Boy with Goat*, Ralph Burke Tyree. 1977, oil on velvet, 18" x 24", Tyree family.

Afterword

I have had the privilege of calling CJ Cook a friend and distinguished colleague for over 20 years. I first met him at the San Diego Zoo Safari Park, where I had the pleasure of working as a zoo and wildlife veterinarian for almost 35 years. He assisted us during a complicated injury in an infant gorilla. Since gorillas are anatomically similar to humans, we had a successful outcome. The clinical management of this gorilla incident was, at the time, an early example of the emerging concept of a One Health approach to problem-solving, where a multidisciplinary group of professionals with seemingly unrelated career focuses work to resolve a health issue of an animal(s) or ecosystem.

Beauty in the Beast is not only CJ Cook and Paige Herbert's tribute to Ralph Burke Tyree's brilliant artwork, but also the curiosity and concern for the culture, tradition, and wildlife conservation in Tyree's art form. Throughout the book, the authors uniquely tell a story of the significance of conservation issues through art, effectively connecting people with wildlife. This message is similar to the goals and objectives of zoological collections or the memorable experiences gained by in situ observations of these animals and ecosystems in the wild.

The authors highlight the historical significance of the African Big Five: the elephant, leopard, buffalo, lion, and rhino. This group of animals attracts worldwide attention due to their tremendous size, potential danger, and trophy status amongst hunters. *Beauty in the Beast* fully provides the conservation significance of less-known species. Still, I would like to boost further attention and importance to a lesser-known group of smaller-sized animals, the Little Five. A more diminutive group compared to their counterparts from the Big Five, this group consists of the elephant shrew, leopard tortoise, buffalo weaver, ant lion, and rhino beetle (figures 12.2–12.6). Although they are less famous, their plight is just as precarious. Because they are not as recognized in the African landscape, they receive less public attention, which increases their populations' vulnerability. The loss of lesser-known species of plants and animals reduces overall biodiversity, disrupting the balance of ecosystems, which is necessary for sustained viability. However, conservationists promote the worldwide principle of using iconic or "flagship" species, like the Big Five, to help improve the status and health of many other species that share the same habitats, including the species of the charming Little Five. In this book, the authors bring attention to an array of unique and endangered species that deserve better than to end up as specimens only found in museums or epitaphs like the extinct Dodo bird.

The book opens with a statement from Albert Einstein that reads, "Our task must be to free ourselves by widening our circle of compassion to embrace all living creatures and the whole of nature and its beauty." Coincidentally, Einstein is one of my real-life superheroes. Through his genius, he made landmark discoveries that have greatly contributed to our understanding of our universe. Who knew

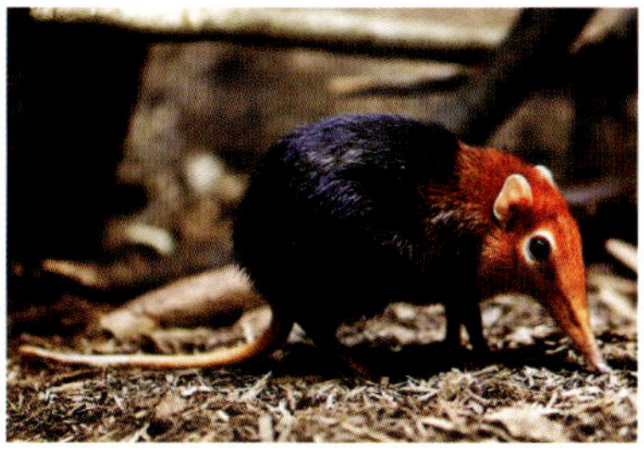

12.2 Elephant shrew (black and rufous). Joey Makalintal.

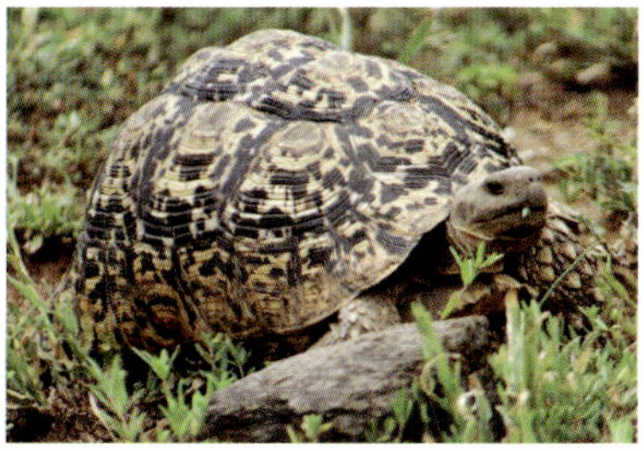

12.3 Leopard tortoise. Katelyn Deelefs (@bushbabyk8), South Africa, 2019, courtesy of the photographer.

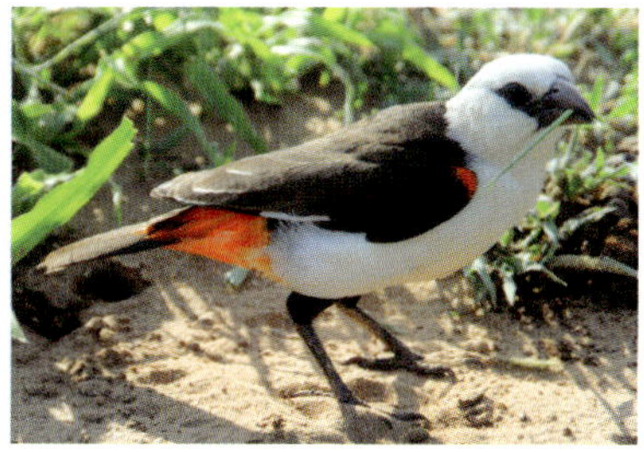

12.4 Buffalo weaver (white-headed). Bob.

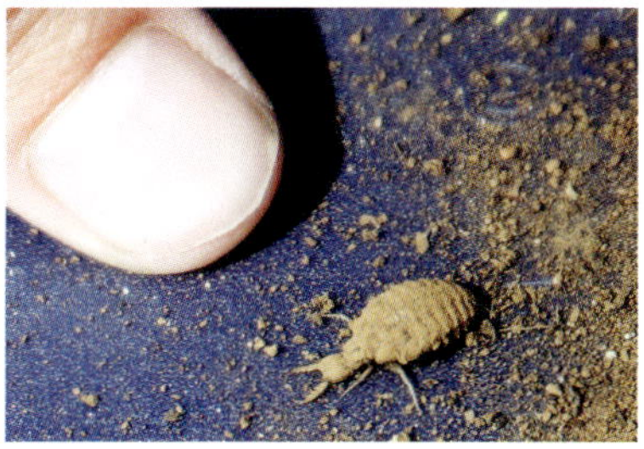

12.5 Ant lion larva. Djzanni.

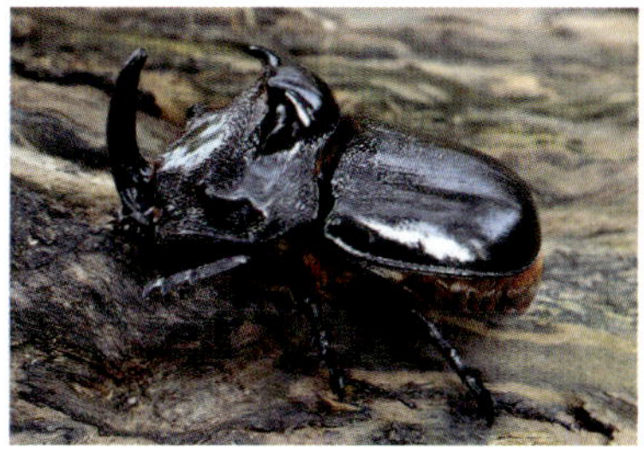

12.6 Rhinoceros beetle. Len Worthington.

that Einstein's thoughts were not only directed toward conserving mass and energy but also conserving species and habitat?

Beauty in the Beast is a unique and magnificently written book. I became captivated and before I knew it, had read through the entire book. The book is not only well done but significant.

—Jeffery R. Zuba, DVM
Zoo and Wildlife veterinarian
San Diego Zoo Global (retired)

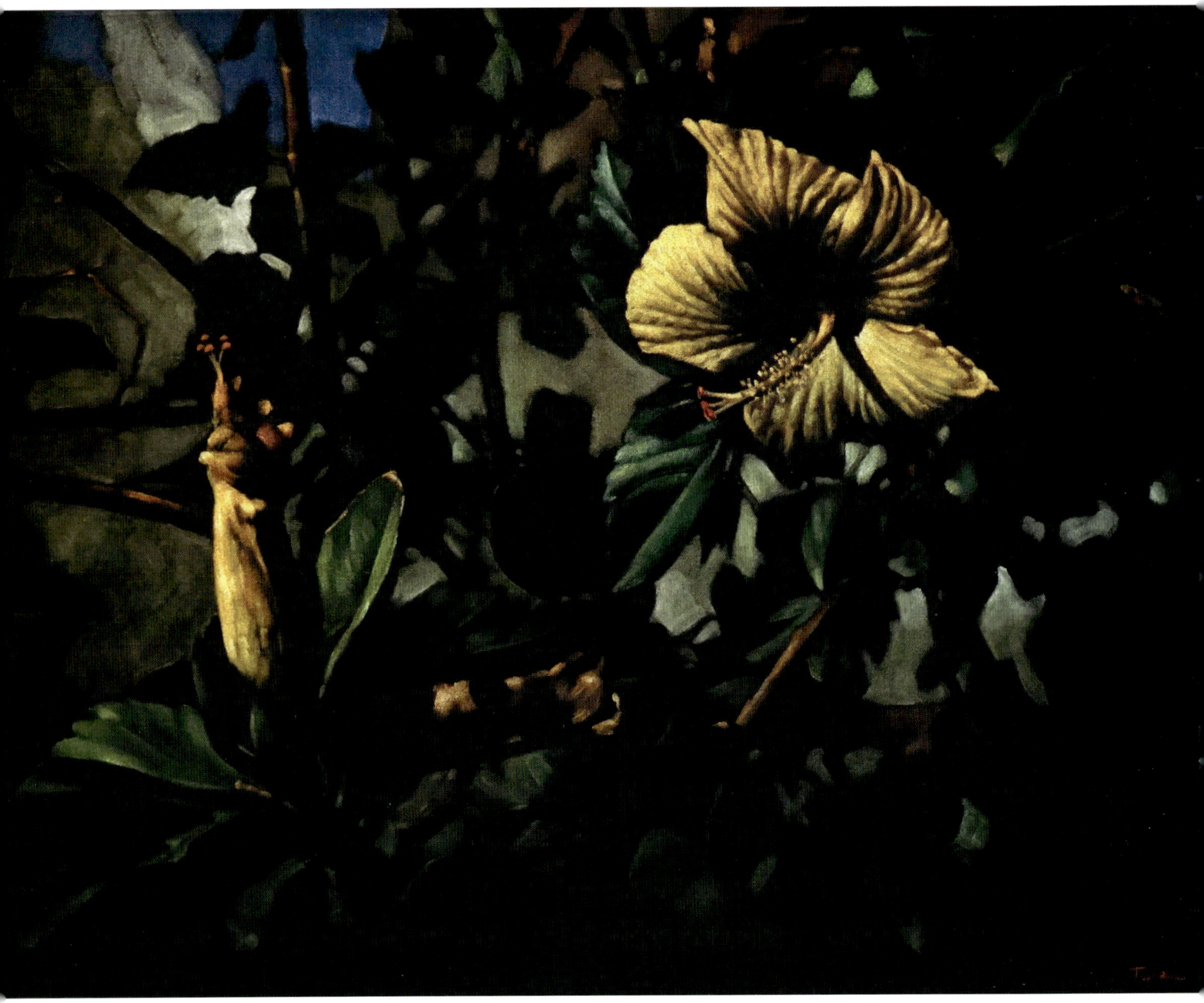

Figure 12.7 *Yellow Hibiscus*, Ralph Burke Tyree. 1968, oil on velvet, 16" x 20", author's collection.

Acknowledgments

I am indebted to the many who helped me with this book.

With this book, as with my first book, Tyree, Artist of the South Pacific, I had the support of the Tyree family in further exploring this fantastic man and artist and finding more of his animal paintings.

Tyree's granddaughter, Paige Herbert (daughter of Marda), is my editor and coauthor. She brought her art history and dedicated research to the book. Besides editing and proofing thousands of times, she made many excellent contributions. Her mother, Marda Herbert, once again, as with the first book, was a reviewer and wrote a perfect introduction. Thank you, ladies.

Thank you to my book designer supreme, Claudine Mansour, who, as with my other books, turned my collection of photos, paintings, and words into a fascinating and hopefully pleasurable tome.

We needed photographs to amplify the animal's description of Tyree's oil paintings. With his nighttime cameras and telephoto lenses, photographer Jim Bremner added some beasts and birds from our neighborhood. Thanks, Jim (www.jfbphotos.com).

Another photographer who contributed many photos was Katelyn Deelefs, a hobbyist wildlife photographer from Zululand, South Africa, where she is spoilt for choice as the area boasts many different Wildlife and nature reserves throughout the region. Kate spends most of her free time in nature, which she loves so dearly. It offers fantastic photographic opportunities for megafauna and tiny birds and bugs, and please see her Instagram @bushbabyk8. Keep shooting, Kate.

Neighbor Brian Caldwell captured a mother Bald eagle and chick (www.lakehodgestours.com). Plus, countless photos from photographers who have open-sourced their creations on Wikipedia Commons and Unspashed.com. Thank you all.

Thanks also to Jeff Zuba, DVM, longtime San Diego Zoo veterinarian and close friend, who provided animal advice, spectacular photos of an African bull elephant vasectomy in the wild (Chapter 8), and a terrific afterword.

And to Rolf Benirschke, San Diego Chargers kicker 1977–86, and author, Saving Wildlife, whose kind words about our book and his father, Dr. Kurt Benirschke, creator of the "Frozen Zoo" at the San Diego Zoo, which led to the birth of rare Przewalski horse.

Also, thanks to Karla Olson, publisher of Patagonia, for her lovely comments on our book. To Annie Toomaru and her dear friend Juliette Mills's assistance with Sir David Attenborough's remarks.

And finally, to my wife, Sandy, and granddaughters, Scotlyn, and Sadie, who were models for the flower chapter.

Mahalo all,
CJ Cook

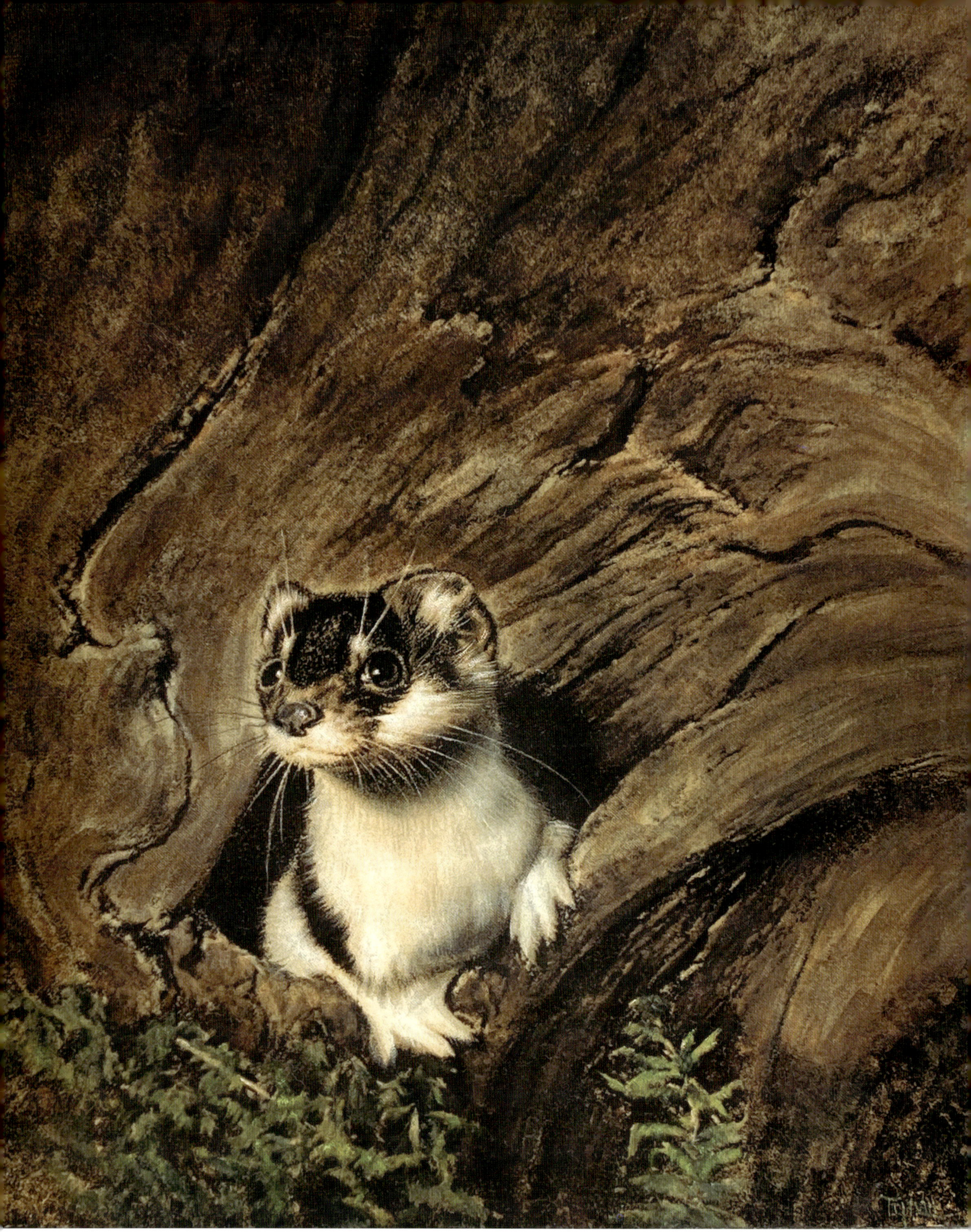

Bibliography

Bell, Catharine. "Be a Part of Ndjia's Team." *ZooNooz*, San Diego Zoo, March 1998.

Benirschke, Rolf, and James Lund. *Saving Wildlife: The Extraordinary Life and Legacy of Dr. Kurt Benirschke*. San Diego Zoo Wildlife Alliance Press, 2022.

Cook, CJ. and Ashley, Michael, *Leeteg: Babes, Bars, Beaches, and Black Velvet Art*. South Pacific Dreams, 2021. www.southpacificdreams.com.

Cook, C.J., *Tyree: Artist of the South Pacific*. South Pacific Dreams, 2017. www.southpacificdreams.com.

Dreamtime. In *Dreaming*. Wikipedia, n.d.

Escalante, Ralph Rugoff, John Turner, Meredith Tromble, Charles Kraft, and Ralph Rugoff. *A Rascal in Paradise: The Velvet Paintings of Edgar Leeteg*, 1998.

Flach, Tim. *Endangered*. Abrams, New York, 2017.

Green, Graeme. *The New Big 5: A Global Photography Project for Endangered Species*. Earth Aware Editions, 2023.

Heath, Jennifer. *Black Velvet: The Art We Love to Hate*, 1994.

Kleiman, Joe. "World's First Cloned Przewalski Horse Comes from DNA Preserved at San Diego Zoo 40 Years Ago," September 4, 2020.

Niyopi, Deepanwita Gita. "The Bengal Slow Loris Disappearing Gently into That Good Night," 2021." *Mongabay Indi*, 2021.

OpexNews. "The French Army Is Using Birds of Prey to Take down Drones to Protect Its Air Bases as Well as to Secure Public Airspace in Case a Drone Poses a Threat." February 12, 2017, OpexNews (@OpexNews) edition.

Wikipedia.com, 2023.

12.8 *Weasel*, Ralph Burke Tyree. 1977, oil on velvet, 20″ x 16″, Tyree family.

About the Authors

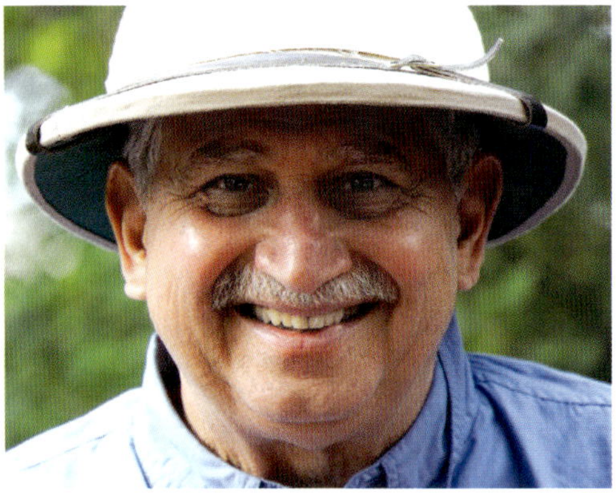

CJ Cook is an author, historian, and biographer of artists of the South Pacific, primarily Hawaii and Tahiti. His first biography about Ralph Burke Tyree, *Tyree: Artist of the South Pacific* (2017), was an incredible success, winning two Gold Awards for Best Cover and Best Biography from the Independent Book Publishers Association (IBPA) in April 2018.

His second book about Edgar Leeteg, the rascal in paradise, entitled *Leeteg: Babes, Bars, Beaches, and Black Velvet Art*, was equally successful, winning a Gold Medal for Best Biography from the IBPA in April 2022.

Cook is a lifelong manuscript collector who has a particular interest in art and the South Pacific. He has written biographical sketches on historical figures. He is a prominent manuscript society board member dedicated to preserving historical documents.

With the help of Tyree's granddaughter, Paige Herbert, he began this book with more to say about Tyree's flora and fauna. Like Tyree, Cook is a plant and animal lover. Many photographs supplement this book from his collection and travels or those of his friends. With the world-famous San Diego Zoo nearby saving endangered species such as the California condor, featured in chapter 6, and cloning first, the rare Przewalski's horse discussed in chapter 9. In 1997, Cook even assisted the efforts of pediatric orthopedic surgeons and zoo veterinarians to correct a severe knee deformity in an endangered infant lowland gorilla. Please enjoy the flora and fauna of artist Ralph Burke Tyree.

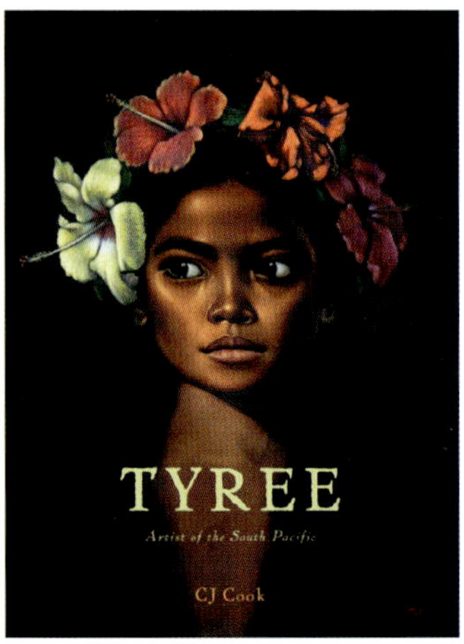

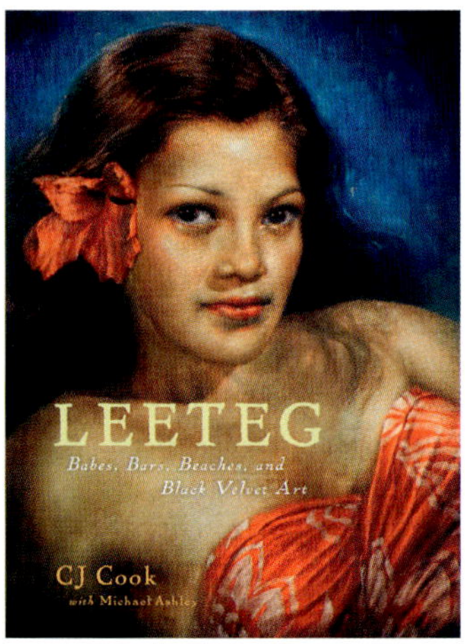

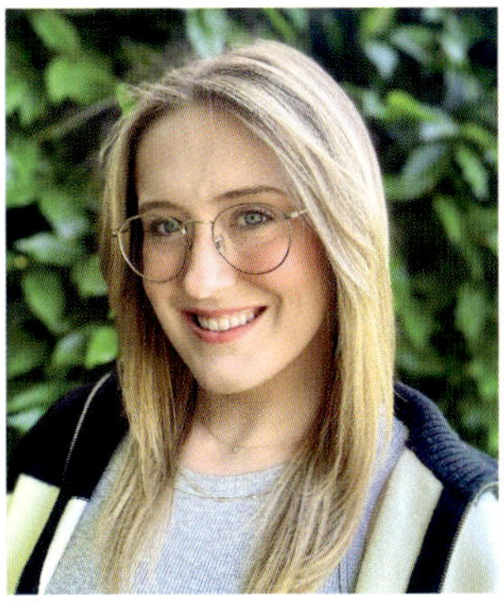

Paige Herbert is an author, editor, art historian, graphic designer, and the granddaughter of Ralph Burke Tyree. The only child of Marda, Ralph and Margo's daughter, she grew up in the California Bay Area, less than two hours from Turlock, where she spent weekends and holidays with her extended Tyree family. Whether a quick visit to see her grandmother, or a big celebration with all her uncles, aunts, and cousins, these times were treasured and foundational. She was born over a decade after Tyree's passing and never knew her grandfather personally, but she learned of him through stories told over the years, which shaped him into an almost legendary or mythical figure. His adventurous and artistic spirit infused the entire family and her childhood, inspiring an appreciation for the creative and daring and a yearning to live boldly, unconcerned with societal expectations.

This love for art and cultural exploration were passed down to Herbert, who traveled with a small group of students from across the US to Australia as an 11-year-old student ambassador and later went on to study Art History at Loyola Marymount University in Los Angeles, graduating *magna cum laude* in 2016 as a Scholar of Distinction. Since she had loved CJ Cook's first book about her grandfather, *Tyree: Artist of the South Pacific*, she jumped at the opportunity to work on a sequel to it that focused on his animal paintings. This process has been incredibly rewarding, allowing her to get to know and connect with her mysterious and incredible grandfather. Today, she lives in Los Angeles with her boyfriend and their dog, working as a writer, editor, and graphic designer, and traveling as much as possible.

(Following painting) Still Life, Hawaiian Fruits (Mango, Papaya, Passion Fruit) and a Hibiscus, Ralph Burke Tyree. 1972, oil on velvet, 16" x 20", private collection.

(Following photograph) Tyree with a finished painting in his studio. Circa 1960, Tyree family.

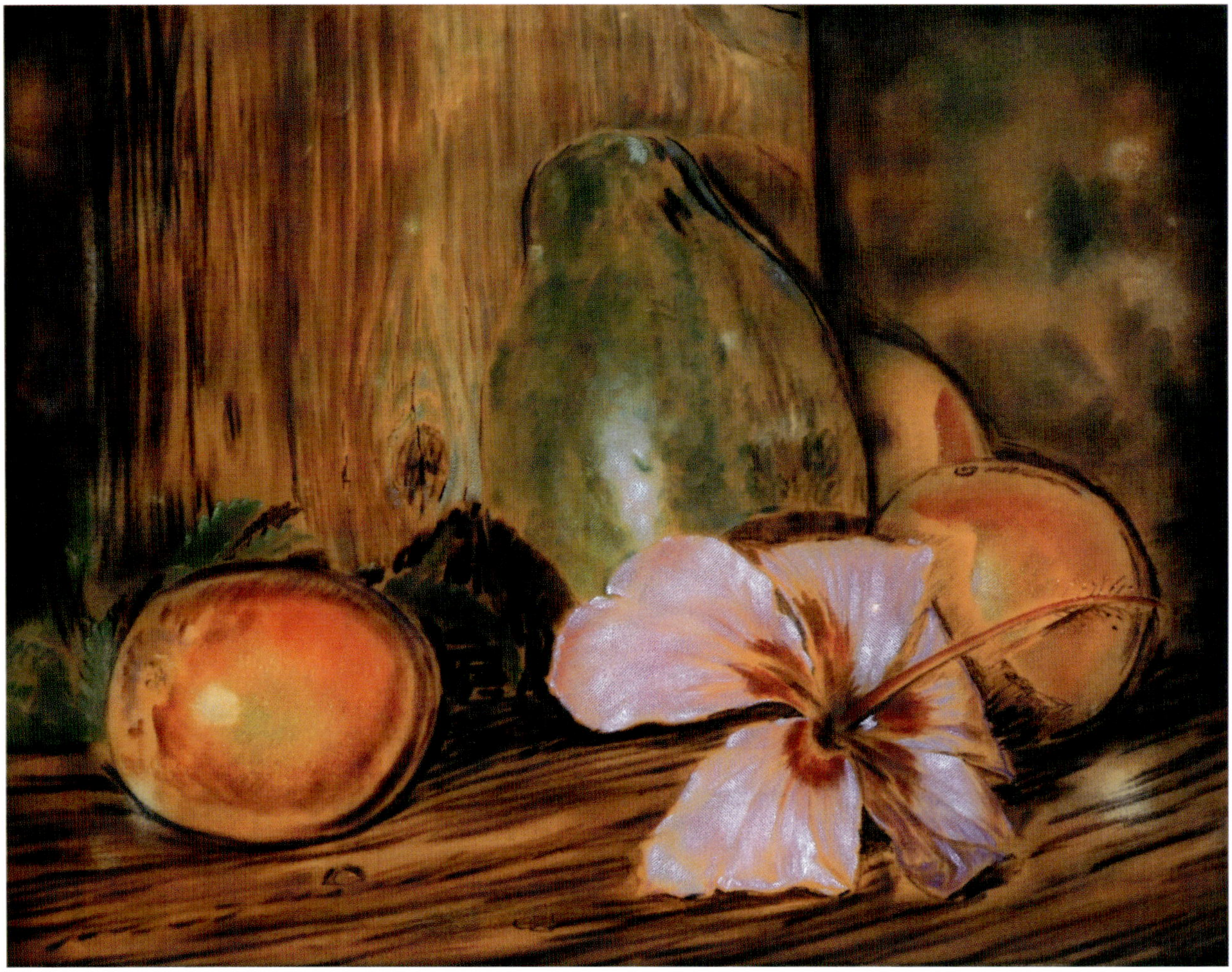

"The earth does not belong to man; man belongs to the earth.
All things are connected like the blood that unites us all.
Man did not weave the web of life; he is merely a strand in it.
Whatever he does to the web, he does to himself."

—CHIEF SEATTLE

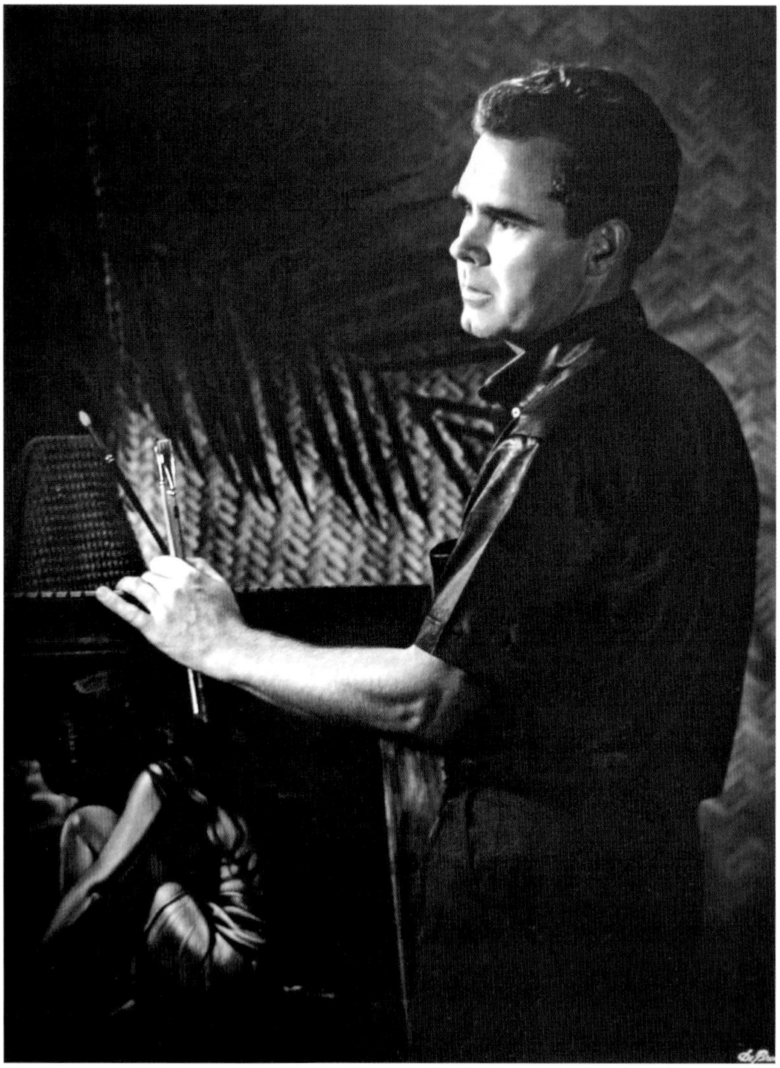